To Harry,
With kind regards
from Dad.
July 7-1909.

H. M. Pope.
This book, written by his "best friend", was autographed by Dad at the Medical Center, Jersey City, N.J. for his son Charles, recalling Charlie's lifelong faithfulness and love for his father as exemplified by many visits, by numerous and thoughtful letters, by generous assistance in times of need and by his keen interest in his father's work.
Allen Pope, July 9, 1950.

THE BULLET'S FLIGHT

FROM

POWDER TO TARGET

THE INTERNAL AND EXTERNAL BALLISTICS
OF SMALL ARMS

A STUDY

OF RIFLE SHOOTING WITH THE PERSONAL
ELEMENT EXCLUDED, DISCLOSING THE
CAUSE OF THE ERROR AT TARGET

ILLUSTRATED WITH ONE HUNDRED AND EIGHTY-EIGHT PLATES
SHOWING THE RESULTS OF OVER THREE HUNDRED RIFLE EXPERIMENTS
PERFORMED AND CHRONOLOGICALLY ARRANGED

BY

F. W. MANN, B.S., M.D.

MEMBER OF THE CORNELL UNIVERSITY ALUMNI, BOSTON UNIVERSITY
ALUMNI, AND LIFE MEMBER OF THE MASSACHUSETTS
RIFLE ASSOCIATION

1980
WOLFE PUBLISHING CO., INC.
PRESCOTT, ARIZONA

COPYRIGHT, 1909,
BY FRANKLIN W. MANN.

Copyright © 1980
WOLFE PUBLISHING CO., INC.

Manufactured in the United States of America
Reprinted July, 1980
ISBN: 0-935632-04-2 Hard Bound

Wolfe Publishing Co., Inc.
P.O. Box 30-30
Prescott, AZ 86302

Respectfully Dedicated

TO

BROTHER T. H. MANN, M.D.

WHOSE AID IN PREPARING THIS WORK
HAS BEEN INVALUABLE

AND TO

BROTHER W. E. MANN, ESQ.

WHO SO MATERIALLY ASSISTED AND ENCOURAGED
THE AUTHOR IN THE EXPERIMENTS
RECORDED

This replica edition
is dedicated
by the Publisher
to
HARRY MELVILLE POPE

FOREWORD

To the best of our knowledge, this is the fourth printing of *The Bullet's Flight* by Dr. Franklin W. Mann. The first three, published by Munn & Co., Standard Publishing and Ray Riling, are now out of print and eagerly sought by collectors.

Early in 1972, Neal Knox (then editor of *Handloader* and *Rifle* magazines) and I learned that a copy of the first edition once owned by famous barrel maker Harry M. Pope was owned by Pope's son, Charles. And best of all, this particular book was full of marginal notations, most of which were hand-written by Harry M. Pope, and some by Dr. Mann.

We called "Charlie" in South Miami to see if he'd grant permission to reprint his copy. "Sure," he said, "but I won't let it out of my sight. It's too valuable to send in the mail." Consequently, Neal Knox, (tape recorder under his arm) and our photographer, Walter Schwarz, flew to Florida. On April 1, 1972, the following conversation was taped between Neal and Charlie Pope:

Knox: Let's talk about your Dad. What was his type of personality and general character. How would you describe him?

Pope: As far as I was personally concerned, I found him very easy to get along with. But the bane of his existence were people who came into the shop and wanted to tell him how good they were, using up his valuable time. So, I would think that a fair answer to that question would be that he was sometimes crotchety. He was very well liked by the people who knew him; knew his abilities. And some of that rubbed off on me — the abilities I mean. Many times I have been called an expert, which I'm not — not by a long shot. I just happen to be "gunny" — I like guns. But I never had the ability that Pa had for fine work on a lathe or with hand tools, of the type required in gun work. I would say Pa was a recognized expert in the rifle barrel business, and he knew it; consequently he had short shrift for what he used to call "cold tuna and hamburger mechanics," which so many so-called gunsmiths seem to be. But as I said, he probably falls into the general classification of all the gunsmiths that I ever knew: they are crotchety when they get to be old men. And that just about covers it.

Knox: About what time in his life did he have to quit shooting? He was such a good shot in his younger days.

Pope: I'd say within a few years after the First World War because after I came home from France, I went with him to the Seventh Regiment Armory in New York City to a rifle match, and he shot at that match. Now that would be approximately the middle of July of 1919. I know he used to go to other matches after that, but whether or not he shot in any of those matches, I can't say.

Knox: How about Camp Perry?

Pope: Yeah, Perry and Sea Girt both. But he belonged to several shooting clubs up and down the Hudson Valley. Because I was away, either at sea or on some construction job, I had few opportunities to accompany him on any trips. The last one I remember was at Sea Girt, New Jersey, and I don't recall the date. I would say it was 1928 or 1929 because I was working out of a New York office. That particular shoot was on the Fourth of July, but the exact year I don't recall.

Knox: Did you shoot with him?

Pope: No, I never shot with him. I was never a match shooter. Most of my shooting was done in the Massachusetts Militia or in the service. We had a small team on the destroyer I was on and we used to shoot a little bit out on the West Coast at different clubs. But I was never a real top-notch man in the field.

Knox: Did your Dad ever give you any coaching in off-hand rifle shooting?

Pope: No. When I was growing up and became a gun bug, was at the time that Pa lost everything in San Francisco. From then on most of my shooting was what they call plinking nowadays.

Knox: You have a range right here on your place and you have some fine guns that you have shown us, including some Pope-barreled guns. This one — this Colt Lightning: What year did you say he made it?

Pope: He made that for me when I was 15 years old. Let's see, that would be in 1911 because I was born in '96. But he told me the barrel he put on that rifle he made before I was born. And I will be 76 next week.

Knox: Charlie, you mentioned he had a basement window that he shot out of.

>***Pope:*** That was in Hartford. He shot from the cellar of the house out to the shop. The targets were out at the shop. It was underground, in a tunnel of sorts. I don't recall ever being in that tunnel. Of course, I was only a sprat then, but I don't think it was a sort of tunnel you could walk in. Possibly it was a large pipe underground — concrete or ceramic perhaps — something of that type. That was when he had his shop at 59 Ashley Street in Hartford.

Knox: That was right before he went to Stevens?

>***Pope:*** I don't know exactly when he started in business for himself, but he was in business for himself as long as I can remember — that is as far as the Hartford times were concerned. That would be at the most four years.

Knox: Do you remember hearing your Dad comment on the newer firearms designs coming out before and during the war when the boys all started going to jacketed bullets instead of the lead bullets used in Schuetzen guns? Did he prefer the old Schuetzens, or did he like the new . . .

>***Pope:*** I don't believe I can give you a definite answer on that, but he made a lot of barrels for Krags, which used jacketed bullets. And he designed some lead bullets for the .30 caliber. Lyman still makes the moulds for them. He made a lot of Springfield barrels. Also, one year he made the barrels for the Olympic team. That year I think I still lived in Springfield. I had a chum by the name of Pierce whose father was a colonel in the Ordnance Corp and was commandant at Springfield. We had the run of the armory. Of course the boss' son, you know, could go where he wanted to. My favorite spot was the museum. It was just across the hall from the colonel's office. We had been in the museum one afternoon, and coming out, we heard people talking about Pa. So we eavesdropped a little bit. They were discussing the first contract to make those heavy barrels for the Olympic team. They were Springfields.

VIII

Knox: I wonder if those were the barrels ordered in 1-10 twist, which was standard for the .30-06. Your Dad made them 1-12 twist. He said, "I'll make them on the outside the way you want them, but I'll make them inside like I want them."

Pope: That could be. I don't know. But I do know this: when we closed his shop, among the barrels was an unfinished .30-06 barrel which was the one Roy Dunlap put on an action for me. That was 1-in-14 twist. It was stamped on the breech. It was 32 inches long, and Roy cut it off to 28 — made me mad as hell. But he did a good job.

Knox: In that shop in the Colgate building, you said there was old equipment — presses, tool-room equipment — that they were running on belts and you mentioned that this caused a big vibration which gave him trouble.

Pope: That was because the power unit that drove the belts was on a reciprocating engine, which I never saw, but every time the crank would lower, there was a thump. From what I know about that stuff nowadays, I figure it wasn't bolted down very tight. Probably would lift a little bit every time she would come over. That was when he was on the first or second floor at 18 Morris. When they moved him upstairs to the fourth floor, that address became 22 Morris Street. It was the same building, just further up the street. To get into that building, you went up one of those enclosed fire escape things. They took the belt-driven machinery and power plant out and electrified the whole establishment. The last time I was in the shop, Pop was the only tenant. All the rest of the building was the Colgate warehouse and storage facility.

Knox: This book, Dr. Mann's book, *The Bullet's Flight,* stayed in that building then, in his shop and was pretty well thumbed over it looks like, for all those years.

Pope: Yes. People would come in and go over and sit in his chair or on the couch and that book is about the first thing they would reach for. I guess practically every real rifleman of those days who was in the shop had handled the book.

Knox: Well, there are thumbprints here of undoubtedly no telling how many famous shooters, then.

Pope: No doubt about it.

Knox: Do you know or do you have any idea at about what time Harry M. did the writing in this book? Do you think it was over a period of time, or did he do most of it shortly after the book came out?

>*Pope:* There again, I don't know exactly, but I would assume that it was done right after he got this copy. Probably sometimes as he read through the book, or after a conversation with other shooters, a thought would enter his head and he would check back and write about what he thought on a particular subject.

Knox: Dr. Mann and your Dad were good friends for many years, weren't they?

>*Pope:* Yes, from way back when we lived in Springfield. When Pa was with Stevens, Dr. Mann was at the house several times. One particular instance I remember quite vividly. I think I was probably six or seven years old. In the dining room we had what most people had in those days — a big round oak table you could put leaves into and stretch it out. At that time my brothers, Joe and Allen, were away at college, so there was Ma and Pa and Agnes and myself and Dr. Mann at the table at one time. Of course, I wasn't allowed to say too much.

Knox: What do you remember about Dr. Mann?

>*Pope:* Very little, except that he looked like the picture in the book, and was an affable sort of person with me.

Knox: A child has a pretty good sense about kinds of people. Do you recall that you liked Dr. Mann?

>*Pope:* My Dad liked him.

Knox: Did you?

>*Pope:* Oh, yes, I liked him — from what I knew of him, yes.

Knox: Those times he came to the house, I assume that he was quite an honored guest and friend.

>*Pope:* Yes. And in those days, a man's wife never called him by his first name in front of company. It was always Mr. Pope. I remember Ma saying, "Well, Mr. Pope likes so and so, and so and so." Words to that effect. She was a good hostess and a heck of a cook. Do you want to take pictures of that old Colt?

Knox: What I would like to do, if it's not still raining, is to walk out there and shoot a few rounds through it.

Pope: Sure. I have a little portable shooting bench. I'll put a target up and you can shoot as much as you want.

Knox: Let's go.

It should be noted that both Charlie and his wife, "Tommy," were very reluctant to discuss their personal relationship with Harry. But Charlie later confided to Neal that Harry became almost a recluse after the San Francisco fire, and after moving back East, he buried himself in his gunsmithing. He had very little contact with his family, and in fact did not attend Charlie and Tommy's wedding which took place only a few miles from his shop in New Jersey.

Before and during the above interview, and the target shooting session in Charlie's back yard, Walter Schwarz, with great difficulty, captured on film each page of the book on which Harry had penciled notations. On extremely dirty pages, several exposures were made. Luckily, we ended up with a legible print of every page with handwriting. But the hard work was yet to come.

Old Harry's "pencilmanship" was certainly not the equal of his gunsmithing ability. Art Director Dave LeGate, in his spare time, began the tedious task of clean-up — removing smudges from around and "in" the handwriting, and filling in some letters in words where Harry's pencil had become dull. After several years, only a few pages were completed. And for several years more, the project was in limbo.

Early in 1980 an additional artist joined the Wolfe Publishing staff — genial and talented Mark Harris. His first assignment: get the Mann/Pope book ready for the printer. During three months' back-aching work at the light table, runny eyes and all, Mark completed the job.

The result, we believe, is a book that has historic significance for all gun enthusiasts. And to Dr. Mann and Harry Pope, wherever you are, we sincerely hope our efforts have pleased you.

Dave Wolfe

THE HUMAN QUALITIES OF DR. MANN

It is with a mixed feeling of pleasure and sadness that I write these words for the Second Edition* of my Father's book. To appreciate fully the precision and quality of his experimentation, it is, I believe, desirable to know something of his personality. A few of his friends are still alive who can recall experiences and associations with Father and who can reveal the many sides of his lovable character. It is possible here, however, to illuminate only a few of these facts, the ones which sparkled in the eyes of his family.

From Father's picture you can see that he was a fine looking man. In his fifties when he wrote "The Bullet's Flight," he was still young. Of medium height, thin but muscular, he was agile and graceful. Father's pet personal vanity, if he possessed such, was his feet — feet with high insteps, straight toes, and very strong. He could stand all day without tiring; he ran easily, which came in handy in tending the target. His hands, which Mother admired so much, were strong but delicate in movement, the hands of a surgeon and scientist. Father's hair always remained dark brown and thick, but his moustache and side whiskers greyed slightly. To hear him speak, to look into his clear grey eyes you knew he was a man to be trusted.

Father loved animals, especially cats, and provided us with almost every kind of a pet. There were the two parrots who cackled like Mother's hens and commanded Aunt Kate to go home, an old circus donkey with a broken leg, white rats, guinea pigs, plain and angora rabbits. Billy Angora was a grey rabbit who lived in the house and ate out of Father's hand at the table. Our horned dorset sheep, in true feminine fashion, swooned at the noise of Father's first automobile and had to be put to sleep under the apple blossoms. We had horses, of course, but never a dog; Father didn't like the odor of them. Nor did we have a cow; I guess Father had to milk too many of them when he was a boy.

Father loved fairy stories. Each time he read us a story, he would mark it with the date in his small, neat hand. One of our favorites, Snow White, had six dates after it. He liked flowers and music, too.

XI

* Refers to the 1942 Standard Edition in which this section appeared.

XII THE HUMAN QUALITIES OF DR. MANN

Among the clippings in his rifle scrapbooks, pictures of pretty girls occupy the focus of attention on each page. We used to tell him that he must have married Mother because she was the prettiest girl in Windham County, and he never disputed our conclusions.

Father loved people and people loved him because he didn't allow his erudition to be a barrier to common folk. He was modest and retiring, and never too busy to lend a bit of help and advice. Many was the time when he would stop in the middle of an experiment to mend our dolls.

Time was of inestimable value to Father. He knew how to save it in many little ways. He bought lead pencils by the gross, his lined paper pads by the quire, common pins by the ten pound can. Once when Bowker, the clothier, had black French felt hats weighing one ounce apiece, he anticipated future needs by buying a dozen and kept them in his closet. His tailor made black suits with an original cut six at a time. Was he afraid of moths? "Oh, no," he would reply confidently, "those moths can't stand worsted 100% saturated with my brand of old cigar smoke."

Father did the work of three men, except when he had one of his headaches. At such times he was so unlike the Father we knew ordinarily. On these occasions he would lose his temper at the first thing to present itself. Perhaps we children heard a dish smashing against the wall. "Hadn't Mother been told to put all cracked and nicked dishes in the trash?" he would complain. "Surely she realizes that cracks are filled with germs and that nicks can cut the children's fingers. What are a person's nerves made of, anyway?"

As he grew older Father seemed so much more hurried. At night when he came home tired, with face drawn and pale, Mother would ask, "Why didn't you come home and eat your lunch this noon, Frank?"

In an exasperated tone Father usually replied, "Frances, when I come home at noon, I eat. When I eat, I get sleepy. When I am sleepy, I can't work. Then I waste time." After a pause he would continue. "The experiment didn't go well today. It's the seventh time I've had to start it all over again." But he would finish speaking in a quiet tone, because Mother calmed him. Her sympathy and interest in his work gave him confidence and comfort.

During the last six months of his life he did almost no rifle work. He spent most of his time with Mother as in their younger days. She said that he apparently knew he was about to die and wanted to get away from rifles back to the time when they spent so much time together. And she felt that the reason he hurried so and did the work of three men, as any of his friends will tell you he did, is because that he had a premonition of his death and wanted to leave as little as possible undone.

In spite of his accomplishments, Father died a tired, worried, and disappointed man. After the shock of his death had passed, we realized that all his heartaches and worries were over. We couldn't wish him to return except for our own sakes; he was the backbone and mainstay of our family, the constant inspiration of his friends.

The first of the two influences which brought on my Father's worry and consequent premature death was the utter lack of appreciation shown for "The Bullet's Flight" by the scientific world at large. How many times have I heard him say to Mother, "The people do not appreciate my work, Frances. They have given my book hardly any notice. Strange, England and Germany have felt its worth more than our own country."

And Mother would say to comfort him, "Frank, you are fifty years ahead of your time. People as a whole can't understand your "Bullet's Flight." They are not smart enough. Those who could understand it, if they would, don't want to bother with it. You will never live to see your work appreciated." Indeed, of the men who were supposed to be interested in ballistics, only a few ever gave "The Bullet's Flight" a second thought. But with the passing of the years has come a new appreciation for this lifetime of labor, spent in studying the intricacies of ballistics, and the publishers of this new edition of the book are of the opinion that today's public will be sufficiently receptive to warrant the expense of republication.

Father claimed that the expense involved in experimentation did not bother him at all. He said he never dreamed of making any money from it. On the other hand, since he was a frugal man, he must have been disappointed to see many thousands of dollars invested in experiments which didn't even yield returns of academic value.

The second factor which caused Father so much worry was World War Number I. With his sensitive nature and understanding of the havoc which the rifle bullet and cannon shell could effect, the force of the seriousness of war fell upon him heavily. "It seems ironical, Frank," Mother once said, "that a man who has spent all his life perfecting bullets should take any war so seriously." Father didn't answer. His face became grey and his jaw set. For one whose work with rifles was always that of a scientist and experimenter who was motivated by the highest type of intellectual curiosity, that perversion of science against mankind was all the more crushing.

It is impossible for me to present Father in these pages as he really was; you would have had to know him. I am sure, if he were here now, that he would heartily welcome you readers, students, scientists, collectors and shooters, who have recognized the worth of his effort.

(Mrs.) Willard Lewis

Forest Hills
Augusta, Georgia
1942

DR. FRANKLIN W. MANN, THE SCIENTIST

The author of this book was Franklin Weston Mann. He was the son of Levi and Lydia Lurana (Ware) Mann, and was born in the old homestead, at Norfolk, Massachusetts, on the 24th of July, 1856. He was of the eighth generation from William Mann, who came from England; settled in Cambridge, Massachusetts; and sold his farm afterward for the erection of buildings by Harvard College. William's son, The Rev. Samuel Mann, was a graduate of Harvard in the class of 1665; was teacher of the Dedham school; and was the first minister of the church in Wrentham. Samuel had two sons, one of them, Thomas, was the great-grandfather of the distinguished educator, Horace Mann; and Samuel's other son, Theodore, was Frank's progenitor. The members of the family, from the first, were sturdy, active and efficient in their communities. Frank's grandfather, Salmon Mann, bought the village church and held it as private property, to prevent a possible schism of its members and its transfer to another denomination from the Congregational. Frank's father, Levi, was a farmer and mill owner; deacon and choir leader in the church; and was superintendent of the Sunday School. He was also Chairman of the Board of Selectmen, and member of the legislature. His mother, Lurana Ware, was a descendant of the sixth generation from Robert Ware, "The Aged," an original settler of Dedham, whose daughter, Esther, was the wife of The Rev. Samuel Mann.

It was an advantage for Frank that he was born on a farm, where there was also a sawmill, and that, as the years passed, he could take his place in the seasonable activities of them both. Strength of body was gained and quickening of mind. The principles of farming were discussed by all frequently, with the aim of obtaining reasonable rewards from animals and earth. He was trained to realize the thought and labor needed to produce an income, and early in life came to recognize in its expenditure what results were most beneficial. A sense of duty and of personal responsibility, as one of the family, came with the years, and the acquirement of self control, of management, and the knowledge of relative values were absorbed steadily and unconsciously.

The village school was entered in time, followed by the High School at Walpole. The village Lyceum was a center for debates, in those days, and was a strong incentive for the study of history, politics, law and economics. The town meeting was attended by a large portion of the townspeople, where they learned the rights and duties of citizenship, and where ideas and ideals of government were advanced, subjects that were considered by all in their homes. Religion was also a constant subject of thought and discussion. All of the family were members of the church, and attended its services, regularly; were scholars or teachers in the Sunday school; and took an active part in the weekly prayer meetings. In the home, when all were in the dining room, and were seated about the table, the beloved father asked the blessing. No profane or indecent word was ever heard in that home. Alcohol in any form was never used. Those were the influences and surroundings that were conducive to the development of a strong, clean and thoughtful character.

There were other influences in the home that were important in giving a direction to Frank's thoughts. Connected with the house was an ell, the ground floor of which was used as the storage place for carriages, wagons and farming utensils. The second floor of this building was one long room, known as the "shed chamber." In the center was a carpenter's bench, with the usual tools. Nearby was a masonry outfit. At one side was a shoemaker's bench, with its equipment. Around, in racks on the walls, were hubs and spokes of various sizes for the wheels of carriages and carts, that had been turned from lumber at the mill, prepared in stormy weather, and made ready to be sold to wheelwrights. This shed chamber was the place for the repair of anything about the home that was broken; and where each one of the family could apply the skill he had acquired during the years of his growth. But this chamber was also the center for practice with a gun, when unpleasant weather prevented outdoor operations. The gun was a rifle, which was fired out of a window at a target, perhaps 300 yards away, for practice and to test the marksmanship of each one of the boys. But at other times, the rifle was screwed in the vise of the carpenter's bench and aimed at a target at the far end of the room. The results of bullets, fired from a fixed base, at a uniform distance, were inspected and studied by them all. It was in that shed chamber that Frank's research began in his boyhood, which culminated many years afterward in "The Bullet's Flight".

While this young mind was being molded by the influences of his home life and the community impressions, there was developing a yearning for more knowledge. The result was a four-year course of study in the scientific department of Cornell University, and the award, in 1878, of the degree of Bachelor of Science.

"One of his favorite ideas at this time was the scheme of making a top that would spin absolutely true and remain standing until motion ceased. When his friends denied his ability to produce a top that would continue its revolution more than twenty minutes at the longest, he effected their complete astonishment and open admiration, by exhibiting a top that would spin three-quarters of an hour. In Prof. Anthony's laboratory under glass the top spun two hours and fifty-seven minutes, a truly remarkable time. Prof. Moler is present at Cornell today to vouch for the truth of this statement, as indicative of young Mann's speculative scientific tastes."

Still the yearning was for more knowledge, and he decided to study medicine. The Medical School of Boston University was chosen for that training. The degree of Doctor of Medicine was granted to him in 1883, after he had served as interne in the Massachusetts Homeopathic Hospital. While there and learning the practice of his profession, under the supervision of the attending physicians and surgeons, he made tracings of heart beats in various bodily conditions, as a part of the records of patients. Making a record and filing it with the case history was his duty, but not the end of his interest. He collected many of the tracings he had made, showing the characteristic signs of various heart disturbances, and divided them into groups for use afterwards in the diagnosis of cardiac diseases. Here again became apparent the mental traits, whose seeds had been sown years before in the shed chamber.

After graduating from the medical school, in 1883, he began the practice of medicine. A year later, on the 29th of September, 1884, he married Miss Frances Gertrude Backus, of Ashford, Conn., "a lady of French descent from the old Burgevine family, a niece of Gen. Henri Burgevine, U. S. Ambassador to China, who lost his life in a native rising. Two daughters were born of this marriage." Four years were spent in the practice of medicine and care of the sick, but they were not years of satisfaction. His mind, ever alert to new fields, preferred to move in other channels.

In 1887, he entered the machine business at Milford, Massachusetts.

"In the winter of 1888, his attention was called to the food value for poultry of fresh market bones and the raw meat attached, also to the fact that an enormous quantity of this valuable food was a waste product due to the lack of any successful reducing machine. Mr. Mann, disregarding all methods for grinding green bone, hitherto so unsuccessful, very quickly produced an experimental machine on entirely different lines from anything thus far known. . . . And in 1889, he produced his world famous Bone Cutter which has given him his reputation and success. . . . Poultrymen in all sections of the country were quick to recognize the immense value of these machines and the bone cutter business increased rapidly. This necessitated many and rapid changes in its construction and multiplying of sizes. In 1891, the increased demand for the Mann Green Bone Cutter compelled its inventor to seek larger quarters, and, equipped with complete machinery, it was the largest manufactory of its kind in the world. Having, in 1893, a flourishing and increasing business, and a competent partner, Mr. Mann devoted himself entirely to the mechanical side, so that plans and inventions so long considered were carried out."

Now, when thirty-seven years of age, he had attained the goal of which men dream, and for which they strive. He had become a leader in a branch of industry of his own creation; he had won an honorable reputation; and he was financially secure for the rest of his life in the lucrative rewards of his genius and enterprise.

But the problem of the bullet had been the boy's puzzle in the shed chamber. Its solution was deferred necessarily while his mind was developing in school, college and university; while solving other problems; and while earning his daily bread. Still, as he has stated in the preface of this book, the "spare moments of his whole life were given to rifle conjecturing and experimentation." Then through sixteen years of intensive study and large expenditure, he accumulated and condensed the facts that have made this book the fundamental source of information in this branch of science. Specialists came from other countries to confer with him about their ballistic problems, during the remaining years of his life. They regarded him as a Consulting Engineer of Gunnery, an expert in the science of ballistics.

This glimpse of the author's life has been contributed because of my intimate association with him. He was my cousin. My mother, Charlotte Mann, was the sister of Levi Mann, his father. My summer vacations were spent in Norfolk for several of my boyhood years, in his home, when my parents lived in Albany, N. Y. Happy days with Frank were spent in that delightful home, haying, berrying, or taking turns in driving the oxen and holding the plough. Other days, when rain was falling, some or all of those five brothers found occupation for their bodies in the shed chamber, or they found mental stimulus in their attempts to explain the action of the bullets fired from that history making rifle. I saw the variables on the targets.

In the years that followed, he acquainted me with his studies and the results. There was the record-making top that would not stop spinning, or so it seemed. The groupings of smoked-glass tracings of heart beats were informative and stimulating. When visiting the factory in Milford, where the bone cutters were made, he told me about his experiments for determining the most satisfactory metal or alloy for the cutting knives. And then, during the last years, he unfolded to me his studies for solving his lifelong problems of the bullet and the gun. The summation of his trials and errors makes this book in reality an encyclopedia of information; and it will be a lasting monument of his genius. It is fortunate that he was able to complete his investigations, and to preserve the results in permanent form before his death at Milford, on the 14th of November, 1916. A short account of his life was written, several years ago, by The Rev. Frederick H. Danker, a former pastor of his church; and extracts from it have been made and are shown by quotation marks in this brief biographical sketch.

That the author was not a scientist exclusively may be indicated by the verbal snapshots of him made by his daughter, Gertrude, who says: "Of father's love for children, he never was too busy to talk over our problems with us. We went to him on every occasion for encouragement in our varied enterprises. He knew all about flower gardens; provided us with all kinds of pets, from a circus donkey with a broken leg, which he mended, to a monkey, who used to watch him as he moved about the house. Then he loved fairy stories, and he bought seven or eight volumes. We have some of them yet; and he used to read to Agnes and me fairy stories in the evenings. All these little things are so dear. I hate to have people think that Dad was 'all books,' for, indeed he was very human, as well as scientific."

I conclude that when Frank was groping in the dark vaults of the unknown for science's sake, that he truly found his lamp of Aladdin. And the Genie of that discovery has made possible the contribution of the riches of knowledge and achievement which he so generously imparts to others in this volume.

              Nathaniel Emmons Paine (M.D.)

1640 Washington Street
West Newton, Mass.

PREFACE

During the year 1868, when twelve years of age, I came in possession of the family rifle, a 44-caliber, powder and ball affair, with 41-inch octagon barrel and weighing 12 pounds.

Not being satisfied with the way it performed nor the manner in which it had hitherto been managed by various members of the family, experiments were immediately commenced to improve its shooting at 35 yards; beyond that distance its inaccuracy was too great to present any inducements.

The rear sight, being attached to a spring, was screwed to the top flat of the barrel and would easily slip out of place, — a useless error which I undertook to overcome.

One cold Thanksgiving morning — one of the only two holidays in the year for me — I undertook to fasten firmly this spring sight to the barrel by solder. Not succeeding, I walked a mile to the village blacksmith and induced him to lend a hand. After he had expended some time without success I returned home, not at all discouraged, however.

Without any further aid, the soldering was well and firmly completed at ten o'clock that Thanksgiving evening after fourteen hours' almost continuous work during a precious holiday, — a day laid out for rifle shooting instead of rifle repairing.

This same spring sight is still held firmly in its place as left by my soldering, and the rifle occupies its place in my cabinet.

This feature of unaided perseverance is mentioned to indicate my method of hunting for the cause of the x-error at the target in rifle shooting, and partially explains why it has been persisted in with some degree of success to the present writing.

The results of my experiments of the past thirty-eight years, here recorded, have been as persistently and laboriously worked out with an earnest desire, born of a scientific bent of mind, to assist my fellow-craftsmen and add my mite to the world of scientific progress.

PREFACE

Recorded experiments in the field of rifle work, particularly the unsuccessful ones, have been very meager, and no doubt many of the same mistakes here described have been and are being made by thousands of enthusiastic riflemen far and wide.

A detailed and descriptive record of such personally performed, made with one of the most fascinating mechanisms, the rifle, it is hoped will add to the rifleman's comprehension, and prevent repetition after repetition of the same errors.

He is wise who learns by his own failures, and should be able to "cut across" many corners, if the mistakes as well as successes of others are faithfully recorded for inspection; so the tabulation of unsuccessful experiments is here made a prominent feature.

Keeping in mind the conjecturing and theorizing so prevalent in rifle literature, speculations have been omitted in the following pages, except where they may add to the interest of the reader, and only such conjectures have been allowed as are afterwards either proved false by actually recorded tests, or fully substantiated by recorded experiments.

A fortunate combination of events and conditions conspired in the production of this book. There was required a person whose bent strongly inclined toward a scientific education, one whose spare moments for his whole life were given to rifle conjecturing and experimentation, a natural mechanic, with a machine shop of his own, and means enough to perform all needed experiments, a disposition to provide for the expense of these experiments instead of providing for the more usual amusements of life, a knowledge of photography, a keeper of records, and a person with sufficient time and perseverance.

The author has endeavored in these pages to give credit to the persons and sources from which he has received suggestions or help.

My grateful acknowledgments are particularly due my brother, W. E. Mann, of Norfolk, Massachusetts, for his years of material aid and enthusiastic support towards these recorded experiments; also to E. A. Leopold, of Norristown, Pennsylvania, for valuable aid and information.

F. W. MANN.

MILFORD, MASSACHUSETTS, 1909.

CONTENTS

PART I

	PAGE
The Old .44 Muzzle-loading Rifle	1
The F. Wesson .32, Rim-fire Rifle	2
The Stevens Taper-chamber Rifle	3
Some Experimental Bullets	8
Incidental Questions	10
Winchester Ballard, 36-inch Barrel	12
Winchester Ballard, 30-inch Barrel	15
Winchester Ballard, 20-inch Barrel	17
.30–40 Winchester-Krag Ammunition	18
Pope-cut Special .38-Caliber	22
Auxiliary Chamber, .33-Caliber Rifle	26
Shooting Braces	28
Pope-Ballard .28-Caliber	31
A Woodchuck Experience	33
Reflections	37
The Personal Element vs. Mechanical Rifle Shooting	39
Ross-Pope .32-Caliber Second Hand	41
Testing Muzzle Blast	43
A "Shooting Gibraltar"	45
Bullet Press or Nutcracker	50
Testing Bullets. Snow Shooting	52
Recovering Bullets from Oiled Sawdust	59
Short-barrel Shooting	60
Short Barrels become Interesting	66
Short-barrel Experiments Continued	72
Where the Upset Occurs	81
Ross-Pope .32-Caliber Continued	82
The Pope Breech Loader, .25-Caliber	84
Muzzle Loader, Pope .32-Caliber	86
Letter to Dr. Skinner	88
"Medicus"	90
Pope 1902 .32 Rifle. Compliments for Mr. Pope	93
Testing a Brass-base Bullet	95
Dr. Skinner's Shooting Range, and a Disappointment	99
12-inch Barrel Experiments	104
Smooth Bore, .32-Caliber	107
Vented Barrel, Pope	109
Utility of Vented Barrels	112
The Whizzer	115
Pope 28-inch, .28-Caliber Barrel and Fixed Ammunition; 1903	117
Reflections upon Black Powder and Cast Bullets	120
Laflin & Rand High-pressure Sharpshooter Powder	123
Telescope Mounts; An Invention	125
Accurate Fixed Ammunition Difficult	132
.28-30 Pope 1904 Rifle	132
Trouble with Smokeless and Rifle Bores	134
Ruined Rifle Bores vs. Smokeless Powder vs. Primers	135
.28-30 Pope Continued	136
A .28–9 Barrel, 1904	140
Discarding Two .28–8, 1905 Barrels	141
Remodeling Rifle. Experimental Shells	143
.30-Caliber, 21 & 8 Twist, 1904 Barrel, Jacketed Bullets	145
Reflections; Pipestem Rifles and Jacketed Bullets	149
Ammonia vs. Primer Acid	151
Accurate Ammunition Difficulties, 7 mm. Caliber Rifles	152
24-inch 7 mm. Barrel	159
A Shooting Match	160
.25–36 Marlin Factory Barrel	164
Metal Jackets, Short Barrels, .25-Caliber	166
Short Barrels and Full-mantled Bullets, .30-Caliber	170
Special .25-36 Marlin, 14-inch Pitch	174
Mirage vs. Telescope	177
Space covered by Cross Hairs of Telescope	180
Distance Measurement with Scope	181
Conveniences	182
Superiority of Bore-diameter Bullet Discussed	183
Flight of Bullet; Screen Shooting	185
Comparative 100 vs. 200 Yard Butts	202
Plank and Screens	205

CONTENTS

	PAGE		PAGE
Unbalanced or Mutilated Bullets	212	Trajectory Deflection	245
Gyration and Oscillation	221	More Reflections	247
A Spinning Bullet	224	Cause of x-Error Located	249
Tipping Bullets Deceptive	229	Cause of y-Error Located	253
Bullet Tip Scale	232	$x + y = 80\%$. The Rifleman's Rainbow	258
Correcting Measurements	234		
Flight of a Bullet	236	x and y Epitomized	260
Measuring Wind Drift	239	Unbalanced Bullets, how Produced	261
Motions executed by Normal Flying Bullets	242	Difficulties with Rifle Twists	267

PART II

	PAGE		PAGE
Verification	268	Illustrated x-Error	328
y-Error disclosed by Plank Shooting	270	Jacketed Bullets throw Melted Lead	332
Cause of y Illustrated	276	Plank Shooting, Service Rifle	334
Cause of x and y disclosed at Muzzle	280	Paper Plank Experiments. Tube Shooting	336
Success comes; x-Error stands alone	319		
Bullets oscillate about Center of Gravity	321	Determining Rifle Twists	338
		Plank Shooting, Spherical Bullets	343
y-Error stands alone	322	Cause of Excessive Tip Disclosed	344
Cylinder Bullets do Stunts	323	Retrospect	347
Driving Tacks with Bullets	326		

PART III

	PAGE		PAGE
Mathematics of x and y	350	Kinetics of Spin	368
Heavy Ordnance	358	Stripping the Grooves	371
The Spitzer Metal-cased Bullet	359	A List of Constants	376
The Spitzer Bullet straightens up	365	Table of Metals and Gases	378

INDEX 379

Translation of marginal notes 385

NOTES

Pope Muzzle-loading System. The Author does not wish it understood at any place in this book that the bore-diameter bullet or the two-cylinder bullet, loaded at the breech, is as accurate for target work as the Pope muzzle-loading system. He simply tabulates a few tests which indicate that under certain conditions and for a very limited number of shots, without cleaning, the bore-diameter bullet did outshoot the muzzle-loading system as operated at the time.

The Five-shot Group. The adoption of the five-shot group in my rifle work, instead of the ten-shot group, was an immense saving of time and it was well adapted for experimental purposes. The five-shot group in most cases answered the question at hand. The system of using the five-shot group was not devised for competitive but for experimental rifle work. Any tests recorded in the book where the five-shot group does not seem to be conclusive may be repeated and supplemented by as many trials as is found necessary. This work is going on continuously (1910) at the homestead-range.

The "Lucky" Group. The Author does not claim that a very small five-shot group gives a basis for judgment or proves that the rifle and ammunition are working properly, but he does claim that a spreading five-shot group proves that they are not working properly.

ERRATA

P. 71. 5th line, should read .32-40 instead of .30-40.

P. 115. Substitute the following for the first paragraph under the heading "The Whizzer."

During Aug., 1902, in order to determine if the powder gases at the muzzle of a rifle traveled faster than the bullet which had been propelled by them during its flight in the bore, and to determine if the grains of unburnt powder were shot past the bullet while the latter was leaving the muzzle, Mr. Leopold suggested a machine, which, after its completion, was christened "The Whizzer." The machine and its mode of operation is well represented (in Fig. 51) on the following page. This machine did not throw much light upon the question for which it was designed.

P. 115. 20th line, should read choke out instead of choke-out; omit hyphen.

P. 116. The word As at the beginning of the last paragraph should be omitted.

P. 128. 8th line, should read 1906 instead of 1806.

P. 159. 15th line, should read triangulated instead of triangled.

P. 164. Under Test 110, 3rd line from bottom, should read .258 instead of .285.

P. 219. 1st line, should read (Fig. 88) instead of (Fig. 86).

P. 233. Bottom line on page should read tip instead of trip.

P. 303. 1st line, should read 7½-inch right twist instead of 10-inch.

P. 303. In Figure 124, top illustration, should read (6 m.m.) instead of (7 m.m.).

THE BULLET'S FLIGHT FROM POWDER TO TARGET

PART I

The Old .44 Muzzle-loading Rifle. DURING the first four years of experimental work with this ancestral, .44-caliber, muzzle-loading rifle, at 35 yards, it was firmly and persistently believed that the ball did not always fly straight from the muzzle; that is, in the direction which the bore of the rifle pointed.

In discussing this matter with my immediate comrades, it was suggested that the errors in shooting were caused by the blast from the muzzle acting to deflect the bullet, and it was planned to cut six long slots, longitudinally, through the barrel at its forward end near the muzzle, thus allowing the major part of the gas to escape after imparting the proper velocity to the ball before it left the muzzle.

Although cutting these slots, being impossible with our limited facilities, was never attempted, it was pretty firmly believed at the time that if a few inches of the muzzle were so slotted, it would enable us to put every shot into the same hole, every time at 35 yards.

Thus at between 12 and 14 years of age attempts were commenced to place the sights of a rifle in their true place and reliable condition, and to cause the ball to fly straight every time from the muzzle of a rifle, or know the reason why it would not. The following photographic illustration (Fig. 1) represents the first attempt, with plaster of paris, jackknife, and lead, to make a bullet that could be entered concentric with the bore of the old rifle.

2   THE BULLET'S FLIGHT FROM POWDER TO TARGET

The bullet standing upon one half of the plaster mold was first made from a simple chunk of lead by passing it back and forth through the bore of the rifle, swaging repeatedly and jackknife whittling, until the rifle grooves were comparatively perfect upon its sides, oval point centrally balanced and base

FIG. 1.

at right angles to the body. Then the plaster-of-paris mold was made by aid of the first bullet and several others were successfully cast in it, one of which is represented at the lower right-hand corner of the illustration.

The F. Wesson .32, Rim-fire Rifle   This fine-looking sporting rifle came to hand during 1879, and was a finely finished breech loader, made special for an older brother, and he prized it very highly. It was a fair sample of the best of its kind made 40 years ago, and occupied my rifle experimental days for the next two years, though proving as unsatisfactory in its performance as the old muzzle loader, making even larger errors at 35 yards.

There was no satisfaction in being reminded from time to time that my inaccurate shooting was due to errors of aiming or holding because the theory that the difficulty could only be explained by assuming the bullet did not fly from the gun in a true line with its bore, was too firmly rooted to be shaken by anything less than a practical demonstration.

A visit was made to Mr. Wesson, the maker, at Worcester, and the matter discussed with him. A special bullet mold was designed and made with the hope of causing the bullet to enter the bore from the chamber, at the time of discharge, with its long axis true with the axis of the bore of the rifle, claiming that accuracy of flight could not be expected if the bullet did not stand point on with the bore.

No improvements were made or satisfaction obtained, however, from this rifle; the average size of groups made at 50 yards, with regular factory ammunition, being 1.70 inches.

The Stevens Taper-chamber Rifle. It was 15 years after experimenting with the Wesson rifle, or during July, 1894, before owning my first rifle, the intervening years being given to student life at two universities, and work connected therewith. This first rifle was a Stevens .32-40-caliber, 26-inch barrel, Mogg telescope, and with a special chamber cut by myself.

Two or three years were spent in designing this chamber, discarding one design after another, before one was completed and sent to the Ideal Co. The sole object of it was to compel the bullet to enter the bore immediately in front of the rifle chamber true with the axis of its bore.

So far as known at the time no form of ammunition or rifle chambers, regularly made, complied with sufficient accuracy to this requirement, all being made rather to assure an easy and uncomplicated insertion of the cartridge. This is not intended as a reflection upon manufacturers, for it is true they had made great improvements over the old muzzle-loading, round-bullet affair, and commercial needs required that the arm and its ammunition should be easily and safely handled by the class of marksmen to whom they catered.

The final design under which the chamber was made was simple enough. The chamber and the steel swage for the bullet, as also the swage for reducing the Ideal .32 shell, were all made with the same taper reamer, .008 taper to the inch. The bullet, after being swaged tapering, was dropped into the taper chamber, and the loaded, tapering shell was dropped in after it, and this is termed "front seating" in contradistinction to fixed ammunition.

The front end of the bullet was .315 inch, the same diameter as the bore of the rifle, and the large or rear end was .320, corresponding to the groove diameter, while the mouth of the shell fitted up snug to base of the bullet.

The rifling or grooves in bore of a rifle are usually cut about .003 inch deep, but vary somewhat. The grooves being thus cut would make what is termed the groove diameter .006 inch larger than bore diameter, the latter being the size before grooves were cut.

It was thought that the bullet resting in the throat of the rifle, as explained above, on a true taper with the lands left by the tapering reamer in the chamber,

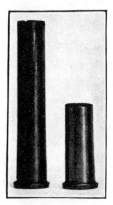

Fig. 2.

must be in position before discharge to lay straight and central with the bore. The bullet seated itself by its own weight. It had not been deformed by forcing it into the shell or into a rifle bore smaller than itself. It had not been deformed at its base by pressure from a bullet seater. Being formed independent of the shell, in a tapering swage identical with the taper of the chamber, it must lie in the chamber, taking the same position that it did in the swage. This was a mechanical certainty and (Fig. 2) is a correct illustration, exact size, of the two different lengths of the taper shells which were produced from the Ideal .32 shell by swaging, and they fitted the taper of the chamber.

This special chamber was also designed to make front seating more practical by abolishing the cumbersome bullet seater, and doing what a bullet seater will not do, that is, to enter the bullet central and straight with the bore.

My several rifle friends rather smiled at the over-particular nicety displayed in this matter, averring that the bullet was entered central and straight with the bore, either with a bullet seater or when shot from a shell, and it required several years' controversy, with material proof tests in evidence, before a single rifleman showed signs of being convinced that such was not the case.

Still unconvinced of the necessity of absolute correctness in this matter, they failed to comprehend the utility of so many weeks and months spent in trying to overcome so slight an error as could possibly exist in the ordinary fixed ammunition and commercially made chamber. The only reply that could be made to such a criticism was through the Yankee way of answering a question by asking another, viz. Why are you so particular in your attempts to make an accurate shot or a good score?

A special feature of this chamber was its design to take a tapered, ungrooved, non-lubricated bullet; and if a soft one was found to lead the bore, another of antimony and lead was to be used. If trouble came from this, a copper-plated one would be tried. To prevent the blast from the muzzle throwing the bullet wild, a spherical base bullet suggested itself, reasoning that such a hemisphere in front of the blast would offer no sharp edges for its leverage.

Four varieties of bullets were made for this particular ammunition and chamber, each of two lengths: one oval point, one sharp cone-shaped point, one flat base, and one hemispherical. The adjustable mold, swages, and chambering tools for these eight bullets were comprised in one order to the Ideal Co., while the dies for reducing the Ideal shell were made in the home shop. The Mogg telescope was for the express purpose of testing with certainty the accuracy of this particular ammunition and chamber.

A cast-iron forestock, weighing six pounds, extending from the rifle action to its muzzle, was made to increase its weight if desired, fitted to the barrel with melted lead. But no practical method was found of attaching it immovably to the barrel; screws were of no value; any amount of waxed twine would yield after several shots, allowing the iron forestock to slip forward; besides this tight binding would spring the barrel because the lead casting, upon cooling, did not conform itself perfectly to it, therefore the rifle was not tested at target with this extra forestock.

Test 1. — August 8, 1894, with this tapered-chamber rifle, a tapered, ungrooved, swaged 151-grain sharp-pointed lead bullet was seated in the tapered chamber in front of the shell, seating itself by its own weight and mechanically fitting the front end of chamber.

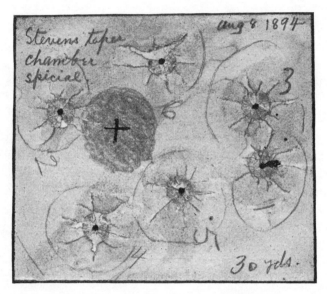

FIG. 3.

After the first shot a group of six was made at 30 yards, shown by (Fig. 3), using 10 grains powder, spiral spring muzzle rest, telescope sight, no wind. Size of group was 2.25 inches. A second group of five was made under the same conditions, only using 25 instead of 10-grain charge, making a group of 1.75 inches. It was a surprise, however, to find that the center of this latter group was three inches directly under the first, a fact which was carefully noted because it opened up a new problem.

There was no leading during these tests, nor were other tests made at this time, as the affair was considered useless. Some time later, however, an amateur made 55 consecutive shots, without cleaning, with these non-lubricated, ungrooved, oversize bullets, and no leading could be detected though carefully searched for; but it was concluded this non-leading was due to some peculiar condition not connected with the peculiar taper chamber.

Five years later, before this barrel was rechambered and disposed of, it was put upon a Pope machine rest during a perfect day, making 10 shots at 100 yards with same kind of ammunition as in the above test. Two of them missed the paper, while the remaining eight formed a 6.25-inch group. Multiplying the spread of the 30-yard group, shot five years previous, by 3.33, and 6.66 results, very similar to the group made at 100 yards from machine rest. This might have been a 14-inch group at 200 yards.

8   THE BULLET'S FLIGHT FROM POWDER TO TARGET

Some Experimental Bullets. The plate (1) on opposite page well represents some of the bullets experimented with during about 10 years, commencing with the old .44-caliber round ones. It will be found somewhat interesting by comparing one with another.

Figures 1, 2, 3, and 4 were cast from a home-made steel mold 12 years prior to experimenting with those represented in the upper row, and were made to use in F. Wesson's .32-caliber rim-fire rifle.

Figure 5 is one of F. Wesson's best bullets for .32 rim-fire.

Figures 6 and 7 were made for the old muzzle-loading, .44-caliber of 1868.

Figure 8 was made for the same rifle by means of a home-made plaster-of-paris mold.

Figures 9 and 10 were the regular round bullets for same rifle.

Figures 11, 12, 13, 14 and 15 are some .32-caliber cylinder bullets made for experimental purposes.

Figures 16 to 24 inclusive were designed for the special Stevens rifle with its Mogg telescope of 1894.

Figure 25 is a hollow-base bullet into which brass screws represented by Figure 28 were placed and swaged in.

Figure 26 is a hemispherical base bullet of 1902.

Figure 27 is the Figure 25 bullet completed with the screws shown in Figure 28.

PLATE 1.

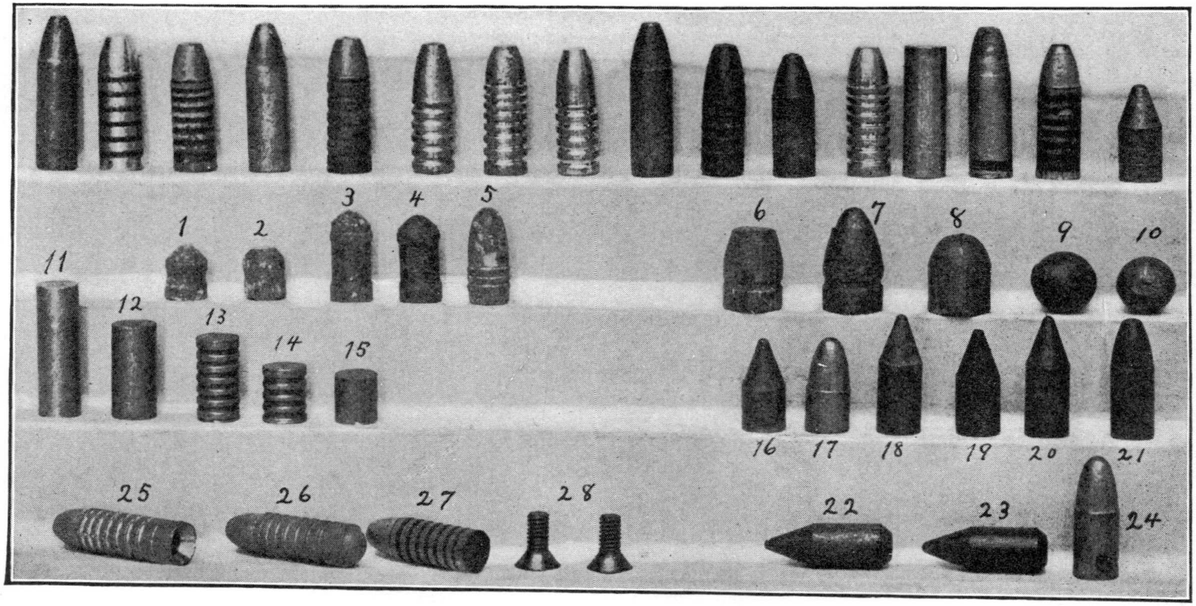

Incidental Questions. At this early stage of experimenting I had no acquaintance among riflemen with whom to compare notes or from whom to obtain suggestions; had never seen a telescope rifle; knew nothing of any governmental experimenting, almost nothing of the refined methods in use upon different rifle ranges, and had read but few books, papers, or articles upon the subject. True, in a general way, I knew of rifle tournaments and competitions, but the erratic shooting so often displayed savored more of chance than science.

It was therefore intentional on my part to investigate in my own way, unhindered by previous experiments of others, all of which seemed to have overlooked the vital error. This peculiar disposition not to follow the crowd kept me alone and emphasized a disregard for the conclusions as stated by others unless such were self-evident.

To one who is delving and plodding after the reason for things, the unexpected frequently happens while expected or sought-after results elude the grasp. Testing a theory experimentally has been found, however, a pretty tedious process compared with the rapid work of brain in compounding it.

During my student days repeated trials were made to produce a spinning top that was perfectly balanced, and many of different sizes, shapes, and of different metals were turned in the lathes supplied to the students of Cornell University.

A top so perfectly balanced that it would keep its standing position after coming to a standstill was the desideratum; though known to be as impossible as perpetual motion, the very fascination of the idea kept pushing the attempt further and further. Some beautiful tops had been finished, to the amusement of friends, when the subject was dropped, partly on account of other pressing duties.

Ten years later, however, with improved machinery, material, and new experiences at hand, the manufacture of tops and gyroscopes was again commenced, the work becoming more and more refined. Finally, by submitting the finishing of the hardened tool-steel top to Brown & Sharp, of Providence, mechanical perfection was undoubtedly obtained, yet the spinning top would tremble and hum, showing that its balance was not perfect.

Mr. M. might have saved much time & learned faster had he mixed with really expert shooters. Hrm

This statement may be would be modified, if Mr. Pope or Mr. Mann knew what in reality Mr. Mann was trying to accomplish. (Frank)

Without knowing at the time, and without the rifle question coming to mind in connection with balancing of this almost perfect steel disk, it was found to have a close relation to the bullet in rifle shooting, as will be noticed later. The negative from which this half tone (Fig. 4) of the spinning top was made

FIG. 4.

was exposed for 20 minutes while it was spinning, and the top continued spinning for 35 minutes. Photographers can readily understand how well balanced this top must have been to make so perfect a print. Of course the block at left of cut, from which the top was spun, was motionless, as was also the string which spun it, and the similarity presented in the minute detail of the several figures would suggest that the top was also stationary.

The amount of labor and time spent in experimenting to produce a disk that would spin like this need not be detailed in this connection, except to draw attention to the relation of this spinning top and the slight motion of its spindle to the spinning bullet upon the air, and the impossibility of obtaining perfection.

Turning from this digression regarding spinning tops back to rifles: before this Stevens special chambered one was thrown aside an old timer, a long-range telescope man showed up who said that two errors appeared in the mountings of the Mogg 'scope: the tube would take two different positions in its front ring, and the binding screw in the rear mount held the tube so firmly that if tension was placed on the center it would not return to its normal position.

This discovery divided my rifle work into separate lines for several following

12   THE BULLET'S FLIGHT FROM POWDER TO TARGET

years: one to overcome the inaccurate flight of bullets, and the other to construct telescope mountings which should combine reliability with the necessary accurate methods for adjustment.

From this time, during the following 12 years, all my practice was group shooting, for reasons which must now be obvious, and all at 100 yards; during this time also all groups made, with all rifle barrels, some 30 or more, have been filed and preserved for purposes of comparison and study later.

No doubt it is safe to say that no rifleman in America has kept every target made for 12 years, as has been done in this case, usually keeping, showing, and publishing only their best scores. With my recorded tests the best groups are considered only as fortunate or accidental and are not the important ones which demand attention.

Winchester Ballard, 36-inch Barrel.   Some time previous to 1895 a special .32-caliber was ordered, but after waiting until patience ceased to be a virtue a .32-40, 36-inch (their longest), No. 4, full round barrel, factory made, was ordered from the Winchester people that shooting might be commenced with the club.

A commodious house was built two miles out of the village of Milford, Mass., with a shooting annex and a range of 200 yards for our club of six. Five of the club invariably shot at 200 yards, while my own was as invariably at 100, and until the advent of high-pressure smokeless powder all recorded tests were made with Hazzard's F. G. powder. In all breech-loading work with front seating a wad was used, that is, where the bullet was not inserted into the shell, except when the Pope muzzle system was employed; and after 1901 the oleo wad obtained from E. A. Leopold excluded all others.

Test 1 a. — This was made April 5 and July 16, 1895; unlubricated, non-grooved bullet, making six 5-shot groups; the smallest was 1, largest 2.12, average 1.56 inches. With several groups there was no leading, but with some there was. It was not certain that all lead was removed between shots, but most of it was.

THE BALLISTICS OF SMALL ARMS 13

These 214-grain bullets, 29 grains heavier than the twist was cut for, showed only normal tip, no greater than the regular ones.

With this shooting 45 grains of powder were used, a thin felt wad, 214-grain, medium long-pointed, lead bullet, swaged cylindrical to .315 inch; non-grooved, non-lubricated, seated ⅛ inch beyond chamber; Ballard action rifle, double rest, telescope sight, and the barrel swabbed after each shot with kerosene and machine oil, half and half, the oil being partially wiped out after each shot.

This is next to the largest of the six groups and measures two inches. All groups made and recorded are measured in this manner, that is, the distance

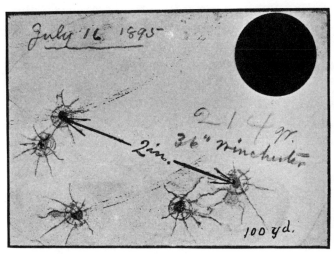

Fig. 5.

between centers of the two shots widest apart. This two-inch group (Fig. 5) indicates that some error exists, aside from entering the bullet central and straight in the bore before discharge, because these bullets were so entered.

The close similarity between the above and the then famous Chase-patch system of the Massachusetts Rifle Association, Crank's Corner, Walnut Hill, omitting only the paper patch and substituting thin oil for water, will be remarked. The .315-inch cylindrical bullet seated itself in the .32-caliber bore "like old cheese," an expression originating at Walnut Hill. So far as could

It has been well known for many years before this by me ('84 with my first 32/40 Ballard, 20" twist) that a smooth bullet will shoot in a closer twist than a grooved .185 patch in a bore printing Jupetty while 1895 keyholed flat, experiments to correct this showed that a fine point bullet also showed much less tip due to less air resistance, a 175 gr nearly sharp pointed bullet was adopted, which tipped less than a 165 gr Marlin.

be learned the above test was contrary to all rifle ideas at the time or since. Although practically identical with the Chase system, that is not why it was employed, but rather due to a conviction that a bullet larger than bore diameter could not, with any certainty, be entered straight or central with it, either with a bullet seater of any construction or a charge of powder.

This cylindrical, bore-diameter bullet, therefore, was so made that it should be possible to enter it point on in the bore ready for discharge, and there can be little doubt but it upset into the grooves sufficiently for all demands of rotation. There is no question but the Chase-patch system which gave a bore-diameter bullet, first suggested to the writer the theory of its superiority, though from the beginning a patched bullet was a thing that must be eliminated, else a target rifle would not be practical in the hunting field.

Test 2. — April 9, 1900, and May 26, 1898, a test was conducted like the preceding one except the ungrooved bore-diameter bullet weighed 159 grains instead of 214, and was lightly swaged into a new rifled bullet seater before entering into throat of the rifle. From the double rest, the smallest group was 1.25, largest 3, average 2.31 inches. Prints of the bullets were good but not perfect. This new seater was made by turning down a three-inch piece of new Winchester barrel to fit the .32–40 chamber, and so arranged that the grooves in the seater and in the barrel properly followed each other. It was used in many tests and always behaved itself. The bullets were not swaged in the seater to insure their taking the grooves, but to insure that they should lie in the bore, before discharge, with point central with it, and with their base at a right angle to it. If swaged to fit the grooves snugly, it would have been impossible to push them out of the seater or into the bore.

Test 3. — July 16, 1895, and April 27, May 5, 10, 13, and 26, 1898, an experiment was conducted like the preceding except the bullet weighed 185 grains and was grooved, lubricated, and carefully swaged without any taper; smallest group 1.25, largest 2.37, average 1.62 inches.

Test 4. — April 8 and 9, 1898. This was like the last except the 185-grain bullets, lubricated and swaged, were lightly swaged in the bullet seater; smallest group was 1.25, largest 2, average 1.62 inches. These last four tests indicate the x-error is not eliminated by placing the bullets, whether long or short, central with the bore before discharge.

Test 5. — This was made May 5, 1898, for purpose of comparing a paper-patched Winchester bullet, front-seated, with special bullets as used in the preceding and following tests. Selecting the Winchester factory-made, paper-patched, regular 185-grain bullets, seated $\frac{1}{8}$ inch beyond the chamber, they were tested in the same manner as in the preceding; smallest group made was 1.50, largest 2.25, average 1.85 inches, — not so good as bore-diameter bullets. With this longest and selected Winchester barrel, at double rest and fine weather conditions, there seems to be a constant spreading of groups at 100 yards with all forms of ammunition tested.

Winchester Ballard, 30-inch Barrel. After accidentally ringing the 36-inch barrel about five inches from the muzzle, a 6-inch piece was removed, leaving it with a length of 30 inches, and this also removed the original choke in the barrel; then many of the tests made with the 36-inch were repeated with this, for comparison.

Test 6. — June 18, July 16, December 26, 1898, and April 17, 1900, under similar conditions as test 5, with 185-grain bullets, lubricated, and swaged, double rest; smallest group was 1, largest 2, average 1.54 inches. One of the July 16th groups is shown in (Fig. 6), measuring 1.70 inches, the one print standing off is a very common occurrence, many groups being spoiled by this one in five or two in ten off shots, which has caused unending discussion.

Test 7. — This experiment was made July 16, 1898, with the same ammunition as test 1 a, except a 30-inch in place of a 36-inch barrel was used, shooting but one group which measured 1.09 inches. Perhaps the absence of choke had its influence here, or it may simply have been a fortunate group.

Test 8. — December 28, 1898. In this F. J. Rabbeth's bullet, Chase-patch system, water cleaning, same as at Crank's Corner, were employed with great care and a new swage was required; otherwise the same as other tests. Two groups were made, largest 1.68, smallest 1.50, average 1.59 inches. The error in shooting still persists, and it is not due to faulty sighting or holding, nor because bullets were not seated central with bore from chamber.

Fig. 6. (See page 15.)

Test 9. — May 23, 1900. This was interesting, with the 30-inch barrel, lubricated bullets, bore diameter and front end as square as the base, that is, swaged to perfect cylinders; 45 grains powder, double rest; smallest group 2.25, largest 6.50, average 4.50 inches, and the prints made by all the bullets appeared perfect, looking like holes cut with a sharp steel die.

Test 10. — This was the same as the 6th test, except this 30-inch barrel had been badly ringed about 8 inches from its muzzle, occurring accidentally by shooting when there was a second bullet near the muzzle. Four 5-shot groups were made, however, the smallest being 2.2, largest 5.25, average 3.75 inches. Evidently the ringed barrel put in its influence here.

Winchester-Ballard, 20-inch Barrel.
After accidentally ringing the 30-inch barrel and cutting 10 inches from its muzzle to clear the defect, one of only 20 inches was left and with this the following tests were made: —

Test 11. — June 23, August 25, and October 18, 1900, with 187-grain bullet, lubricated and swaged to bore diameter, the smallest group was .85, largest 2, average 1.06 inches. It seems the shorter barrel made an improvement in groups, and six years later W. E. Mann still finds this one of his most accurate rifles.

Test 12. — The conditions in this experiment were the same as in previous one, except the front rest was placed 12 inches from muzzle instead of four. The result of a 5-shot group was two inches lower than the last, clearly the result of changing position of rest.

When this barrel was ordered it was the longest furnished by the Winchester people, and at present its length is only 18 inches beyond the chamber. Other tests were made and enough shooting has been done with this short barrel to make certain that its groups are as small or smaller than during its 36-inch days. *Better muzzle just as likely to be an improvement due to gradual improvements in detail of which Mr. Mann was ignorant, also better swages or better methods in swaging bullets or moving other details —*

.30-40 Winchester-Krag Ammunition. While experimenting with the .32–40 Winchester in 1898 the 1895 model Winchester-Krag, made for the .30–40 government ammunition, was ordered to gratify a desire to test the great power of this, then celebrated, ammunition.

The first interesting experiment was made May 28, 1900, to test the force of its recoil, and (Fig. 7) presents the results. The bullets standing on each end of the block were shot from the .30–40 Winchester through $41\frac{1}{2}$ inches of dry chestnut wood, while the middle bullet was tied to the butt plate, leaving the rifle free to recoil when discharged. The force of recoil drove this center bullet into the soft pine plank half an inch, as shown. The explanation of these two forces acting simultaneously, or the same force which sent one projectile through $41\frac{1}{2}$ inches of wood and the recoil from it, was explained to the writer by his old professor as follows: —

"Cornell University, Ithaca, N.Y.

"My dear Mr. Mann: The exhibits you sent are very interesting. I have never seen anything to illustrate so clearly the difference in the *energy* imparted to the ball and to the gun. I will try and explain the apparent anomaly.

"Force by definition is measured by the product of the mass it puts in motion, by the acceleration it imparts to that mass. Action and reaction *are equal*. The mass of your rifle is approximately 250 times that of the ball. The same force, therefore, imparts to the ball 250 times the acceleration as to the rifle. Since it acts upon both for the same time, that is the time required for the ball to be expelled, the force acts upon the ball through 250 times the distance that it acts upon the gun, supposing both equally free to move.

"Now the work done by the force is measured by force × distance through which it acts. Hence, the force does 250 times as much work upon the ball as upon the rifle, hence imparts 250 times the energy to the ball. Hence, the ball should do 250 times the work on being brought to rest.

"Yours very truly,
"W. A. Anthony."

Fig. 7.

The next experiment attempted was to ascertain the action of this powerful projectile upon confined water. An ordinary steam pipe, 14 inches long and two inches inside diameter, was plugged at each end with wooden stopples one-fourth inch thick, and filled with water. At 20 feet distant a shot was made directly end on, and this Krag bullet, by a lucky hit, entered the pipe at center of the wooden plug. As the bullet flew from the muzzle the water in the pipe returned in the path of its flight, and back over the line of sights, dowsing the shooter's eye in quite a startling manner. The steam pipe was split its entire length, and for half the distance the opened seam was a finger's width.

Seven tomato cans, filled with water and balanced upon each other, thus forming a perpendicular pile about 40 inches high, were shot into from a position directly above, and the bullet opened all the cans by splitting down their sides, through the sudden pressure transmitted to their contents. So great and sudden was the pressure that the circular sides of several of the cans were straightened out flat.

Plate (2), on next page, is a reproduction from a photograph of a bullet hole made through common window glass as it was firmly supported in the sash, by the regulation government charge at 20 feet distance, enlarged to a size and a half.

The bullet which made this hole was shortened from its front end to weigh but half the regular one, thus leaving its front end as square as its base. The negative from which this plate was made appears like the window glass through which the bullet was shot; so close is the imitation that by placing each in like frames it would puzzle one to select the negative from the mutilated glass without a close inspection.

A curious desire to ascertain, if possible, the length of time required by one of these Krag bullets to make its perpendicular flight and return to earth, instigated the next experiment. From the end of a boat landing, reaching out into a pond that was about 440 by 200 yards in size, half surrounded by a grove of pines, during a perfectly calm day, the test was made.

Plumb lines were attached to outstretching limbs above, the butt of rifle being placed upon the landing and so held as to cause the barrel to assume as nearly perpendicular position as possible, by sighting from the lines, in the effort to cause the returning bullet to drop into the placid lake.

Eight shots were made from this carefully plumbed position, and any returning bullets would surely make a splash in the water that could be seen, or if striking in the grove of pines the day was so still they could be heard. With all these precautions, however, the experiment was devoid of results, as not a splash was seen in the water nor a sound heard of any one of the returning eight bullets.

PLATE 2.

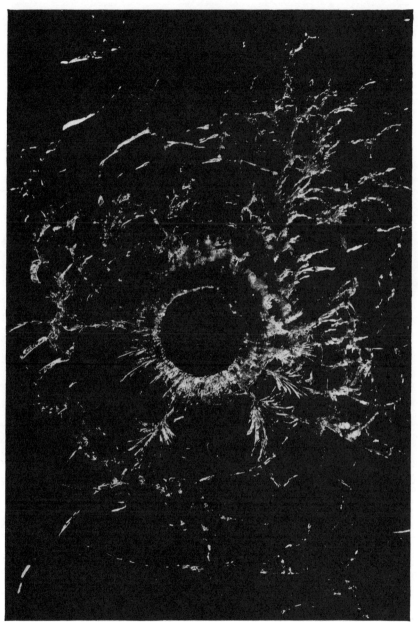

30 Krag 3 ovl load 2000 ft vel + (lighter bullet)
1½ times size   @ 20 ft
Sq end bullet

Another experiment was made to test the penetrating force of these Krag bullets, and Plate (3) on the following page indicates the results obtained.

Figure 1 is a slab of iron one-half inch thick through which the metal jacketed bullet was shot at close range.

Figures 2 and 3 are blocks of lead into which the bullets were fired; the cavities made are evenly lined by the lead of the bullets, while their metal jackets are crumpled up and left loose in the holes.

Figure 4 shows one of the unshot bullets which, like the whole plate, was photographed to exact size.

Pope-cut Special .38-Caliber. During May, 1898, a special .38-caliber Winchester No. 4, full round 32-inch barrel was received. It was rifled by Pope with eight narrow lands, $\frac{1}{32}$ inch wide, bottom of the grooves forming a true circle, — "lands" in rifle parlance meaning the spaces left between grooves in the rifle's bore, surface of the lands constituting the bore diameter.

The special chamber was cut $\frac{1}{4}$ inch longer than the shell and to exact diameter of bottom of the grooves; that is, .386 of an inch. The .38–55 shell was reduced to make it more tapering, which made its outside diameter at mouth .386 inch, or groove diameter. It was a beautifully cut rifle and tests were immediately commenced.

Test 13. — This was made June 25, using a bullet that was nine parts lead and one part antimony, weighing 290 grains, ungrooved, unlubricated, swaged cylindrical to bore diameter except $\frac{3}{16}$ inch of base, which was left with a diameter exactly fitting front of the chamber; using 55 grains powder and felt wad; rifle bore cleaned and slightly oiled between shots; double rest and telescope sight, — this style of ball being almost identical in principle with that used in modern cannon although made of different metal.

The bullets were purposely made hard, intended to be beyond the upsetting limit of the 55-grain powder charge, so it was impossible to swage them by usual methods. After catching several bullets in soft snow, neither upsetting or slipping grooves was observable.

PLATE 3.

The results of this test of three shots at 100 yards were: one missed the paper target, while the other two printed 4.50 inches apart, but the prints were very good if not perfect, as indicated by (Fig. 8), a reproduction from the original target, and notes thereon explain themselves.

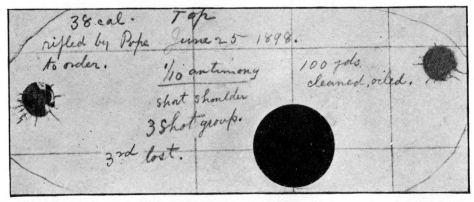

FIG. 8.

This experiment was made with bullets so hard that their bases could not be made oblique by pressure of exploding powder, and so shaped that they could be entered straight and central with bore. It should have put the three shots into the same hole at 100 yards, but they went wild. Why?

Seven rifles of different calibers were ordered at intervals, chambered on this principle, making front seating always necessary, because the base of bullet is same diameter as outside of the shell at its mouth; and leaving no appreciable shoulder at end of the chamber, it does away with a bullet seater, besides making easy swabbing.

This Pope-cut special barrel seemed to fill requirements for test 13 that had been studied and planned for several years and for which rifle, swages, and reducers were exclusively made. No particular account was kept of the expense, but that, however, has little bearing upon the results obtained.

Test 14. — This was similar except the bullets employed were an alloy of lead and tin, 20 to 1, which would upset and take the grooves, and was conducted in the same manner and during same day as the preceding, two groups of five being shot; the smallest measuring 1.50, largest 1.75, average 1.62 inches, thus sending us back to the same tiresome group dimensions made with the Winchester .32–40 barrel.

Test 15. — June 30, five days following the last test, Mr. William Dougherty, one of six club men who was making the best scores at 200 yards, was engaged and paid for leaving his work as a machinist, to go to the range and make two 100-yard groups. None of the others could be induced to shoot at this distance, in their estimation not requiring much skill, but Mr. Dougherty promised not to vary his ammunition or sighting while making the groups.

He made the two groups without reference to where they printed, according to agreement, and the originals were photographed as (Fig. 9).

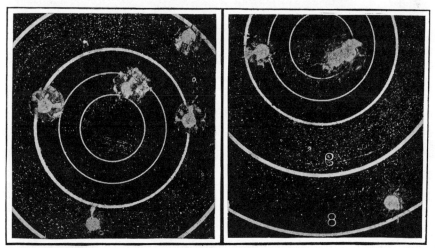

Fig. 9.

The smallest was 2.12, largest 2.25, average 2.18 inches. The shooting was done with his own .38–55, No. 4 full octagon Winchester rifle, muzzle and wedge rest, Chase-patch system, — the whole outfit was supposed to be the most accurate in this section at the time, and the weather was fine. Here again we have the constant error at the target exemplified.

My own attempts to obtain accurate shooting not being satisfactory, the above plan was adopted to see how matters stood with others, with above results. Although Dougherty, by previous agreement, made no excuses and had no opportunity to throw his targets away, it could be truthfully said that if he had shown his best 200-yard scores, those carefully preserved, he would be ranked first class.

It should be fully comprehended at this stage of experimenting that my rifle work absolutely discards fortunate scores or groups which are not due to the mechanical accuracy of the rifle and ammunition, attending only to such as fairly indicate it.

Auxiliary Chamber, .33-Caliber Rifle. During April, 1899, after a year's delay, the .33-caliber, Pope-cut special, was received. It had a chamber $1\frac{1}{2}$ inches longer than the .32 Ideal shell, and $\frac{1}{4}$ inch larger in diameter. Into this large chamber a steel shell or bushing was fitted and doweled; afterwards the barrel was reamed, rifled .33-caliber, and chambered for the .32 shell as though no auxiliary chamber existed. The extractor, acting upon a flange, threw out this steel chamber as a shell, after which it could be swabbed, oiled, and loaded, either as a breech or muzzle loader; thus the cross patch invented by Dr. S. A. Skinner of Hoosick Falls, N.Y., could be used, or the linen patch of bygone days.

By using a primed shell it would take a charge up to 55 grains, and bullet could be seated with or without air space, while a lubricated bullet could be firmly swaged into the grooves, thus leaving the point central and base at right angle to bore. The bullet then received the explosion in the swage which formed it and where it perfectly fitted, instead of being driven out by a punch and reëntered into another hole of, perhaps, a different size, by a bullet seater.

This device of an auxiliary chamber was planned after several years' experimenting in an attempt to send the bullet out central with the bore and point directly on, and to demonstrate or disprove a theory, long held, that the cause of inaccuracy in shooting is at the breech and must be corrected there.

After all the trouble and labor required to obtain a rifle with this auxiliary chamber and necessary tools, plenty of difficulties were encountered. As the bullet passed into the rifle proper from the rifled chamber, it threw thin lead foil into the joint between them, which required time and patience to remove. By exercising great care, however, a few shots were made, though very disappointing in not showing improved accuracy.

Then a grain of unburnt powder, or some other hard substance, was caught in the joint at front of the chamber, and upon closing the Ballard action, metal was forced or upset into the bore of rifle, making it necessary to ship the whole thing to Hartford, where Mr. Pope succeeded in working it out, though it gave him one of the problems of his life.

Only a few groups were made before the barrel was cut off and re-chambered especially for Dr. Skinner's red fiber-clad bullets. It seems that "Iron Ramrod," Reuben Harwood, some time before this designed an auxiliary chamber similar to this, but never attempted its construction, a fact unknown to me at the time, only being confessed on one of his last visits to my shop.

Test 16. — This was made May 2, 1899, with this auxiliary chambered barrel, and the result is seen in (Fig. 10). The smallest group was .87, largest 1.87, average 1.50 inches. This seems to be good shooting, so why should this auxiliary chamber be condemned? Simply because this is good shooting only by comparison. Swaging the bullet into the rifled chamber by an accurately made plunger must have caused it to lie straight in center of the bore, since the chamber was thoroughly cleaned and oiled between each shot. The third group, however, of the four made happens to be the average, as shown in accompanying cut — no better than from other and normal rifle chambers.

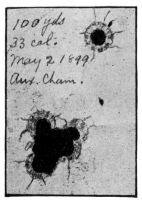

Fig. 10.

Test 16 a. — June 3, or four weeks later, a 6-shot group was made, swaging five of them very uniformly in the auxiliary chamber and the other one very much harder. It had been stated by riflemen of experience that it might be difficult to swage bullets uniformly hard in a contrivance of this kind, but (Fig. 11) of the original 6-shot group speaks for itself. The shots did seem to fly up and down, but hard swaging on the center shot does not indicate that it was due to irregular swaging, if its position in the group can be taken as proof.

Still something further is required for accuracy than entering a swaged bullet true in the bore before discharge.

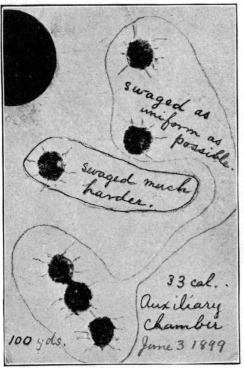

Fig. 11.

Shooting Braces. Of front rests for use at different shooting ranges, their variety indicates the conceptions of many minds in regard to what constitutes a practical article, and the illustration (Fig. 12) brings another into notice.

This "bob-sled" affair is designed to overcome its theoretical jumping and clattering which occurs in the usual form of brace as it slips backward on the rough shooting table at the time the bullet is traveling through the bore, before its exit from muzzle. If the marksman depends upon having the plank made true and smooth for the bottom of his front brace to slide upon, he may find himself against hard lines when shooting a match from some other club's shooting table.

Again, the dampness and sun may affect his own table, if not more fortunate than most shooters. Such conditions are often occurring with careful riflemen, as indicated by the multitude of measures employed to overcome them: some have made heavy, broad, flat bases to their front rest, hoping to overcome the trouble, but this being immovably attached to barrel will not seat itself upon the table only when elevation is in line with bottom of the base, a condition which, theoretically, will rarely occur; or if this broad bottom is made convex, such a device reverts back to the old style which presents the error already mentioned.

Fig. 12.

To those who are using these various devices and who desire accuracy, this " bob-sled " brace is suggested, and the illustration tells its own story to the glance of any rifleman. Attention may be directed, however, to the brass bushings contained in the ring, which make it fit a No. 4 barrel, while the clamping screws on either side will make the brace firm. Cardboard inserted will cause these bushings to fit a No. 3 or round barrel, or other brass bushings may be constructed. The V-notch in lower edge of ring is made to act in conjunction with the Pope machine rest, or may be used against a steel pin at muzzle and shoulder rest. The two pivoted pieces are about one and a half inches long, and owing to their freedom to revolve, will find an even bearing upon any plank, be it warped, rough, or smooth. These runners or pivoted pieces are split at one end, and the clamping screw takes up all looseness at the pivot, a small but practical necessity.

30 THE BULLET'S FLIGHT FROM POWDER TO TARGET

The exact amount of error introduced by a heavy unbalanced device attached to muzzle of a rifle barrel has never been tested, to the writer's knowledge.

Fig. 13.

As all riflemen know that the barrel recoils about one-tenth of an inch before bullet leaves the muzzle, they will readily recognize the desirability of a front

Fig. 14.

brace which will properly care for this sliding motion, which occurs at its maximum just as bullet is leaving the muzzle.

The late Horace Warner recommended that the arms of any front brace should invariably project from the body on a line that passes through the rifle bore. Combining this recognized principle with our "bob-sled" device will give us a balanced bob-sled front brace as here illustrated (Fig. 13).

After all our devising and experimenting to present an accurate and convenient front brace to riflemen, Dr. S. A. Skinner steps in with a simple block of wood, screwed to the rear end of the butt stock, see (Fig. 14), which applies to double-rest target work with wedge, and eliminates all our front brace business.

In his shooting with this butt brace the barrel simply rests in a wooden V placed six inches from its muzzle.

Pope-Ballard, 28-Caliber.

This .28-calibered barrel, cut by Pope of Hartford from a regular 22 Ballard, was received during September, 1899, being the first step taken for anything smaller than a .32. It was advised by Pope and never regretted by me. The use of a Pope flask throwing nitro priming and Hazzard's F. G. was also commenced at this time, gauged to fill the shell, leaving only space for oleo wad. *a decided cause for bad shooting.*

For many years, in fact from boyhood, it had been our ideal to find or construct a rifle that would hit and kill a crow at an attainable distance, but during this year that long-cherished object was abandoned. The size of groups made in the past, as also those made by other riflemen, did not encourage the anticipated fun of hitting crows whenever desired.

Perhaps not all riflemen are acquainted with the actual size of a crow's body when compared with its apparent size as that bird is seen in full winter feather. It surprised me, after a lucky shot had been made at short range with an old Queen's arm, heavily loaded with double-B shot, and after a close examination had been made, to discover whereabouts on its anatomy the killing shot struck. The question of crows was becoming a mooted one among agriculturists, some claiming they were beneficial to the farmer, while others retained the old grudge against their corn-pulling propensities. Anyhow, crows were losing their interest, since they could not be hit often enough to be exciting; besides, the 25-cent bounty hitherto paid for a dead crow had been discontinued. Henceforth attention was turned towards woodchucks, which, by common consent, were a nuisance; and this rodent offered a larger target without the feathers, therefore future efforts were directed towards a " gilt-edged " woodchuck gun.

By scores of tests, not tabulated here, it had been shown in a satisfactory manner which bullet and what loading would make the smallest groups at 100 yards, though contrary to the generally accepted theory and practice that a lubricated bullet must be as large or larger than groove diameter of the rifle. No explanation of these tests attracted attention among the constantly accumulating number of rifle friends for they were as positively convinced regarding the truth of their theories as the writer was skeptical.

Test 17. — This was made September 28, with this .28-caliber rifle, lubricated lead bullets, swaged cylindrical bore-diameter, no base band, front-seated, $\frac{1}{8}$-inch air space, Pope shell charged full and nitro-primed, Leopold's oleo wad, brass shell with wooden plunger for bullet seater, Pope machine rest, 12-pound trigger pull, dirty shooting. Four groups were made; smallest being .75, largest 1.75, average 1.09 inches. (Fig. 15) gives the fourth group, not so much to show accuracy of the rifle as to enable the reader to size up other groups by their listing in inches. This one measures .90 inch. Place a " nickel " on its center.

Fig. 15.

These small groups, however, can be attributed to that form of bullet which foregoing tests had indicated to be the best, and to a new rifle barrel. The commonly accepted theory that inaccurate shooting from machine rest is often due to a heavy trigger pull has never been accepted by the writer, only as one of the 64 reasons given by others for poor shooting in general.

Test 18. — This experiment was made October 13, 1899, in much the same manner as the preceding, except a Pope 1 to 30 muzzle-loading ball was used at breech, being considered by many as superior for breech loading. It measured .279 at its front end and .290 inch at base; smallest group made was 1, largest 2.37, average 1.81 inches.

Test 19. — This was made September 24, with same .28-caliber rifle, muzzle and shoulder rest, regular fixed ammunition which was specially made for this rifle by an expert .28-caliber man, his woodchuck load. One 10-shot group was made, measuring 3.25 inches, and is illustrated with others under the head of " A Woodchuck Experience," page 34.

THE BALLISTICS OF SMALL ARMS

A Woodchuck Experience. During September, 1901, a woodchuck hunt was undertaken with a well-known rifle expert. He criticised my .28-caliber rifle because it required so much time to put the bullet and shell in separately, claiming that one might not have time to load by this system while hunting, a rather thin excuse, because this front seating might take 10 seconds longer than with fixed ammunition.

Our hunt lasted 13 hours, during which he shot only five times, at five different chucks, each one sitting still, getting three out of the five tails, and there seemed to be plenty of time for front seating in this case. The two chucks missed were less than 75 yards away, and the hunter was positively known to be a fine shot of 30 years' experience. His rifle was a .28-caliber Pope with telescope sight, and with it he was induced, before the hunt, to make a group of five shots at 100 yards from a good shooting table and his fixed ammunition. The size of group made indicated clearly why he lost two chucks out of five.

Test 19 a. — A year later, on "Medicus" woodchuck preserve at Hoosick Falls, one of the best shots of Crank's Corner was persuaded to make a 5-shot group at 100 yards, double rest, with his .28-caliber Pope, fitted with a 12-power telescope, specially made fixed ammunition with two kinds of powder and nitro priming, cartridges being carried each in a separate aluminum case in his belt, his special woodchuck outfit.

The size of the group he made was three inches, and Plate (4) gives it as photographed from the original paper. His shots are numbered 6, 7, 8, 9, and 10, the black paster being the mark aimed at, and his rifle was intentionally sighted for less than 100 yards. This same ammunition was then tested in my own rifle, double rest, same cut barrel, using a new target. In plotting the shots, using the same black paster for mark, the black-faced prints, marked 1, 2, 3, 4, and 5, as will be seen, made a 4-inch group. Five shots were then made with my front-seated ammunition, aiming at the left-hand cross, and the group marked "group 3," inclosed in pencil lines, was made.

On the same plate is also shown the group of test 19 made by another expert rifleman's ammunition, from my .28-caliber, the right-hand cross

PLATE 4.

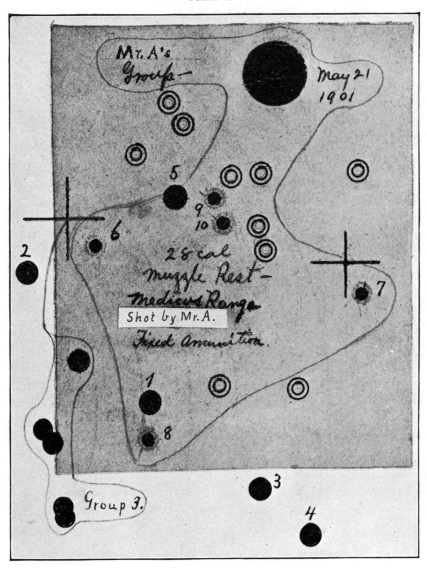

showing position of the paster, and ten double circle prints give the test 19 group.

As before mentioned, to front seat a bullet takes a few seconds longer than fixed ammunition, but in chuck hunting there is usually plenty of time for loading, the necessity for rapid manipulation only indicating that the first shot was too rapidly loaded to stop the chuck.

Looking at this kind of hunting as a rifle exercise rather than for the extermination of the animal, experts would not trust a shot over 75 yards with a rifle

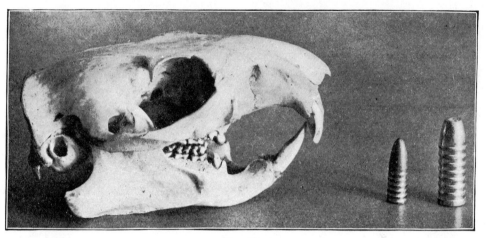

Fig. 16.

making a group shot of 3 inches at 100 yards, under the most favorable shooting conditions, and 50 yards should rarely be exceeded with ammunition of this character.

The usual target the marksman has offered to his skill when hunting this sly animal is well shown by (Fig. 16), which is the skull of a full-grown woodchuck, photographed to exact size. It was prepared by the late Dr. S. A. Skinner, of Hoosick Falls, N.Y., with great care, and is a perfect specimen. It is well known that the chuck's head very often offers the only target, and that it is usually found sitting at the mouth of its hole. If its tail is to be captured, the first shot must kill in its tracks, because a badly wounded one will crawl into its hole out of reach.

After this woodchuck experience some further tests were made with the .28-caliber, Pope-Ballard rifle, to show its buckle at different rests, also with various methods of loading, and experiments with some cylinder, flat-end bullets.

Test 20. — September 25, 1899, Homestead range, with the .28-caliber, breech load, 138-grain bullet, telescope sight and double rest; with front rest on solid plank, the group made was 2.87 inches higher than with front rest of a woolen cloth folded several thicknesses. Changing the rest of a rifle from one thing to another, or to no rest, demonstrates the buckle, as it is termed, and indicates one reason why shots print differently when the rest is changed from a fence rail to the side of a tree trunk or to arm's length. A rest must be chosen that is uniform.

Test 21. — August 25 and September 28, 1899, using same rifle loaded at muzzle with a Pope muzzle bullet through a false muzzle made by E. A. Leopold, a well-made affair. The smallest group was 1.62, largest 1.87, average 1.74 inches. Here again it will be noticed that larger groups resulted than from bore-diameter bullets.

Test 22. — On June 8, 1900, we experimented with the same rifle, 138-grain, lubricated, grooved, .280-inch diameter cylinder bullet; that is, the front end had same shape and size as base. One group of five was shot, making 4.50 inches, and the prints were as perfect as if made by a sharp steel cutter, corresponding to test 9, page 22, where the same form of bullet was used. Both this and test 9 were unusual and odd, making unexpected prints.

Test 23. — On the same day as last a like test was made, except the rifle bore was clean and dry, and lubricant was wiped from grooves on one side of this cylinder bullet. Only one shot was made, and it struck 8.50 inches from center of the foregoing group, keyholing badly, while a pointed bullet treated in the same manner, by actual test, shoots almost or quite as well as the normal close lubricated one. Seldom have such tests as 9, 22, and 23 been made or recorded, and they serve to open a question by presenting facts which will be explained later.

Test 24. — December 12, 1900, a test group was made with same rifle, regular charge of 23 grains powder and three grains nitro priming, front seated, muzzle and shoulder rest, telescope sight, very strong 9 o'clock wind varying to nothing, over a broad open plain. An 8-shot group was made, printing 1.30 inches perpendicularly and all the way from the inch paster to 5.50 inches horizontally, 3 o'clock, thus showing that some of the bullets were driven 5.50 inches or more by the wind.

Reflections. During two years' work with this re-bored .28-caliber rifle, bullets were tested with four different-shaped points besides those given in foregoing tests; a hemispherical, Leopold, ogival, and true cone-pointed. Bullets were tested with different alloys and different diameters, made in seven different steel swages; special tools were made for preparing fixed .28-caliber ammunition; and many other tests were made which space would not allow for recording: some made a better showing and others worse but enough are tabulated to enable the reader to form intelligent conclusions. All the experiments made with the .28-caliber rifle were for the purpose of perfecting a gilt-edged woodchuck affair, for which purpose Ballard set triggers had been added.

The most accurate bullets used in group shooting, starting from a clean bore, were without doubt the 138-grain, soft lead, lubricated and swaged lightly to a cylindrical, bore diameter, these making the finest groups of the 13 varieties tried; after 7 to 10 shots, however, with this bullet the black powder, even with nitro priming, was liable to cake a little in throat of the rifle, where the ball lay before discharge, so the woodchuck load was devised, the only one used with this rifle for that game, as follows: —

The bullet was swaged bore diameter to within $\frac{1}{16}$ inch of its base, which was left with a diameter of .290 inch; then it was entered point first into a $\frac{5}{8}$-inch piece of the rifle barrel which had been cut from the muzzle; when the base was flush with the die it was shaved with a sharp knife, thus cleaning it up, then pushing it entirely through, thus rifling the base band of bullet.

These bullets with rifled bases were easily entered at the muzzle of rifle with the fingers and sent down through the barrel to their place with a rod. In thus sending them through the bore any large collection of dirt was scraped down into the chamber, where it would do no harm — the Pope system adapted to hunting. It was the only black powder system used by the writer in chuck hunting, and in no single instance was a chuck lost because the rifle could not be loaded quick enough.

After this rifle, however, all woodchuck work was done with Laflin & Rand high-pressure smokeless, and the advent of this powder threw down all my rifle work of the past, including 6 Ballard actions, 17 rifle barrels, swages, and special tools which had been gathering for 12 years.

Before going further it may be well to explain some terms in common use by riflemen; though familiar to them, this may be read by some who are not so well up in rifle parlance. Most factory ammunition is made with bullet as large as, or a little larger than, the groove diameter of the rifle for which it is intended, that is, about .006 inch larger than bore diameter; thus made, it is supposed, to assure that bullets take the grooves sufficiently to shut off the escape of gas when discharged. The Chase-patch system used a swaged, ungrooved cylindrical bullet, smaller than the bore and made up to bore diameter by one thickness of thinnest patch paper, folded in at the base. This was one of the most accurate shooting bullets, but required water cleaning before each shot.

By "point on" is meant where the bullet does not tip in the least, and a tipper is one where the base does not quite follow the point in its flight. When the bullet tips very badly, it is termed keyholing, a perfect keyhole being made when the bullet strikes the target with its flat side.

The Personal Element vs. Mechanical Rifle Shooting. In writing an exclusive personal experience it is not easy to entirely eliminate the egoism of it; but the writer hopes, under this heading, to dispose of it to a large extent, and begs the reader's forbearance while doing so.

During all the years of my rifle experimenting, one to five shots was often a day's work; 10 to 15 was the usual number; rarely as high as 30 were made. It was my habit to lay out the experiment and prepare the ammunition before the day's shooting, sometimes requiring months. Before making a test group everything is made ready on the range, quiet days selected, flags set, etc. All original groups were carefully filed away and accurate copies of such will be seen here illustrated in part and tabulated in full.

I have no system to uphold or defend, my search being simply and solely after mechanical accuracy in rifle and ammunition, and thus far no system has been found too laborious to test that held out any promise of producing results or accurate groups; thus all the work might be termed mechanical rifle shooting, personal element being absolutely eliminated as far as possible. My whole work has been alone and carried out on my own initiative, regardless of systems or competition with others; it being an axiom with me that when competition enters scientific investigation exits.

During the two years' work at the Milford range, where competition ran high with the other members of the club, it had no influence upon me; my own shooting butt of 100 yards they had no use for. Under these methods it was soon discovered how difficult it is to find what one wants to know.

It is easy to sit about the fireside or under the shade of the home trees, after a day's work at competitive rifle practice, and talk over the causes of bad shots, and its good fellowship's pleasures are not to be denied; but it is not so easy to prove by repeated and, maybe, costly experiments that our fine theories are correct.

After 1899 another range was built on the homestead farm, 16 miles from Milford, for during that year the Milford rifle club was given up, disbanded, and the fine shooting house was presented to a blacksmith for a shop.

It was about this time I wrote Mr. E. A. Leopold, Norristown, Pa., and his first letter showed that he was no ordinary rifle student; correspondence with other riflemen amounted to very little because they seemed to follow entirely different lines of investigation, or none at all. Mr. Leopold was original in thought and methods of investigating in many respects, and our correspondence became quite voluminous. Looking over his letters recently, I was surprised to find there were 286, extending over about six years, which contained questions and problems relating to rifle experimenting which were worth preservation.

If the personal factor in rifle shooting can be wholly eliminated, as has been conscientiously attempted in these experiments and tests, the opportunity would then seem wide open for discovering defects in the rifle and its ammunition, or the cause of the x-error, the constant and varying deflections of the bullet from a bull's eye. If mechanical accuracy can be attained, the personal element required will be that of pure skill in handling the arm. If the obstructions to the accurate flight of a bullet are such that they cannot be scientifically overcome, are they not worth discovering, so the rifleman may approximate a distinction between these and his own skill?

If this aspect of the rifle shooting question is clear to the reader, I am confident that the years spent in the effort and the multiplicity of tests and experiments made, to discover the cause of the x-error, and what the rifle can mechanically be made to do, will be useful.

Target shooting and shooting from a fixed rest such as I have used in all these experiments are so different from each other that a word regarding the meaning of the term "mechanical rifle shooting," by way of explanation, is offered.

Target shooting by sighting with rifle held in the hands, butt against shoulder, or with rifle held on a rest, is universally a method the whole object of which is to hit a given point or target. This is rifle shooting in which the personal element enters to a great extent. Mechanical rifle shooting eliminates at once the personal factor as far as possible, for the rifle barrel rests upon a fixed base.

But the main distinction, which should be clearly understood, lies in the fact that in this mechanical shooting no attempt is made to hit anything like a mark, the object being to have the shots print on the paper wherever natural laws throw them, and to have the line of fire remain the same from shot to shot.

THE BALLISTICS OF SMALL ARMS 41

With few exceptions all experiments here recorded are those of strictly "mechanical rifle shooting" as understood by the definition the writer has endeavored to make clear.

Ross-Pope .32-Caliber, Second Hand. Returning again to the business of actual tests from apologies or explanations; during September, 1900, was received from Ross one of his No. 4 octagon, Pope muzzle-loading barrels which had been shot from 10,000 to 15,000 times. It was in perfect condition, never having been abused or suffered accident. It was purchased to be looked over, handled, and to experiment with the Pope system before ordering a new special Pope muzzle loader. *Was not perfect, was breech worn.*

Test 25. — An experiment was made September 20, 1900, with this second-hand barrel, from Pope machine rest, telescope sight, 200-grain 1 to 30 bullet, muzzle-loaded, 45 grains powder nitro-primed, making 15 shots; smallest group .69, largest 2.19, average 1.37 inches.

Test 26. — On the same day this test was conducted like the other except breech-loaded instead of muzzle, with bullets weighing 187 grains, swaged to bore diameter and front-seated, oleo wad; and the three 5-shot groups resulted as follows: smallest .56, largest 1.62, average 1.25 inches. One of the three groups, which happens to be the average size, 1.25 inches, is shown in (Fig. 17). The smallest, or .56-inch one, being less than half this size, would look quite small in comparison.

FIG. 17.

Test 27. — September 29, another experiment was conducted like the last, making three groups of five; smallest .81, largest 1.25, average 1.03 inches.

Test 28. — On the same day, one group was shot under same conditions as the first (test 25) with this barrel, muzzle loading; it measured 1.56 inches.

Test 29. — Fifteen consecutive shots were fired under same conditions as tests 25 and 28, muzzle loading, and on the same day; smallest group 1.81, largest 2, average 1.93 inches.

The above five tests do not properly represent the Pope system as operated with a new rifle, but do show how a second-hand one acts with the two bullets in question, and as a test between a breech load with bore-diameter bullet and the regular Pope muzzle loader. Having been one of Pope's disciples for some years, it may be taken for granted that the Pope system was well understood. These five tests also seem to show that the breech load is superior to muzzle with this particular barrel and bullet. *It apparently was not as no attempt was made, apparently, to discover best load, or trouble)*

Test 30. — October 15, ten consecutive shots were made without cleaning, under the same conditions as test 26, that is, breech-loaded; smallest group .87, largest 1.12, average 1 inch.

Test 31. — On the same day 20 consecutive shots were made, similar to tests 26 and 30; smallest group was 1.50, largest 1.75, average 1.56 inches. Comparison again shows the breech load with bore-diameter bullet to be best.

Test 32. — November 8, with Pope muzzle load, same as tests 25 and 28, two 5-shot groups were made; smallest 1.75, largest 1.87, average 1.81 inches.

Test 33. — On same day, with breech load, tests like 26, 30, and 31 were shot; smallest group 1.06, largest 1.37, average 1.25 inches, and again the bore-diameter bullet shows superiority at breech load over muzzle.

Test 34. — October 5, 1900, two 10-shot groups were made, one with muzzle load and the other breech-loaded, conducted same as the nine foregoing tests; size of group made with muzzle load was 1.62 inches; with breech load it was 1 inch.

Comparing these tests made with this old barrel, it will be seen that though a limited number were made, they were sufficient to demonstrate that while muzzle-loading, or Pope system, may keep up its accuracy for a large number of shots without cleaning, the cylindrical bullets, breech-loaded, are well adapted for some methods of hunting, and are even superior in accuracy for the first seven to ten shots without cleaning.

Testing Muzzle Blast. No doubt these compiled tests will interest those mechanically inclined more than competitive rifle shooters, where a large number of shots are required without cleaning; but the latter class are reminded that if mechanical accuracy can be obtained, competitive shooting will quickly follow in its wake.

Test 34 a. — This barrel seemed to be doing such good work it was subjected to several tests in an attempt to discover the power or influence of the muzzle blast upon flying bullets. An oak plank six inches wide was placed directly under the course of bullet at the muzzle and lying against bottom of the barrel as it lay in the machine rest, thus causing the bullet and muzzle blast to pass for 6 inches only half an inch from upper surface of plank; upon discharging, the bullet printed in its normal group; plank exerted no influence.

The plank was then raised and placed so the bullet and blast passed only one-fourth inch above it; again discharging, bullet printed in its normal place on the target. Several times the plank was thus raised until bullet flew across as near its surface as possible without touching, and prints still kept their normal position.

A trough was built six inches long, its two sides at right angles, and placed tightly against the muzzle, causing bullet to pass through only $\frac{1}{16}$ inch from the two surfaces, and still it printed a normal group at target, notwithstanding the fact that the heavy muzzle blast was closely confined for six inches by two adjacent sides which, one would suppose, should blow the bullet widely out of line. No matter how the plank was arranged at the muzzle, the blast seemed powerless to deflect the bullet and, in fact, the best groups were made when shooting along the angled bottom of this trough.

This indicates that general conclusions obtained among rifleman, " that the muzzle of rifle should be kept away from near obstructions when shooting," are erroneous, and that no error can come by having muzzle, when aiming, close to plank of the shooting table.

Test 35. — On November 8, 1900, with the same rifle which shoots smaller groups when its muzzle was penned in by two adjacent sides, W. E. Mann was interested to make further tests on the same line; so a square box was fitted and driven on over the muzzle end, thus retaining to a certain extent the blast on all sides. Then mandrels were carefully made and Babbitt metal poured over them and the barrel's muzzle; withdrawing the mandrel, a bell-shaped muzzle was formed, $4\frac{1}{2}$ inches long, from which all surplus metal was removed to make this bell muzzle as light as possible. The bullet, flying through this muzzle, surrounded by the blast, printed four inches higher, at 100 yards, than when shot without this extra $2\frac{1}{2}$ pounds added weight; otherwise it made normal groups. This higher elevation was simply due to less buckle in the rifle and not to any influence of the muzzle blast as disclosed by tests immediately following.

This barrel, without metal end added, at muzzle and shoulder rest, front brace two inches from muzzle, made a group $4\frac{1}{2}$ inches higher than when brace was placed 20 inches from it.

During September 2, 1901, with the same rifle at muzzle and shoulder rest, it printed four inches higher with front brace at extreme end of barrel than when placed 10 inches back. The groups, using bell muzzle of November 8, with front rest at extreme end, were 3.75 inches; with bell muzzle removed and front rest at extreme end, the group size was 2.25 inches, both printing in same place.

A 10-shot group was then made with bell muzzle and front rest 10 inches from it; size of group was 3.75 inches. Again removing bell muzzle, leaving front rest in same position and group size was 2.75 inches, printing four inches lower than the bell-muzzle group. Driving on the bell muzzle again, one shot was made which printed in center of group which had just printed four inches above normal, line of fire being kept always the same.

THE BALLISTICS OF SMALL ARMS

There is no reason for believing that the muzzle blast had anything to do towards making the group an inch larger, nor are there reasons for supposing that the Babbitt metal casting was rigid with the barrel, so movement between them may have caused the groups to grow an inch. This bell-shaped muzzle test was not of much value because the $2\frac{1}{2}$ pounds attached to the muzzle brought in other errors, the influence of which could not be fully ascertained.

Such tests as these and many others, which we were constantly devising, were not valueless. They caused our minds to dwell upon the bullet's flight. They entertained us and held our minds in a studying mood. We had not discovered for what we were searching and had no idea of the form of the things after which we were looking.

We prided ourselves that we were not over particular as to what form an experiment should take, provided its results were novel. Our desire was to keep our minds constantly alert by surprising experiments.

A "Shooting Gibraltar." During the fall of 1901 it was found advisable to construct a machine rest and shooting pier for more accurate tests and experiments, and several illustrations of the pier, V-rest, concentric actions, and range are shown to help out the following description. After it was finished Dr. S. A. Skinner christened it a "Shooting Gibraltar."

This pier, with its V-rest and 100-yard shooting range, was built in a growth of mostly pines, on the north side and foot of quite a hill, in fact running along the most sheltered strip upon the homestead farm, Norfolk, Mass., and conveniently near the commodious farmhouse with its roomy sheds and workshop. (Fig. 18) on the following page was reproduced from a photograph of the general location of range as seen when provided with its muslin covering, but before its cement pier was built; as will be noticed, a very sheltered place.

The pier, taking the place of wooden one shown in figure, was built of gravel and Portland cement, standing 26 inches above ground and extending below its surface 40 inches into a gravel bank that showed no signs of having been disturbed since the glacial period. A heavy iron bed, well ribbed underneath, was made to cover the top of this pier and, besides being laid in cement, was bolted with $\frac{5}{8}$-inch stay bolts extending 14 inches into the masonry. The iron brackets and V-rest were heavily bolted to this bedplate. Plate (5) on opposite page shows the pier, iron bed with its iron brackets, V-rest, and recoil block better than it can be described.

This heavy cast-iron V-rest was machined above and below and on its sides in a modern planer, and allows the concentric rings on the rifle barrel to find

Fig. 18.

an easy position, giving barrel perfect freedom to recoil backward against the block provided, or to rotate without changing its line of fire by a hair. No springs, stops, or weights are needed except the wooden block mentioned, loosely set behind to catch recoil.

The concentric rings on telescope and on barrels were turned to same diameter, 1.635 inches, bringing its line of sight covered by cross hairs in telescope and line of fire or center of rifle bore into one and same place at shooting stand and target, when barrel with its concentric rings replaces the telescope in rest. Each ring on the scope is retained by six large headless screws, rings being so

Plate 5.

adjusted that if the scope is revolved in V-rest the cross hairs always remain on the tack head in the 100-yard shooting butt; and concentric rifle rings are arranged to serve barrel in the same manner. For three years, during summer and winter, no movement of the line of this V-rest has been observed.

(Fig. 19) illustrates the manner in which the rest retains telescope or rifle barrel, showing the scope inclosed by its concentric rings lying within it, and by

Fig. 19.

its side lies the rifle barrel, also fitted to its concentric rings, ready to replace the scope in rest; tilted against the barrel is a concentric action which is screwed to chamber end of barrel for firing.

This is not an adjustable machine rest, but having line of fire and line of sight identical and uniform from month to month, always ready for immediate use, is a marked advantage. In short, the good qualities of this V-rest and concentric actions when mechanical rifle shooting is to be performed are self-evident. It takes the buckle all out of a barrel, for the heavier the charge the higher it prints, very different from the normal muzzle and shoulder rest. The rotation test from this rest indicates whether dealing with a well-straightened bore or otherwise.

Not to individualize the many other machine rests that have been devised and utilized to more or less extent, it is claimed for this that by removing the action and stock from rifle and substituting a concentric action, a long step has been taken towards a successful construction of a strictly mechanical rest.

As an illustration of action of the rifle barrel in the rest, which will be appreciated, may be mentioned the fact that a four-inch steel spirit level balanced across the top flat of a .32–40 barrel, when ready for discharge, is not disturbed or jarred when the barrel is discharged, recoils to the block, and rebounds again. The level rides on the top flat as if nothing had happened.

A description of this V-rest would not be complete without the following caution: The line of fire is supposed to be the continuation of a straight line drawn through center of bore of the arm to be tested. This definition is simple enough if bore of the rifle is straight from end to end. Rotation tests with this V-rest, however, indicate that the bore of a rifle is seldom straight, often

Fig. 20.

far from it; the line of fire from an ordinary arm, therefore, is not easy of definition. In a curved barrel this line might be a straight one drawn through the center of that part of the bore which imparts the final direction to projectile before leaving the barrel, therefore it is not certain that line of V and line of fire lie always in the same direction; and this fact has been a serious handicap all through different experiments attempted at this range.

The concentric actions so intimately associated with this Shooting Gibraltar and its V-rest are too important to drop without further notice and illustration. (Fig. 20) shows three concentric actions on the right, with hook for drawing

50   THE BULLET'S FLIGHT FROM POWDER TO TARGET

bolt and wire which acts as a trigger. This wire trigger acts in open slot, shown at top of two actions, which are represented cocked ready for firing; giving the bolt a slight turn, but before it reaches the primer, the trigger is separated from it. This whole action is unscrewed from barrel to insert charge, and no appreciable time is lost in the operation. Several bushings are also seen in this cut, which adapt the actions to the threads of various makes of barrels. These actions have been used five or more years without breakage or accident.

Fig. 21.

This description of our ideal regarding a range which did accommodate itself to solving the questions self-proposed cannot be better closed than by the presentation of a photographic illustration (Fig. 21), showing its location and appearance, with its muslin cover, as seen from north side of head of the millpond which lay between it and the homestead farmhouse.

Bullet Press or Nutcracker.   The universal method employed by riflemen to swage their bullets with a hammer did not seem accurate enough to meet demands of the V-rest, concentric action shooting at the range, and something must be devised. This necessity, like many others which kept presenting themselves, was finally met as shown in (Fig. 22) on opposite page.

The press is but 14 inches high, weighing only 20 pounds, yet is capable of exerting a pressure of one and a quarter tons and, as may be seen, will take any kind of tool in place of the plunger. The lever arm may be set at any position to exert any desired pressure, from an ounce to a ton or more, by one's hands or by weights, and under cap screws seen at base any mold or device desired may be fastened. It furnishes power to swage bullets of all calibers and do it better than any device known, as the plunger moves in a straight line with a slow motion of great power.

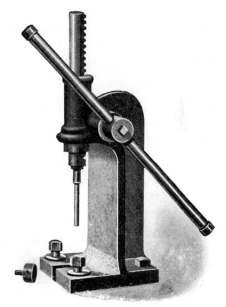

Mr. H. M. Pope, of Hartford, among other things says: "I doubt if many people realize the difficulty of swaging bullets in the old way, with any approach to uniformity. Any form of blow tends to deform the bullet after it is swaged, and in the old way, with a hammer, the bullet must be driven out after it is formed. This will, to a dead certainty, deform the same, and it is not possible to strike so uniform a blow as to have the deformation a constant quantity, therefore the bullets vary in size and shape. Again, it is difficult to strike uniformly enough so bullets do not vary somewhat in hardness. In this press this condition is much changed for the better, it being possible to get very nearly the same pressure in swaging, and instead of the bullets being driven out of the swage, they are pushed out without deforming the same, and the result is much more uniform work than is possible in the old way. To one who uses swaged bullets, it is simply invaluable."

It will take the place of a bench vise for many purposes, and is a first-class nutcracker, a thing most riflemen need in more ways than one. It recaps shells with absolute uniformity, and proves itself extremely useful in many kinds of experimental work. For example, suppose it is desired to swage a lot of bullets with a pressure of 960 pounds; as the plunger pressure is 32 times the pressure applied to lever, hook a spring scale on the lever and pull 30 pounds, or a 30-pound weight may be attached. In this manner the exact amount of pressure wanted may be uniformily obtained. It proved to be one of the handiest tools about the premises, being utilized in all kinds of ammunition and rifle experiments.

Testing Bullets. Snow Shooting. As might be imagined, snow shooting, or gathering unmutilated bullets that have been shot into light, soft snow, is cold, laborious work; if done at all, it must be when snow is plentiful and dry or only slightly moist. A day must be selected when snow fills the above conditions, ammunition perfected for required tests, and if group shooting is to be done a quiet day is necessary. Days that fill all these conditions during a New England winter are rare indeed.

Such days were found, however, during the winters of 1900–1 and 1901–2, and during the latter season 43 different tests were made, mostly at 100 yards, though the soft lead bullets of the .32–40 charge must be recovered at 200 yards. Many of the tests were made on the lake which lay alongside the range when ice and snow were in suitable condition, and five and a half pounds of bullets were secured, 206 in number, from the snow.

Our brains were not left behind while doing this work from day to day, and if too cold out on the range with its snow butt, or upon the lake, it was not so in brother's upstairs den, which was not far away, and from which most of the shots were made and all records kept.

The bullets to be tested were stamped on their points and side of points with steel die figures, thus enabling a correct record to be kept of the great variety of experiments made with different rifle bores, ammunition, and methods of loading, some of which are interesting. Tests were made with the .28–30, .32–47 and .33–47 calibers, with fixed ammunition, front seating, Pope's muzzle-load system, and front ignition system; bullets of different alloys of tin and lead, some smaller than bore diameter, standard size; bullets

THE BALLISTICS OF SMALL ARMS 53

with cylindrical bodies, some with proud base edges and others with slightly beveled base edges; brass base bullets and some with Babbitt metal bases, with many others.

Soft lead bullets were usually mutilated, more or less, at 100 yards, and as these were the most important for furnishing the desired information, a bank of snow was made at 200 yards to receive them. As these bullets were picked from the snow, one by one, the records obtained, though seemingly slight, were important, and as experiments progressed they become more and more intelligible though better understood a few years later.

There was no bullet tested that did not take the rifle grooves sufficiently for all purposes of rotation, and none were gas-cut beyond the first base band, though some which were tested were .005 inch smaller than bore diameter, that is .012 inch smaller than standard factory made, and these were not gas-cut in the least when a wad of any kind was used. After a few shots, front seating without cleaning, grooves were often found the whole length of recovered bullets that at first were mistaken for gas-cutting. They looked like an extra land mark, but it was soon discovered that they were caused by small spots of powder residue left in throat of the rifle, and marking showed that this residue did not collect in throat in uniform positions. By touching throat with a damp swab before loading these markings always disappeared.

The above information was obtained incidentally, but is not why the tests were made. The real object of these varieties of loadings and bullets was to determine if base of the bullet was oblique to the bore at time of its exit from muzzle. All bullets before loading had bases at right angles to their cylinder sides, made so by proper swages, except those of the Pope system; these, being tapering in their body and unswaged, could not well be tested in this respect. Since Mr. Pope makes his own bullet molds, however, it may be assumed that bases were square; he does his work that way. These bullets when recovered from the snow (Pope's) however, were found to be cylindrical, made so by the rifle's bore, with no sign of their former taper, and their bases could be tested for obliquity with certainty. Of the 206 bullets recovered in good shape, 122 were so perfect they could be put in a lathe and bases tested.

The tools employed in testing these recovered bullets are here illustrated in the foreground of the pile of recovered bullets. Figure 1 is the chuck or bushing that fits the lathe head center, containing the .32-caliber bushing.

Fig. 23.

Figures 2 and 3 are the .28 and .33 caliber bushings, a new bushing being made for each set of tests and not removed from lathe center until tests are completed. The lathe tool, represented by figures 4 4, has a square cutting edge $\frac{1}{32}$ inch wide and is brought up to the revolving bullet, cutting a channel $\frac{1}{64}$ inch from edge of its base, and channel is deepened until the tool begins to touch the short side of base.

This (Fig. 24) shows a simple device for measuring depth of the channel in

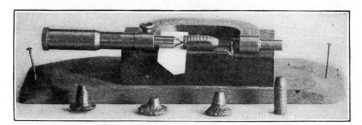

Fig. 24.

high side of the bullet's base. The paper under base of the bullet acts as a mirror, and measuring is done by a sharp-pointed micrometer, the meter swinging on a knife's edge at its other end while using a magnifying glass to follow

THE BALLISTICS OF SMALL ARMS 55

its point. Two persons are required to operate this device, one watching the graduations, while the other observes the micrometer point and swings it over the high side of base into bottom of the channel which represents the low side of base, thus giving a direct measure of its obliquity.

Measurements of obliquity of the 122 perfect bullets, in decimals of an inch, are as follows: Two bullets were 0; 5 were .00025; 6 were .0005; 12 were .001; 14 were .002; 22 were .003; 4 were .0035; 11 were .004; 10 were .0045; 4 were

FIG. 25.

.005; 3 were .0055; 8 were .006; 3 were .007; 5 were .008; 1 was .0085; 1 was .009; 1 was .010; 1 was .011; 3 were .0115; 1 was .012; 1 was .013; 2 were .014.

Twenty per cent of these bullets, one out of five or two out of 10, have their bases over .006 inch oblique, and this is the usual proportion of off shots in regular careful target practice. This one off print in five-group shooting has been markedly noticed in all lead bullet work of the writer, and two such shots in 10 have been remarked by a great many target men all over the country.

(Fig. 25) here shown exhibits, photographically, 39 .32-caliber bullets, se-

56   THE BULLET'S FLIGHT FROM POWDER TO TARGET

lected from 206 which were caught in the snow, each one being numbered and its condition tabulated. As will be noticed by mark of the lathe tool, all in the pile were placed in special chuck and their bases tested for obliquity. Of all the 122 bases thus tested only two were found with bases square with body. The fourth row from bottom is interesting because they were shot from front ignition; that is, the primer charge was carried through powder by means of a brass tube to front of the shell, thus igniting powder in front. It will be noticed that, apparently, the primer charge through the brass tube blew a hole into base of the bullets; how this was done has not been explained as the primer,

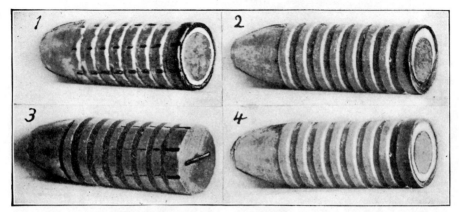

Fig. 26.

in itself, has no power to do this under any condition in which it can be placed. Precautions were taken to keep all powder free of the tube by sealing it with oleo wad prior to filling the shell with powder.

A glance indicates which bullets were shot with Leopold's oleo wad, while the bases which show markings from powder grains were fired without wad.

In (Fig. 26) figure 1 shows one of the two bullets mentioned above, in which the base was not made oblique during its passage through rifle bore. On base of this bullet, as also on figures 2 and 4, is seen the groove of the lathe tool, also shown in pile represented in previous cut. Figures 2, 3, and 4 have not been shot, but were greased and swaged. Figure 2 has a base .0015 inch oblique, owing to an imperfect swage, while figure 4, from same swage, using a new and properly made base plunger, has a square base. Figure 3 shows a lathe-made unshot brass base bullet, described elsewhere.

This snow shooting demonstrated that a bullet which enters the breech of rifle with a square base comes out the muzzle with base oblique 120 times out of 122, and the two which were not oblique happened to be those with Babbitt metal bases and spike. Two of the five that were only .00025 oblique, and the one .0005 oblique were also Babbitt metal bases, the five being all the Babbitt bases shot.

Working backwards and using bullets with bases swaged oblique before discharge, in a close-shooting rifle, tests were made; one of which is given in test 133, group 2, page 214, where the .012-inch oblique-swaged base was entered to come out up at muzzle and made a 10-shot group 1¾ inches from center of a normal group in left-hand twist barrel, printing at 4 o'clock.

June 11, 1903, two normal groups were shot with the Pope "Bumblebee" rifle, and size of groups were .87 and .75 inch respectively. Then one group was made with same rifle, of four shots, using bullets swaged with .006-inch oblique bases, entered to come from muzzle on the quarters, and its size was 1.75 inches. Another 4-shot group with .012-inch oblique bases, also entered on the quarters, gave 2.25 inches.

Turning further on, to test 71, page 110, with another Pope rifle, where bases .012 inch oblique, all entered to come out of muzzle up, made a 10-shot group 1⅜ inches from normal. If the bullets in that test had been entered on the quarter, size of the group would have been doubled, giving 2.75 inches, same as above with Pope 1902 rifle. This is not a selected experiment, being the only oblique base bullet test with Pope 1902 rifle, and test 71 was the only one made with the new muzzle-loading Pope vented barrel, these being the only two fine shooting barrels in which these tests were made.

It is not strange that the rifle used in oblique test 71 should show that oblique base bullets make about the same size groups as normal bullets, only in different places, because with a gilt-edged rifle, bullets with .012-inch oblique bases show their error in a distinct manner, while an error of this magnitude could hardly be surmised with a rifle whose normal group measures 1.75 inches or over.

Mr. Bashforth, of England, in his scientific work, proves that the powder blast acts on base of the bullet with sufficient force to constantly increase its velocity for 25-caliber lengths, or eight inches, from the muzzle. If the bullet with its oblique base up at the muzzle did not rotate the powder blast would, by its glancing motion, force it at 6 o'clock on the target during this eight inches flight from muzzle.

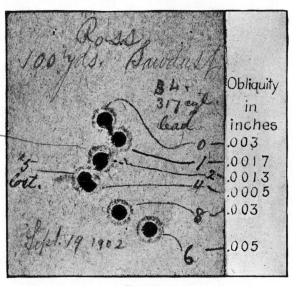

FIG. 27.

The bullet rotating towards the left, in the Pope-cut rifle, causes an oblique base to change its position to the blast, and at seven inches from muzzle its position is reversed; the blast being greatest at muzzle and rapidly diminishing the result of its glancing power seems to be at about 4 o'clock on the target, or when bullet is $2\frac{1}{4}$ inches from the muzzle.

To make a more rigid test of action of the muzzle blast upon an oblique base bullet, a long-desired experiment was successfully made with the Ross-Pope rifle, September 19, 1902, in presence of W. E. Mann and Mr. Leopold. A 7-shot group was made from V-rest, using carefully prepared bullets whose bases were made square; and six of them were secured in oiled sawdust in good condition, the fifth one not being found.

The group size, as will be observed by (Fig. 27), was 1.25 inches. Each bullet was numbered on its point and print of each shot numbered to correspond. The lathe was fitted, channels cut in bases, and with Mr. Leopold's assistance

THE BALLISTICS OF SMALL ARMS

the obliquity was measured, as given in figures at right in cut. The obliquity of bases of the six bullets was in close proportion to the distance of their several prints from center of group. This was the only successfully carried out test of the kind made to this time, and this remarkable proportion of obliquity to the error can hardly be pronounced a coincidence.

Recovering Bullets from Oiled Sawdust. Snow could not be used in warm weather for catching bullets and search was made for some material to act in its place, something practical for warm weather work. Inquiry by correspondence availed nothing; meal, bran, and shorts were tried and found wanting, while soft cotton mutilated bullets worse than a pine board. One day Mr. Leopold said, "Let us try some sawdust," having some in his hand at the time. A box was soon arranged and filled with sawdust that had been used in the hardening room of the shop for drying oily knives, and immediately tested; to our surprise the bullets were recovered without serious mutilation. Not another hour was lost in preparing a box of fresh sawdust with sufficient quantity of thin oil to allow desired experiments to go on.

Dry, wet, and oiled sawdust were severally tested (Fig. 28), showing three .32-caliber, 187-grain Zischang soft lead bullets from a powder charge of 47 grains; the right-hand one caught in dry, the center one in wet and the left-hand one in oiled sawdust. It will be observed that the one from oiled sawdust is perfect as can be detected with eye or magnifying glass, and

FIG. 28.

hundreds have been caught in same manner to perfection, though the reason why oiled sawdust so acts is not easily explained.

The correct method of preparing is to sift through a No. 12 mesh sieve fine maple or birch sawdust and mix with thin machine oil which will not gum. The dust takes much oil, and sufficient must be added to thoroughly saturate without dripping; then resift through a No. 6 or 4 mesh, and if properly prepared it will not cake but will fall back into the furrow made by a bullet, as coarse dry sand will act.

At 100 yards the .32-40 bullet often penetrates this sawdust four or five feet, the distance varying with its speed, weight, and hardness, and whether it keeps point on or travels end over end. To determine many things respecting rifle bullets it is essential that some means should be utilized to recover them unmutilated at point and base after being shot, and this was a lucky find. (Fig. 29) is from a photograph of experimenters in the act of picking bullets from a box of oiled sawdust at shooting butt of homestead range.

FIG. 29.

Short-barrel Shooting. About the time tests were being made by shooting bullets into snow the question was raised regarding what place in the rifle the bullet upsets, and several schemes were devised to determine this, the most promising being to utilize barrels of different lengths, from one with a bore only $\frac{1}{4}$ inch long from chamber to muzzle, up to five or six inches or more in length. So far as could be learned, this question regarding upsetting of the bullet had never been determined or even mentioned except as a mere supposition.

In all these recorded experiments, extending over nearly 40 years, it will be observed they have generally followed new lines, never repeating the experiments of others which have been logically carried out. It has been rather

Fig. 30.

experimenting by exclusion, and this, as will be readily recognized, entails much original labor and no little groping in the dark, so some of these short-barrel experiments were both laborious and amusing, also interesting.

During 1902 pieces were cut from a new .32-caliber Pope barrel, unfinished outside, $\frac{3}{16}$, 1, $3\frac{5}{8}$, and $5\frac{3}{4}$ inches, besides proper lengths left in each for a chamber. Pieces were also cut in the same manner from a new .32-caliber smooth-bore barrel, of $\frac{5}{16}$, $\frac{5}{8}$, and $2\frac{5}{8}$ inches, and from a Winchester .38-caliber,

No. 4 barrel, $\frac{1}{4}$, $\frac{5}{8}$, and 10 inches, while there was found on hand a new piece of a .40-caliber Stevens, 3 inches long, thus making an outfit of 11 short barrels.

(Fig. 30) on page 61 shows all but one of these barrels, the right-hand figure being the concentric action and wire trigger for firing, into which the respective barrels are screwed.

Figure a shows the unloaded shell, $2\frac{1}{8}$ inches long, and so placed that the several lengths of barrels may be compared, as may also be done by bullets in foreground. Figures $n\ n$ show nuts for connecting one short barrel to another.

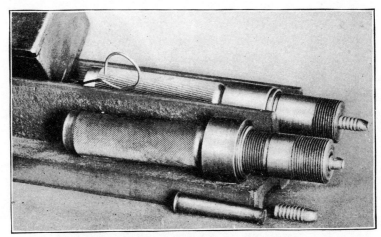

Fig. 31.

The above cut exhibits a short barrel when placed in its concentric action and lying in V-rest, recoil block behind, ready for firing. Another barrel attached to another concentric action is also seen ready for firing when transferred to the V-rest; in the foreground are the .32–40 shell and bullet. The barrel with its concentric action occupying V-rest is $\frac{5}{16}$ inch long, the bullet issuing from muzzle being seated nearly to the shell in its chamber, and projecting from end of muzzle $\frac{3}{4}$ inch.

After turning all these barrels on new arbors to fit an inch reamed cavity in an iron tube, six of the .32-caliber and two of the .38-caliber barrels were chambered for regular .32-40 and .38-55 shells. This iron tube was for the purpose of making one barrel out of two or more by holding their ends together in a true line within; thus a $\frac{5}{16}$-inch barrel could be made into $1\frac{5}{16}$ inches by addition of the 1-inch barrel, and other innumerable ones by different combinations.

That the results of an experiment do not always coincide with a theory, was demonstrated with a vengeance during the first few that were tried, to make two of these short barrels into one within the iron tube. The barrels were cemented in tube with a hard black wax, used to hold arbors from turning in a rifled barrel instead of driving them, which would injure the rifling. In the second shot from this piecemeal barrel the explosion sent the forward barrel with bullet in one direction, and the other with its attached action in another. An attempt to secure these barrels within the iron tube by set screws proved a failure, the explosion was so powerful that if any gas leaked into cavity of the tube the ends of these $\frac{3}{8}$-inch set screws were sheared off. It being evening and in the basement of a dark shop when this first experiment was made, the explosion furnished an entertainment and quite an element of danger which put Fourth of July into the shade.

Proper security was finally obtained by cutting threads on the ends of each barrel and joining by steel nuts (shown by figures $n\ n$ of cut, page 61); and by venting joints with holes bored through the nuts, safety was assured.

The first real experiment was made with $\frac{1}{4}$-inch barrel laid in V-rest; placing a box of snow, with the thin cardboard which formed one end, in front of and 24 inches from muzzle, the first shot was made January 15, 1902. The bullet printed in center of target, point on, making a clean half-inch hole and was recovered from snow completely flattened by the muzzle blast, thus accounting for large print through the cardboard. It is hardly possible for those not present during these tests to appreciate the condition of affairs which are here indicated and recorded for the first time.

Experiments were made with these short barrels, extending over a year, with different lengths, different combinations, with various alloyed bullets, with nitro powders both coarse and fine grained. Eighty-nine 187-grain bullets

This upset be battering up of short bbl bullets suggest that there may be some such action with revolvers & pistols Try it some time.

64   THE BULLET'S FLIGHT FROM POWDER TO TARGET

were secured in the same shape as they left the different muzzles. It was noisy sport and necessitated stuffing cotton into one's ears for protection. This medley of tests was carried on at the range with concentric action and V-rest and multiplied out of curiosity. Perhaps the accompanying cut (Fig. 32) will be of interest as it represents the 187-grain unshot Zischang bullets and a number fired from short barrels.

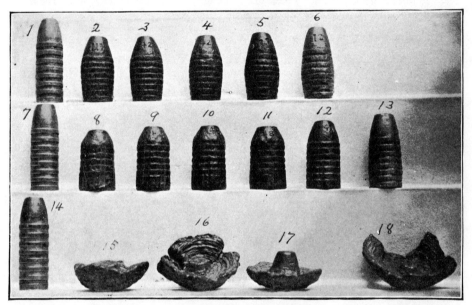

Fig. 32.

Figures 1, 7, and 14 are unshot bullets; 2, 3, 4, 5, and 6 are bullets fired from one short barrel into the muzzle of another, entering into and emerging from the .32–40 chamber of the second barrel. Figures 8, 9, 10, 11, 12, and 13 were fired from a $\tfrac{5}{8}$-inch .32-caliber barrel through a 3-inch .40-caliber which was firmly screwed to the former; figures 15, 16, 17, and 18 were remarkably upset with 21 grains sharpshooter powder, with no air space and .32–40 shell. All these bullets were loose-fitting or bore diameter.

With figures 2 and 8 the charge was front ignited; figures 6 and 13 were

*2 – in double chambered 32/40.
8 – 5/8 – 32 into 3" 40 cal rifled*

an alloy of 1 to 30, tin and lead; figures 15, 16, and 18 were shot with 21 grains powder from a .32–40 shell, bullet entered half an inch into shell, leaving no air space, and extending .06 inch out of muzzle. The apparent tears seen in bullets were produced by a sharp steel prong, placed at varying distances from muzzle and also varying distances from line of low side of the rifle bore.

This prong was accurately adjusted by a screw and by it was determined at what position from muzzle many of these bullets commenced to upset, and at what position they completed their upsetting. It was found that many of them commenced to swell slightly at .06 inch from muzzle, and all that were tested received their full upset during first inch of flight of their bases from muzzle. Figures 15, 16, and 18 pretty well illustrate W. E. Mann's putty-plug theory.

Figure 17 was fired from a .62-inch barrel, .32–40 shell full of Du Pont's .30-caliber high-pressure powder, leaving no air space, body of bullet extending .06 inch from muzzle and no prong interposed to obstruct.

It will be noticed that lead bullets, figures 2 to 6, and even the 1 to 30, after being driven through a 3.12 inch, .32-caliber bore, were expanded into the .32–40 chamber, completely filling it. These bullets continued to enlarge as they passed into the larger portion of the still enlarging chamber until their diameter reached .386 inch for lead, and .379 for the 1 to 30, showing that they filled the chamber for two-thirds its length where its diameter is .388 inch; all this with a normal charge behind a loose fitting bullet.

It will also be observed that front ignition, as in figures 2 and 8, does not do away with upset, as has generally been supposed by some well-known riflemen. Shots represented by figures 8 to 13 inclusive, being bore diameter, not only upset into the .32-caliber rifling, but after traveling 3.12 inches they again upset and, with the lead bullets, completely filled a .40-caliber to bottom and corners of the grooves. The putty-plug theory is here again confirmed, but does not encourage the theory that bullets must be larger than bore in order to take its grooves.

Before snow-shooting and oblique-base experiments were made, or short-barrel tests attempted, brother William claimed that a lead bullet in a rifle bore was like shooting a plug of putty from a popgun. A little later, while working on the range, this idea was again brought up when he asserted, still more emphatically, and seemed to show how soft bullets would stick in the bore, being pushed out sidewise, either side first as might happen, by powder blast behind, like pushing a plug of putty from a popgun, and it was decided to thoroughly test the matter.

Measurements taken from bullet bases which were recovered during snow-shooting experiments, seemed to bear out W. E. Mann's theory, and short-barrel tests showed up his theory so thoroughly that they were multiplied until it was substantiated.

Short Barrels become Interesting. These short barrels opened a large field for investigation and experimental purposes, and about every conceivable test was instituted that could be imagined, or that suggested the slightest bearing upon upsetting of the bullet. Plate (6), on opposite page, will help to form some idea of what was attempted, by showing a part of the .32-caliber bullets, fired from various short barrels and recovered in oiled sawdust.

When examining these bullets, which present such a variety of mutilations, bear in mind they were not mutilated by striking any target, but that the oiled sawdust gives them up in practically the same shape it receives them.

The five figures at left of Plate (6) represent eight grooved, 187-grain, unshot Zischang bullets, made bore diameter, used in these experiments; the top figure is the .32–40 shell which was usually filled with Hazzard's F. G. powder, nitro-primed, for discharging the various bullets represented below it; five figures occupying first row, immediately under the shell, are unshot bullets which had been subjected to various pressures, 250, 416, 520, 930, and 1300 pounds, as marked on plate, by placing between flat-surfaced steel plates. They are interesting only as showing action of different weights upon the cohesive resistance of .32-caliber lead bullets.

All other figures in plate represent mutilated bullets that were discharged from barrels of various lengths, passed through a cardboard target into oiled sawdust which was placed 30 inches from muzzle; all were numbered on their

PLATE 6.

points, thus enabling a careful record to be kept, and each had its influence towards determining an answer to the problem in question.

Plate (7), showing mutilated targets, is interesting only as an exhibit of a few prints made by short-barrel bullets through pasteboard placed 30 inches from muzzle. These peculiar prints should be compared with the bullets which made them, as shown on Plate (6); the occasional number placed over mutilations

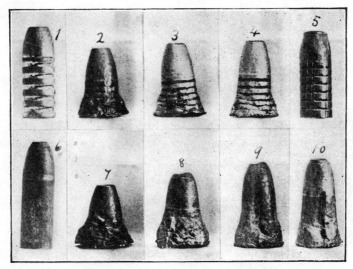

FIG. 33.

there shown correspond with its print on Plate 7, and recognition will be easy. In lower left-hand corner of Plate (7) is a print made by one of the short barrels passing through.

These various and unusual illustrations are introduced not only to assist an attempt at describing curious phenomena, but to aid the reader in understanding what the writer is endeavoring to make plain, and (Fig. 33) well illustrates the behavior of muzzle blast upon the .38-55 bullets when shot from barrels of different lengths.

Figure 1 is a .38-caliber, grooved, unshot bullet; 6 an unshot, .38-caliber, Chase-patch one; figure 2 was fired from barrel only $\frac{5}{16}$ inch long; figure 3 from

PLATE 7.

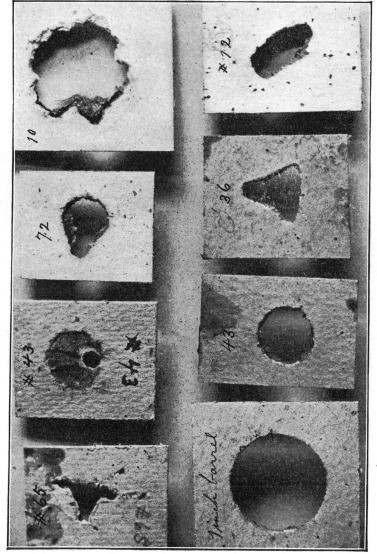

one ⅝ inch long; 4 from one 2⅝ inches long; 5 from an 11-inch barrel. Figure 7, an ungrooved bullet shot from $\tfrac{5}{16}$-inch barrel; 8 from ⅝-inch barrel; 9 from 2⅝-inch barrel, as was also figure 10.

It was discovered that the shortest barrel which could be used, to leave base of the bullet its normal size after shooting, was between 10 and 11 inches, as illustrated by figure 5, which was shot from 11-inch barrel. These eight

FIG. 34.

mutilated bullets (Fig. 33) were fired from a charge of 55 grains black powder, and represent short barrel upset with the standard .38–55 ammunition and chamber, front-seated, bore-diameter bullets.

Turning now to the action of high-pressure, smokeless powder, in low-pressure rifles, the above cut (Fig. 34) well represents it, where the .32 caliber bullets are magnified one and three-quarters times.

Figure 1 represents bullet before shooting; figure 2 was shot from ⅝-inch barrel, .32–40 shell and 12 grains sharpshooter powder; figure 3 from same barrel with 15 grains powder; figure 4 the same with 17½ grains powder; figure 5 the same, but with shell filled with Du Pont's .30–40 powder, presenting end view of the recovered bullet; figure 6 shows reverse side of a bullet shot from same charge as the 5.

It was an interesting discovery that bullet represented by figure 2, which has same velocity with sharpshooter powder that it has from .30–40 shell filled with black powder, has the same upset in same length of barrel, while figure 3 has greater speed and greater pressure than from any amount of black powder that could be placed in a .32–40 shell. When bullet, represented in figure 5, was fired from $\frac{5}{8}$-inch barrel, with the shell full of Du Pont's government powder, instead of a deafening report as from all other charges, it gave a sharp snap only slightly increased over the snap of a primer alone.

Another incident occurred while making tests in connection with those above which was a little surprising: If a bore-diameter bullet was placed in muzzle of the $\frac{1}{4}$-inch barrel, and shell filled with Du Pont's powder behind it, the powder would not explode, while the bullet, being propelled by primer, simply dropped from end of the barrel, the burning powder following behind but leaving considerable powder in the shell, partially melted and cemented lightly to it. A nitro primer in an empty shell propelled this same bullet to the 21-inch strawboard target and nearly perforated it. With a $\frac{5}{8}$-inch barrel, only $\frac{3}{8}$ inch longer, the result as shown by figure 5 was obtained, by same charge and by seating bore-diameter bullet in precisely same manner.

These interesting phenomena, along with other questions, was submitted to H. Pettit, of Belmont, Ontario, for solution, and a part of the correspondence resulting will be found a few pages further on.

Not to be wearisome, but because it is believed this multiplied variety of mutilated and upset bullets will be of some interest, (Fig. 35) on the following page is introduced, enlarged to show markings and mutilations from longest to shortest barrels in regular order, magnified one and three-quarters times.

Figure 1 represents an unshot .32-caliber bullet, and remaining six were swaged to bore diameter, shot from usual charge but from various lengths barrels, as marked in inches on their several points; the amount of enlargement of base in upsetting is indicated below each figure in thousandths of an inch. In figure

2 the bullet as recovered from sawdust has same appearance as if fired from a normal barrel, although shot from one only 11 inches long; figure 4, shot from an 8¼-inch barrel, takes on quite a different appearance.

A comparison between upset of figure 6, shot from 45 grains black powder, and figure 2 as shown in preceding cut shot with 12 grains sharpshooter powder, should be made, because upset was practically the same in both instances, and trajectory would have been the same if shot from full-length barrels.

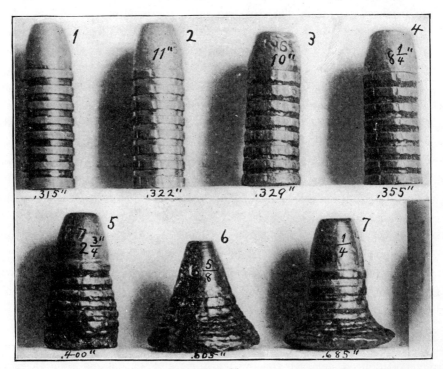

Fig. 35.

Short-barrel Experiments Continued. Not satisfied with ordinary experiments with upsetting bullets and short barrels, unusual experiments were tried, and several were instituted by shooting two bullets from same barrel at the same time, with varying air spaces between; and the magnified view from curiously mutilated bullets, shown on Plate (8), will give some idea of the fun obtained.

PLATE 8.

The figure in left-hand upper corner represents two bullets swaged together, resulting from being fired together from an 11-inch barrel, one bullet being seated in its ordinary place at chamber, while base of the other was extended $\frac{1}{4}$ inch into the muzzle, leaving an air space between the two of $8\frac{3}{4}$ inches.

The figure in right-hand upper corner represents two bullets swaged together as in the other case, except base of the front bullet was entered two inches into the muzzle of 11-inch barrel, leaving an air space between the two of seven inches. The two figures in lower left-hand corner show reverse side of the one in right-hand upper corner after the chamber-seated bullet had fallen away from one entered at muzzle.

The right-hand lower figure is the reverse of the left upper figure.

This No. 4 octagon barrel, 11-inch, from which these bullets were fired was badly ringed in the second instance, and position of ring in the barrel was carefully measured with reference to position of the bullets at time of collision, but without any positive results. It showed, however, that the center of the ring was between the bullets at point of impact where compression of air would naturally take place. Either the lead, the air, or both acting together, produce a pressure No. 4 barrel could not withstand.

(Fig. 36) on opposite page is introduced to illustrate several experiments.

Figure 1 shows a pair of bullets swaged together by being shot at same time, from same barrel, with several inches air space between. Figure 2 is that of a bullet fired from short barrel and entered only length of its body into the muzzle, leaving its point out and with base $2\frac{1}{2}$ inches in front of regular black powder charge, leaving that amount of air space. Figure 3 was entered only $\frac{1}{4}$ inch into muzzle, leaving air space of $2\frac{3}{4}$ inches. Figure 4 was treated similar to figure 1, using two bullets, and in both instances the barrel was badly ringed.

Figure 5 represents the front end of same mutilated bullet shown in upper left-hand corner of the preceding cut. Figure 6 was shot from a $\frac{5}{8}$-inch barrel which took about the length of body of bullet; in front and touching point of this bullet was a dry hickory stick, $1\frac{1}{2}$ inches square and 18 inches long; bullet shows the shape taken when point is retarded by a fairly firm obstruction added to its own inertia.

The peculiar shoulder observed on figure 2, where its point joins the body, indicates clearly that when powder pressure came from behind, the body of bullet was sent forward before point moved, thus leaving its perfect shoulder exactly where it lay at muzzle; in other words, the lead composing body of the bullet must have been shot forward into the point with such rapidity that the point first started at right angles and took impression of sharp edge of barrel's muzzle.

FIG. 36.

A number of theories have been advanced by riflemen to account for the large upset of bullet after exit from the muzzle of a short barrel, the most definite one being that instead of direct pressure it was due to the hammering of unburnt powder grains shot with high velocity from the muzzle, a peening effect of the separate grains like many strokes of a light hammer.

This peening theory was tested by lathe-made brass wads, tightly fitted to the bore and against base of the bullet so as to annul this action, but it was found bullets upset as much and in same shape as those not so protected; and the following cut (Fig. 37) shows this better than any description, all the six figures there represented having been shot with brass wads from barrels of different lengths.

Figure 2 represents a bullet where the brass wad has fallen from its base after being shot; 3, another one with brass wad still adhering after being shot. Figure 5 was shot from a ⅝-inch barrel and shows still greater upset with print of brass

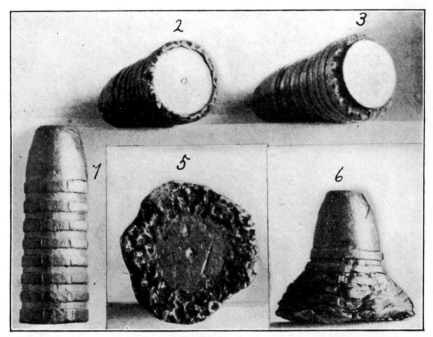

Fig. 37.

wad upon its base, figure 6 being a similar bullet placed in upright position. This test makes it quite plain that brass-wadded bullets had practically same shape and amount of upset as those fired without wads from barrels of like lengths. This will be observed by comparing figure 6 in this cut with figure 2 in cut on page 70, which was shot from the same barrel but without a brass wad.

So far as this experiment goes it disproves the theory that unburnt powder grains, acting as little hammers and driven at high velocity, peen base of the bullet into the shape which is assumed in these various tests.

This short-barrel work has given rise to many interesting theories, too many for all to be noticed, but it called out an interesting and extended correspondence with the well-known H. Pettit of Belmont, Ontario, a little of which follows:—

"BELMONT, ONTARIO, May 6, 1903.
"F. W. MANN, MILFORD, MASS.

"DEAR SIR: Yours of April 30 to hand, also box with bullets came yesterday at dinner time, a long time on the road. Upon opening the box and looking at the bullets I concluded I was 'dead stuck,' for sure. However, before night light (I think) dawned upon the whole matter. You are certainly doing some very original work, and on another sheet I give my opinion on the matter immediately required.

"Until the last portion of this unburnt powder left the bore it would act as a gas check and retain the gases and *the pressure;* and (owing to the light weight of the powder) nearly all the pressure would be transmitted to base of bullet, as if the powder were a light but solid plug. Of course the unburned powder would expand outside of the bore but infinitely less rapidly than pure gas. The space between base of bullet and bore widens very rapidly owing to the forward motion of base (due to upsetting) added to the velocity of the bullet en masse. But, until the upsetting is finished, this rapidly enlarging space is filled with unburnt powder pushed forward with the *whole force* of the burning powder behind.

"If all the powder were in a gaseous state when base of bullet left bore or if charge were ignited at forward end, the rapidly enlarging space due to the two causes above mentioned would allow the pressure to be reduced too soon to allow much work on base of bullet. You have certainly opened the gate into a large field.

"Regarding bullets 21, 24, and 21 out of bore $\frac{3}{4}$ inch long, it is evident that the upsetting was done outside the barrel by a pressure against the base of the bullet, the said pressure being very much in excess of the cohesive power of the lead.

"The medium through which the pressure was transmitted to the base of the bullets was *unburned* powder.

"If the powder were *all* in a gaseous state at the instant the base of the bullet parted from the bore the escape of the gases would be so rapid that (I do not think) they would have time to do the work represented by the condition of the bullets. When bullets left bore a large portion of forward part of powder charge was unburned.

"Yours truly,
"H. Pettit."

Immediately upon receipt of this gas-check theory from H. Pettit the short barrels were tried with front or tube ignition, having all other conditions the same as before. In this case the primer fire was carried in a $\frac{1}{32}$-inch tube to extreme front of the charge immediately behind the oleo wad, and it would seem under these conditions that the unburned powder charge, which acts as a gas check under Mr. Pettit's theory, would be in rear of the pressure; that is, the unburnt powder would not be between the pressure of burning powder and the bullet, and so could not act as a gas check.

A surprise, as usual, was in store for us. The upset of bullet from various barrels with front ignition was increased to a very noticeable degree, while the characteristic shape of the recovered bullets from each barrel was retained. After making this report and submitting samples to Mr. Pettit, he feared that the soldering of brass ignition tube might have been imperfect, and that these tests were really rear ignition instead of front. Consequently new shells and tubes were made and fitted so the soldering could not be imperfect, and front ignition tests were repeated.

Since sawdust and some other substances tested could not be made to take the place of powder grains to act as a gas check, dry sand was experimented with by filling the .32-caliber shell within $\frac{3}{8}$ inch of its mouth with powder and the remainder with dry sand. This combination, in a $\frac{5}{8}$-inch barrel, sent the bullet along as usual, but the sand wad was compressed so suddenly by explosion behind that it bit into the brass of the shell and took $\frac{5}{16}$ of its forward end along through the barrel, both sand wad and section of shell following the bullet. The shell was so evenly and sharply cut off that its mutilation was not noticed until again picked up for another shot.

Hardly believing our deductions from such a phenomenon, a new shell was loaded with a like sand plug, and results were precisely the same. The chamber pressure cannot exactly be known, but diameter of the shell containing sand plug was .034 inch larger than bore of barrel, and sand plug was made hard enough to pick the entire end off a new shell, sending sand and all through bore in wake of the bullet. This brass jacket thus left around the sand wad doubtless acted as a lubricant to the bore, preventing more serious disaster. We stopped using sand for wads.

The results of the second trial of front ignition tests were submitted to Mr. Pettit for explanation, calling out several interesting and instructive letters, one of which follows: —

"BELMONT, June 19, 1903.

"F. W. MANN, MILFORD, MASS.

"DEAR SIR: Your last letter and box at hand. You thought them of slight importance yet they have worried me some. Thanks for specimens of bullets showing effect of pressure. Bullets 66 and 68 show that front ignition can produce greater effect than rear ignition, but I think I understand why in this case there is a difference. You of course know that with a bullet free to move the effect upon base depends upon the quickness with which the pressure is applied. In the case of bullet 68 there was a large mass of unburned powder to put in motion so that it could not get in all its work on base of bullet. In the case of front ignition there is no mass of powder to move and pressure is at once applied to base of bullet. Also as bullet moves forward there is soon a space between powder and base of bullet and as each grain of powder ignites it rushes forward towards the choking mass in front, not only adding to the gas-check effect but adding its momentum to the pressure ahead.

"The larger part in the rear portion will have a space well on to two inches in which to get up motion and you have already demonstrated the effect of unburned powder in motion.

"You are aware that shotgun smokeless will produce a very dangerous pressure in a rifle, especially if bullet is heavy and seated firmly against powder. I have already shown you that an exploding mixture burns more rapidly and produces a higher pressure when under solid restraint. In the case of a shotgun cartridge we have yielding walls all round the powder which yield at once as soon as pressure begins and by the time this yielding has ceased the shot is in motion and giving the powder more room and hence the powder has not the chance to develop the high pressure that the solid walls of a rifle give.

"Du Pont's .30-caliber powder used in shotgun would hardly burn at all because the yielding walls would not resist enough to develop pressure enough to explode the powder. Black powder is less affected by the resistance but yet it has that property.

"The sawdust in front of No. 71 although compressed was still a yielding mass (very much more so than powder) and thus prevented the development of so high a pressure besides acting as a soft cushion between powder and bullet. I do not know of any substance that would act as a non-combustible substitute for powder to act as a test for the gas-check theory.

"Yours truly,
"H. PETTIT."

This correspondence, though imperfect and an injustice to a certain extent to Mr. Pettit, is introduced to show important data respecting the action of burning powder in short barrels. It also shows how difficult it is, even by most careful reasoning, to obtain correct conclusions about the mechanical behavior of a rifle and its ammunition before the test has been applied.

The peening theory which has been previously mentioned gave an attempted solution by one of the most painstaking of riflemen for this remarkable short-barrel upset. It calls to mind that either of these theories would have been satisfactory and would have remained an explanation if they had not fallen into the hands of a careful experimenter. The peening theory avoided an admission that the chamber pressure in modern rifles was far in excess of the cohesive power of lead, which our extended experiments had so completely demonstrated.

Mr. Pettit's letters being based on his thorough and experimental knowledge of the action of gaseous substances must command attention. Since the above correspondence Mr. Pettit has visited the writer and closely examined an exhibition of short-barrel work on the homestead range, but his deductions are not in letter form so cannot be correctly introduced here.

The brass wad largely disproved the peening theory. The sand wad and sawdust wad were failures. The front ignition of the charge, at first sight on Mr.

Pettit's part, disproved the powder obstruction theory but not so on further thought. An unintentional test, however, where 12 grains of sharpshooter powder, in the .32–40 shell, produced the identical mutilations as 47 grains black powder in same shell, seems to show beyond question that it is mainly the pressure which flattens the base. In this test there was no sand or brass wad and only one-quarter the weight of powder to act as a gas obstruction, but the flattening was identical in shape and amount because either the 47 grains of black powder or the 12 grains of smokeless produced enough pressure in the rifle to give the two bullets an equal trajectory from a normal barrel.

The above discussion will be appreciated if it fixes in the mind of the reader the relation which the pressure in a modern rifle bears to the cohesive power of lead or its usual alloys. The .32-caliber, lead or alloy, bullet before the 47-grain charge simulates a putty plug, being held in shape solely by the rifle bore which can only support the surface of its cylindrical sides. The base or point, being free to move, can be shoved one way or another, or any old way, as it passes through its first 10 inches of flight in rifle's bore.

Where the Upset Occurs. This short-barrel shooting was undertaken to determine, if possible, where the bullet upsets; but it led us into new fields, from one complication to another, and did not solve the question. Some competent advice, however, was received about this time regarding a method that would assist in solving it, which was to shorten the .32–40 shell $\frac{1}{8}$ inch so that amount of base of bullet might be left in rifle chamber. In this case, if the base upsets before point is started, it would fill the chamber, which is larger than bore, and on being forced forward the bore would draw circumference of the base backward, leaving it cup-shaped.

Such a test was made in a normal barrel and bullets recovered in sawdust, and their bases were found much concaved, showing all signs of having been drawn back by the rifling and bore, proving the bullet upsets before its point is moved forward.

A test to determine where the bullet would cease to upset, if it had space, would be interesting but has only been carried far enough to satisfy the writer, and may be stated as follows: —

The 187-grain bullet, fired from a .32-caliber rifle by 45 grains black powder, will upset .007 inch after leaving the muzzle of a 10-inch barrel, and there can be little doubt that if pressure is great enough to upset the bullet after leaving muzzle it is great enough to upset the same if in the bore. It was found that a bullet failed to upset after leaving muzzle of an 11-inch barrel, with same bore and charge as used in the 10-inch. If there is a slight ring in bore anywhere within 10 inches of chamber, the bullet, when discharged, will upset into it and partially or wholly fill it, but beyond 11 inches it might not. Evidently the bullet is upset into the first space that offers itself, whether back into the chamber, sidewise into the grooves, or into the air when freed at muzzle from barrels less than 10 inches in length.

The conclusion thus arrived at is not now a theory with the writer. The unusual experiments with short barrels have demonstrated the character of a lead bullet and its behavior under confinement, and out of confinement when subjected to the sharp blow that is communicated by the sudden expansion of exploding powder. Though it has taken years and many useless experiments to determine this property of lead, hoping thereby to eliminate an error from the path to the goal of accurate rifle shooting, these experiments had their educating influence, sufficient at least to encourage the writer to solve other questions still presenting themselves.

Ross-Pope .32-Caliber Continued. Test 36. — Experiments and tests from the "Shooting Gibraltar" were commenced September 2, 1901, with the Ross-Pope No. 4 octagon barrel, in V-rest, concentric action, and it was found to shoot $5\frac{7}{8}$ inches higher than at muzzle and shoulder rest, and it was taken for granted that with this V-rest, concentric rings and action there is no buckle to the barrel, therefore this test gave buckle for this rifle using the following load: 45 grains powder with three grains nitro priming and 187-grain bullet. At 200 yards the buckle would be $11\frac{3}{4}$ inches, which will be news to the average rifleman.

Test 37. — This was made on same day, with V-rest and concentric action, no muslin cover, shooting one group which measured exactly one inch. September 19, under same conditions, another group was shot, measuring 1.25 inches.

Test 38. — A rotation test was made on same day, by revolving this barrel in V-rest on concentric aluminum rings, giving a hollow group with diameter of 4.25 inches, showing that bore was not straight. These aluminum rings are concentric with the bore at muzzle and chamber, being turned on new brass centers fitting the muzzle and chamber of this particular barrel.

Test 39. — September 25 and 27, with same barrel, always with V-rest and concentric action unless otherwise mentioned, under cover, was first day's test of this muslin-covered range, illustrated on page 46. The results of six 5-shot groups were: smallest 1.62, largest 3.25, average 2.25 inches. The cover here mentioned was made of muslin, extending whole length of the 100-yard range, entirely inclosing a space 8 × 11 inches square, through which firing might be done. It afforded perfect protection from any air currents that might be moving.

Test 40. — This was made same day as preceding, with breech-loading, 167-grain cylindrical .316-inch bullet, under cover; smallest group was 1.37, largest 2.62, average 1.75 inches.

Test 41. — October 1, same as last, three 5-shot groups were made; smallest was 1.25, largest 1.62, average 1.50 inches.

Test 42. — This was made on same day and by same method as preceding, except 30 grains F. G. powder was the charge instead of 45, base of shell being filled with Babbitt metal, leaving air space the same as if larger charge had been used. Four groups were made; largest 3, smallest 2.12, average 2.69 inches.

Test 43. — On October 5, this was practically the same as 40 and 41, shooting two groups; smallest 1.19, largest 1.25 inches. In all these experiments nothing occurred to indicate that exclusion of air by the muslin cover added to accuracy of shooting, though riflemen had always contended that if shooting could be done in a tunnel where the air was motionless, their perplexities would cease and rainbow caught. Evidently the muslin-inclosed range did not exclude the x-error.

The Pope Breech Loader, .25-Caliber. The simple tabulation of rifle tests will not appeal to the reader in such manner as to teach lessons which impress so vividly those who are making experiments, but illustrations accompanying many of them may assist somewhat towards adding interest. Constant spreading of groups will be recognized, however, from foregoing groups, whether with double rest or V-rest, with Ballard action or concentric action, covered or uncovered range; but the superiority of cylindrical bore-diameter bullets when front seated, up to 10 consecutive shots, is marked. It is a little curious that the Ross-Pope barrel made larger groups under the covered range than in the open; doubtless there is a reason for this, though it can hardly be charged to absence of air currents afforded by the cover.

During early part of 1901 Mr. Leopold kindly offered to lap out the Ross-Pope barrel, to remove most of the choke and clean it up generally. This operation, judging by following tests, materially reduced size of groups, especially when using bore-diameter bullets at breech; and about this time he presented the writer with a .25-caliber Pope breech loader, 28-inch half octagon barrel, chambered .25–21, and tests were soon under way.

Test 44. — October 9, 1901, with .25–21 Pope, under cover, using Leopold's fixed ammunition, oleo wad, 80-grain bullet; first group of five was made, and measured 1.62 inches; second group of three was .56 inch; ammunition gave out.

Test 45. — On same day, 80-grain bullet, front seated, under cover; one group only was made, measuring .81 inch.

Test 46. — This was made on same day, bullet entered into the shell by hand, fixed ammunition; one group only, which measured 2.50 inches.

Test 47. — October 15, like test 44 except a 90-grain bullet was used and ammunition was not loaded by writer; two groups were made, measuring 3 and 2.25 inches respectively. The x-error still is with us.

This closed tests with this rifle, and it will be observed that a fine rifle barrel, with V-rest and concentric action, does not make fine groups unless there is adaptation of ammunition.

One member of the old club, disbanded at Milford, had recently invested in a new .25-caliber rifle and prepared some fine fixed ammunition for it. Thinking no doubt that any rifle would shoot into the same half-inch hole if placed in the V-rest on Mann's "Shooting Gibraltar," after fitting his barrel to concentric rings and action, he journeyed 16 miles with the writer to test it. As this exhibit is of interest because it indicates why test 47 made a group of three inches, also introducing screen shooting to the reader, its methods and results are illustrated.

(Fig. 38) shows a 100-yard group, as made, in black face, and the cross represents the position of a straight line drawn from center of muzzle to 8.5 inches

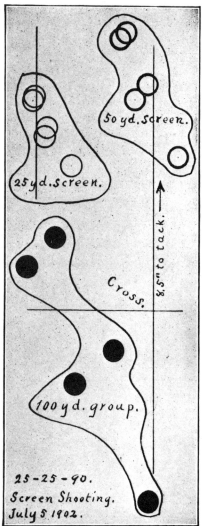

Fig. 38.

below tack on the butt. Screens of paper were placed in line of fire at 25, 50, and 75 yards, through which the bullets passed. The 50-yard group, as illustrated, and its distance above the horizontal line of cross show exactly how much the several bullets were above it. The 25-yard screen, through which the same group passed, together with the perpendicular line of cross, was moved to left to avoid confusion of shots in illustration, but its height from horizontal line is kept true. The 75-yard screen is omitted to prevent confusion, and the 100-yard group, as it printed at the butt, is shown in black faces.

Careful measurements will indicate that the 100-yard group is 3.7 times larger than same group when passing through the 25-yard screen, and 1.75 times as large at 50 as at 25. These screens and measurements show that all bullets in this group followed nearly uniform trajectories throughout their 100-yard flight; proving that the trouble lay in the fact that bullets started off wrong at muzzle of the rifle, as was the case with three-inch group in test 47 using a different rifle.

The old club member who had his rifle tested in this manner went home and thought the matter over for the next two weeks, when he acknowledged what the writer had been asserting for two years, viz. that the bullet starts in the wrong direction from the rifle's muzzle.

Muzzle Loader, Pope .32-caliber. This .32-caliber, muzzle-loading barrel was kindly loaned by Mr. Pope for a few days, and the four following tests were made with it.

Test 48. — October 22, 1901, using the Pope system of muzzle loading, under muslin cover, five 5-shot groups were consecutively made; smallest .62, largest 1, average .81 inch, and the 2d, 3d, and 4th groups are shown in (Fig. 39).

Test 49. — This was made on a cold, bleak day, December 2, muzzle load, Pope system, without cover; smallest group 1, largest 1.56, average 1.19 inches.

Test 50. — The same day and under same conditions, except the bore-diameter bullet was breech-loaded, three consecutive groups were made without cleaning; smallest was .81, largest 1.56, average 1.12 inches.

THE BALLISTICS OF SMALL ARMS 87

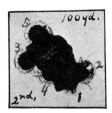  

FIG. 39.

Test 51. — Again on the same day and under same conditions a rotation group of four shots was made, turning the barrel one quarter around in its V-rest between each shot, thus bringing barrel at third shot bottom side up; size of group, as seen in (Fig. 40), was one inch, proving a remarkably straight barrel.

The deductions from these tests may be fairly stated by saying that a fine rifle will do well on a bleak December day while some other, supposed to be equally good, will not do as well on any kind of a day. There is little doubt but this barrel, and not the ammunition, eliminated, somehow, much of the x-error. Riflemen have all manner of excuses for their off days, in shooting, sometimes attributing bad scores to wind, sometimes to temperature, sometimes to mirage, and sometimes they find their rifles will not shoot well on any day; but the above tests indicate that a "gilt-edged" barrel will do pretty good work regardless of all the above excuses.

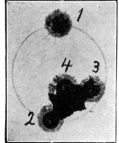

FIG. 40.

It will also be observed that the bore-diameter bullet in test 50 outshot the muzzle load in test 49. The Winchester bore diameter is .315 inch, while Pope's is often .316 or .317.

It is not probable that an oversize bullet can be entered into bore central or straight, either by bullet seater or a charge of powder. It is contrary to any practice obtained in machine shops, and the cylindrical bore diameter was designed to comply with that fact, a fact not generally accepted by other riflemen because contrary to all previous teaching and practice. When it is fully comprehended that a soft lead bullet, with a charge of powder behind, acts in the bore of a rifle like a ball of putty in a popgun, riflemen will have no fear about upsetting or taking grooves. Such being well proved, theory, even without practice, would naturally decide greater accuracy for the bore-diameter bullet, front-seated, than from one that must be smashed into the bore from the chamber, either by bullet seater or a charge of powder.

Letter to Dr. Skinner. At this stage of rifle and ammunition experimenting quite a voluminous correspondence was carried on with Dr. S. A. Skinner, of Hoosick Falls, N.Y., an old rifle crank and expert marksman, where the writer has spent many companionable days at range and "Medicus" woodchuck preserve.

Of all this correspondence this particular letter is selected and inserted as a reminder of the spirit in which these experiments are being conducted, and to put myself in closer touch with my readers.

"MILFORD, MASS., Dec. 19, 1901.

"MY DEAR DR. SKINNER: Your very interesting letter came this evening. The one you wrote Saturday was answered last evening. Now, Doctor, we must be very gentle in trying to solve too many things at once, and very gentle with each other in agreeing what the question is before the house. I like the many thoughts you put forth, and wish I had as delicate an ear for sound as you, but beg you to consider that all my arguments and talk upon rifle shooting have reference to group shooting, which means aiming constantly at the same spot, no matter how the wind is or where the shots go.

"Without this understanding all our mathematics and letter writing will be of no earthly use along the line of progress. I beg of you to keep this everlastingly in mind: A good group made by holding all over the target has no value in the line upon which I am working.

"Now one thing is settled, that when we are talking about the 'off' shots we have sole reference to group shooting. Next, if four shots were made in an absolutely calm day, and one shot made in a terrible gale we should expect one shot out of the five to be wild, but this is not the kind of off shot we are talking about. Again, if a veritable novice was making a five-shot group and he intended to aim at the same point but did not do so within six inches, then we should expect off shots. Again, if an expert tried to aim four times with the bright sun shining on the rifle, and one time when the sun was in a deep cloud, we might expect an off shot; or if he grasped the rifle very firmly for four shots and not at all for one, we might expect uneven shooting, due to the flip of the rifle. If you shot four 200-grain bullets and one 100-grain the five-shot group might not be good. Now, these and a hundred other things of the kind we must not talk about.

"The one out of five off shots still remains when I know that these kind of errors have been excluded. Of course with great irregularities you get poor groups. With lesser irregularities you get better groups, etc. As long as you do not exclude air motion irregularities, and mirage irregularities, you will believe that if these were excluded you could avoid that off shot. When you have, in actual practice, excluded these as I have, you will be obliged to lay the off shot to some other irregularity, but the off shot will come just the same.

"Now put your concentric action on, and the whole shooting match into the V-rest; now weigh your powder and shoot through a tunnel where the air motion is nil; clean your rifle or shoot it dirty,—and the off shot will make its appearance just the same unless, now listen, unless you have a rifle and ammunition which will shoot without this off shot, at shoulder and muzzle rest on your range. It will be difficult to determine the cause of the off shot if we cling to mistaken notions.

"You work at screen shooting faithfully for three years and you will find that an ordinary good shooting ball goes straight in quiet air to within $\frac{1}{4}$ inch, no matter what the humidity or temperature, or how far from or near the ground the ball may fly. The weather conditions which you wrote about affected the ball somewhere between the chamber of the rifle and the first six feet in front of it, and probably, in fact quite surely, within the length of the barrel plus $\frac{1}{16}$ inch out of the barrel, and there comes in the great fact that I wanted you to study out. But really this fact will have no weight with you if you think as most riflemen do that the irregularity comes from the ball flying crooked, or corkscrew-like, or spiral-like in the air.

"The air is not unhomogeneous as you think, remembering we are not speaking about air in motion which carries the ball along with it. You often, or rather sometimes, shoot on your range when the air is nearly quiet, perhaps for not all day or a week, but for a space long enough for five or ten shots, and when this condition does exist you have the advantage of all tunnels in the world, for no tunnel could be better. Do you at such a time avoid the off shot? If you do then you have a good rifle and a good fellow behind it. I am not speaking of a fortunate or lucky group, but am asking if you avoid the off shot always when there is not wind, at which time I know the ball goes straight in the air to within $\frac{1}{4}$ inch, but not in the line of sight.

"I hardly expect that you will at once sit down perfectly contented and satisfied fully and eternally convinced that the ball goes practically straight after it leaves the rifle bore, but until you show some signs of believing this we cannot proceed on our onward journey, because if we determine something by years of patient research, and our conclusions are at once thrown aside, riflemen must go on guessing and dreaming and changing their theories with every wind that blows. If the haphazard guesswork and old notions of the ordinary rifleman are going to be adhered to in spite of cold facts and indisputable scientific experiments, then progress along the line of truth will indeed be slow.

"Yours,

"F. W. Mann.

"P.S. You must not put between the lines of my letter a multitude of things I do not say. When I am talking about an off shot I am not talking about a case where weather conditions are such, or the flight is such that you do not make a group at all but an irregular plastered target. I am trying to combat popular errors, and there are many of them.

"Now if you say that this off shot question is only of minor importance, and that the other and greater irregularities are what interest other riflemen, I will add right here that this off shot is what does interest me, and not the weather conditions over which I have no control. I may be working at an impossibility but not on what I know to be an impossibility, that of altering air motion, mirage, etc. F. W. M.

"P.S. 2d. After you are through with this letter it may be returned to me as I would like to file it away with my rifle work, to be reread some years later to discover how far from the truth it is, and how much progress following years may be able to show. F. W. M."

"**Medicus.**" Reference has been made a number of times to Dr. S. A. Skinner, of Hoosick Falls, N.Y., and his sudden death, which occurred August 15, 1905, calls for a passing word. He was born in Thetford, Vt., graduated from the university and was a general medical practitioner in his native state for 10 years, being appointed a medical examiner by Governor Holbrook, and assistant surgeon of the 7th Vermont volunteer regiment. Owing to ill health, however, he was not allowed to take the field with the regiment, and during 1864 moved with his family to Hoosick Falls, N.Y., where he resided the remainder of his life, one of the most prominent and successful physicians of Rensselaer County, and the leading practitioner of Hoosick Falls.

FIG. 41.

Dr. Skinner was an inventor of considerable note, including surgical instruments, a hospital bed, one of the first stethoscopes, the Skinner oiler used on mowing machines, and many of value and interest to riflemen. He was recognized authority and wrote quite extensively for *Shooting and Fishing*, under the nom de plume of " Medicus," some of his best articles being published during the last few years of his life.

As creeping age debarred him from attending to the practice of medicine, his offices were generally turned into a " Den," filled with material of his own make, largely pertaining to the rifle, its ammunition and paraphernalia, indexing and photographing his very unique and valuable collection.

As has been stated elsewhere, quite a voluminous correspondence was continued with Dr. Skinner up to the day of his death, and very efficient encouragement received through it from time to time. A semiannual pilgrimage was made to his shooting range, by the writer, for a number of years, and to one of the best woodchuck preserves of the surrounding country, which lay in his immediate neighborhood. These pilgrimages invariably proved vacations in the best sense of the term, as the active old-time gentleman always had something new to offer and words of encouragement to give. The "snap shot" on the preceding page is of Maj. George Shorkley, U.S.A., F. W. Mann, and Dr. S. A. Skinner, at "Medicus woodchuck preserve," Hoosick Falls, in the act of preparing the skull of a woodchuck for his museum. The photograph of this skull has already been shown.

Among the many letters received from him, the following is reproduced as a characteristic one: —

"Hoosick Falls, N.Y., Jan. 23, 1902.

"My dear Mr. Mann: I think your remarks on my manuscript, respecting the off shot, are correct and I have summed it all up in a very short synopsis as is here recorded: We will suppose the imperfect rifle is eliminated; imperfect ammunition is eliminated; bad holding, gravity, wind, jump, humidity, a passing cloud and in fact, everything pertaining to, or that can influence the flight of the bullet, is eliminated.

"What we call an 'off shot' is conceived at the breech of the rifle; its period of gestation is 32 inches; its accouchement is at the muzzle of the rifle; its period of adolescence is 200 yards, and of five born four are perfect units, one is not; why? Because of the perversity of inanimate nature. We have not reached or found our zero in rifle shooting. The rainbow is just over the hill and Mr. Mann will ride it as soon as he can get astride it.

"Yours truly,

"S. A. Skinner."

Pope 1902 .32 Rifle. Compliments for Mr. Pope.
After two years of muzzle-loading tests a special muzzle system was devised to overcome a theoretical error in the Pope system, and the new Pope 1902, .32-caliber, No. 4, muzzle-loading rifle was procured at this time for the test. It was full octagon barrel with a dovetail rib running the whole length on the top flat of barrel for attachment of the telescope mountings.

Since experimenting unsuccessfully for three years with many devices for securing these mountings to the barrel, this dovetail rib, which constituted the top flat of this octagon barrel, proved to be the thing.

The special two-part false muzzle ordered with this barrel was an exception to many previously abandoned plans, and was designed to unite the good qualities of the Pope system with the cylindrical, bore-diameter bullet.

To Mr. Pope is due my hearty thanks, and all the good things that can be said of him. During six years, 1899–1905, he completed 21 special orders and contrivances for the writer, without complaint or criticism, and without making an error in any single instance. The goods were received as ordered and as the writer wished them to be, as if he comprehended the bent of the experimenter's mind.

The double muzzle of this rifle was designed to enter and seat an unmutilated bullet central and straight with the bore, and with certainty every time, at the same time scraping down any collection of dirt into the chamber at rear of bullet before discharge. The bore was cut without choke.

This false muzzle, being in two parts, enabled the bore-diameter bullet with base band to be entered point first into rear of its front half; the lead which was drawn back by the rifling being shaved off with a sharp knife against the steel face of the muzzle piece, the two parts then going together, containing the cylindrical bullet with a clean, sharp, and square base, having the appearance and being used as the normal Pope muzzle. Since the grooves are all cut through the base band before being shaved with knife, slight pressure only is needed to seat the bullet to its firing position, hence no deformity would occur, while the dirt wad which is shoved down holds it from falling into the chamber.

This proved to be a gilt-edged barrel, and its shooting qualities were so good that this costly false muzzle, in its two-part form, was never utilized. By a coincidence the barrel was first tested Aug. 12, 1902, on Mr. Pope's birth anniversary, after the long, tedious wait of years to obtain a rifle which would admit of certain desired experiments. The groups which follow should not be classed with the few lucky ones which are selected from thousands and published from time to time.

The 13 groups shot with this rifle extended over a space of four months, but were interspersed among other experiments with the same barrel. These groups were shot to keep tally upon the other experiments, and without this gilt-edged barrel many of them could not have been successfully performed.

All the regular muzzle-loaded groups with this barrel were made with a special swaged bullet, the same as designed for the two-part muzzle had this muzzle been used. It was of pure lead, the body being swaged bore diameter without taper, except $\frac{1}{16}$ inch of the base, which was also a cylinder, but .004 inch larger than groove diameter. No regular Pope muzzle bullets were tested in this rifle, because as was previously known, they could not be successfully seated from the muzzle, since the bore was cut without choke. The bullet without a base band used above in breech loading, was shot in competition with the same bullet adapted to muzzle loading, and not in competition with the regular Pope system.

Test 52. — August 12, 13, 14, 15, 23, 30, September 6, and November 3, 1902, 13 5-shot groups were made, without cover. Six of the groups were made from muzzle loading, using the two-part false muzzle as one piece according to the Pope system; four with bore-diameter, cylindrical bullets, breech-loaded, and three groups not certain whether muzzle or breech; smallest group made was .44, largest 1.25, average .81 inch.

If three of these groups were omitted the smallest would have been .44, largest .93, average .62 inch, and any shot of a .62-inch group will hit a honeybee. By referring to test 48, page 86, the prints made by a .62 and .81 inch group will be seen, and will relieve the reader's imagination.

THE BALLISTICS OF SMALL ARMS

The average size of the six muzzle-loaded groups was .75 inch, and the average of the four breech groups was the same. If it were not for the last group, shot from breech load on a cold November day, or if the three unmarked groups could be tabulated, as they undoubtedly were made, the breech-loaded bullets would have made smaller groups than muzzle-loaded. With that November group, however, the rifle lost its gilt edge.

Test 53. — August 15, a rotation test was made with this barrel, V-rest, concentric action, breech-loaded, making a 4-shot group. The barrel was quarter turned after each shot, and group measured .56 inch, indicating a remarkably straight barrel, approaching mechanical perfection seldom found in rifle bores. Of all the rotation tests made from this V-rest, and nearly all barrels brought here have been thus tested, this proved the most accurate. Other barrels tested have varied in making hollow groups all the way from this .56 inch to 16 inches at 100 yards. (Fig. 42) shows this .56-inch group.

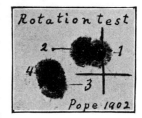

Fig. 42.

While one of the most expert riflemen of Walnut Hill was sighting up his .25-caliber rifle he discovered that when the bore was apparently in line with the 200-yard bull's-eye the bullet printed two inches above it, which was a great surprise to him. A rifle with a perfectly straight bore, under the above conditions, would have printed about 40 inches below the target, owing to normal drop of the ball. The same expert's rifle, put into concentric action and V-rest, would probably have given a rotation test group of 40 inches at 100 yards.

Testing a Brass-base Bullet. **Test 54.** — Aug. 30, brass-base bullets, muzzle-loaded, were shot. This hollow bullet was made in a special Zischang mold, the concavity being $\frac{3}{8}$ inch deep, shaped to fit a $\frac{3}{8}$-inch brass 8–32 machine screw with special shaped slotted head.

Small mill cutters were made to saw a central slot in the flat brass head, giving a good hold for the screwdriver, though the slot did not extend to edge of bullet. This brass base was firmly screwed in and the 187-grain lubricated bullet swaged firmly upon the screw in a regular swage, leaving a base band. Testing the base in a chuck, made for the purpose, showed the base to be oblique to body of the bullet, that is, the swage would not straighten the base. This was gratifying to know, because the brass base was inserted to keep the base from becoming oblique when upset in the rifling by discharging.

If the swage would not set the base square with body, it might be expected that the upset would not set it oblique; and if base was not oblique, would it not fly true to the bud? After swaging the bullets were held in a special chuck

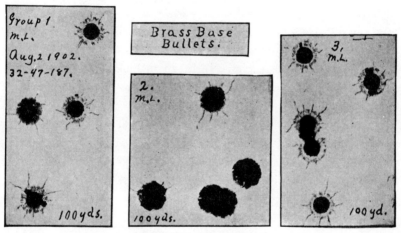

FIG. 43.

and bases perfectly squared in a lathe, then driven point first through the Pope muzzle from its rear end, the lands cutting their way through the lead base band and thin edge of the brass base, thus allowing bullets to be reëntered for loading into the regular Pope muzzle by the fingers, without pressure. Chucking the bullet and squaring base again in the lathe completed an apparently perfect bullet. Other methods had been tried, seemingly easier, but did not accomplish the purpose, and the reader must imagine the days of experimenting before 15 bullets of this character were perfected for the test.

Three 5-shot groups were fired from V-rest, making respectively 1.25, 1.62,

THE BALLISTICS OF SMALL ARMS 97

and 1.87 inches, as illustrated (Fig. 43), and prints showed no tips, but the body of bullets surrounding the screw did not upset into the grooves.

At the same time a test group of five was made with the bore-diameter bullet, under the same conditions as with brass bases, and size of group was only .53 inch. This closed a test which required months for preparation, and it seemed to disprove the oblique base theory of the glancing effect of muzzle blast because we took for granted that the brass-base bullets were not made oblique by the firing charge, though bullets did not catch well in sawdust, which prevented testing their bases for obliquity.

Test 55. — On September 25, with same barrel, breech-loaded, 187-grain lubricated bullet, swaged with hemispherical base, a 10-shot group was made, loading same as with normal bullet which had made an average of .62-inch groups; size of group was 2.75 inches, and bullets tipped very badly. Screens that were placed at intervals along the range showed that they were traveling point on at $6\frac{1}{2}$, $16\frac{1}{2}$, and 88 feet from muzzle.

Test 56. — October 14, with same barrel used as a breech loader, one 10-shot group was made, using three different shaped bullets and all from the 187-grain Zischang mold, lubricated and swaged. Shots 1, 2, 5, 6, 9, and 10 carried hemispherical bases; 3 and 4, deep cup-shaped bases and thin edge with a hole running up $\frac{3}{8}$ inch into bullet; 7 and 8 were normal ones. Size of group will be better comprehended by reference to (Fig. 44) than through any elaborate description. It was 1.81 inches.

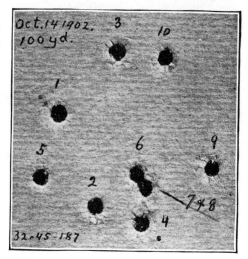

Fig. 44.

(Fig. 44) shows that normal shots, 7 and 8, went into same hole and by referring to the test following this it will be observed that a group made with normal bullets was only .62 inch. It will also be noticed that shots 7 and 8 of this group and all in test 57 would strike a house fly.

This test was made to determine if blast at the muzzle, acting on an oblique base, caused the average spreading of groups. It was theorized that a hemisphered-base bullet would present to powder blast one that would remain equally oblique throughout its whole circumference, even though mutilated in bore, hence would offer to that blast no leverage on one side over any other and must go straight from muzzle and make a small group. The tests indicate, however, that normal flat-base bullets make a 7-shot group of only .62 inch, while those with round bases made almost a two-inch group.

The Pope 1902 barrel was rifled for a 202-grain bullet, had a 14-inch twist, and would carry a 240-grain flat-base one fairly point on for 100 yards, yet the oval-base bullets, which tipped too much, weighed only 187 grains.

As a part of the above test eight paper screens were shot through, each screen being moved between shots. They were placed along the range at 6, 12, 16½, 25, and 50 feet, 25 yards, 88 feet, 50 and 75 yards respectively, and they clearly showed that the oval bases developed a tip at 50 and 75 yards and much more at 100, while the deep cup-base bullets kept point on, though making larger groups. The normal ones also kept point on. These notes are taken from those carefully gathered from each shot by W. E. Mann, who was my constant companion at this range.

Besides following the 10 shots of above group through screens, bullets were caught in oiled sawdust, which preserved them unmutilated. Hollow-base bullets did not turn over in sawdust, but went completely through whole of it, entirely different from hundreds of flat-base ones similarly caught. The body of bullet surrounding center hole in these hollow-base bullets did not upset into the rifling, while solid front part of body and thin edge of base did fill the grooves.

Test 57. — October 14, same day and conditions as the last test, one group was shot with normal breech-loading bullets, to keep sure tally on the brass-base test, that conclusions might be as certain as possible; and this group measured only .62 inch. This is pretty accurate mechanical shooting.

Test 58. — On same day as the last and under same favorable conditions, four shots were made with the hollow-base bullets, same as in test 56, and group measured 3 inches.

Here were three tests made on same day with three forms of bullets, hemispherical, flat- and hollow-bases. The flat base kept point on and made a group any shot of which would have hit a honeybee. The oval-base bullets made about a two-inch group and showed tipping after the first 50 yards of flight. The hollow bases, like flat bases, kept point on, but made a three-inch group. What property existed in these several bases, one over the other as exhibited in the different sized groups, which made one tip and not the other, or kept two point on and not the other, or made the flat base print a small group and the hollow one, which also kept point on, the largest of the three? Affirming that the flat-base bullet was better adapted for this rifle and charge than either of the others is doubtless correct, but such a statement does not explain why. The " why " is the thing we are after, and an explanation of these phenomena is a straight mechanical problem, with no hoodoo or guesswork about it.

Dr. Skinner's Shooting Range, and a Disappointment. **Test 59.** — During September, 1903, the Pope 1902 barrel was given its regular Ballard action and a telescope mount, to be taken on the annual visit to Dr. Skinner at Hoosick Falls, to show an old rifleman what could be done. With the greatest care, on three different days, we made six scores each with the same rifle. Dr. Skinner's six groups averaged 1.37 inches, and my own 1.50 inches.

As usual, each year his scores were better than my own, but the disappointment lay in the fact that this gilt-edged rifle was making groups twice as large as from the V-rest. It was a straight test, to all appearances, between double rest in New York and V-rest in Massachusetts, and my " Bumblebee " rifle would not shoot well in New York. It must be retried in its V-rest and on its native heath.

100   THE BULLET'S FLIGHT FROM POWDER TO TARGET

Test 60. — September 16. After returning from Hoosick Falls a straight test was made with this rifle between V-rest and double rest, 10 consecutive shots from an old discarded shooting table being made by W. E. Mann, using a double rest, resulting in a 1.37-inch group, same as made by Dr. Skinner. This was the first time, as with Dr. Skinner, that my brother had shot this rifle, and he rarely shot any rifle more than five to ten times a year.

After shooting the above group the rifle was taken down and everything made in fine order for the V-rest and concentric action, to prove the oft-repeated

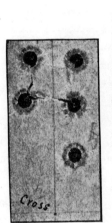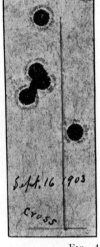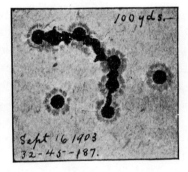

FIG. 45.

statement from other riflemen, " that any barrel would shoot fine if placed upon a 'Shooting Gibraltar.'" The following test, however, indicates that a good barrel and ammunition adapted to it are essential for accurate shooting, whether from V-rest on its Gibraltar or any old rest.

In the V and concentric action a 10-shot group was made on same day in which brother made the above 1.37 inch, resulting in a group of 1.62 inches, showing that this barrel was now shooting better at muzzle and wedge rest than from the V. The cuts show the groups made from both rests. The two left-hand groups from V-rest, when placed together so that the cross on one overlies the other, makes a 10-shot group of 1.62 inches, while the right group of 10 shots, made by W. E. Mann, measures 1.37 inches, same as Dr. Skinner's average.

The V-rest group made with this barrel November 3, 1902, the last one before visiting Hoosick Falls, was 1.25 inches, after making groups that would hit a bumblebee, calling out the remark that "the barrel had lost its gilt-edge,"

FIG. 46.

and the shooting at Hoosick Falls and since has verified that remark and loss. No one has had charge of this rifle but the writer. It had been shot about 200 times. It had not been swabbed between shots. What has changed the average group from .62 to 1.25 inches? It could not have been the V-rest as compared with any old rest, yet this erstwhile "gilt-edged" barrel shot better from the shooting table illustrated above, and from still another equally patched affair.

(Fig. 46) of " Medicus " shooting table, in Hoosick Falls, from which many of the Doctor's rifle experiments were made, is what it looks, a patchwork affair, condemned by himself as a makeshift, yet in a way it served its purpose. It exhibits the boards, chips, wedges, blocks, nails, and any old thing at hand, utilized to prop the rifle into position for shooting, showing how many disadvantages even the expert rifleman seems to labor under when trying to do extra fine work; yet the rifle shot better from this table than from the V-rest on its " Shooting Gibraltar."

Such a patchwork shooting table, however, is an object lesson in displaying the perturbed state of the shooter's mind, as the 60 and more excuses loom before it for that off shot, side shot, high shot, or any poor old shot.

It has taken four years of pretty persistent and persevering work to eliminate the 60 odd reasons advanced by riflemen for their errors in rest target work, and to demonstrate that good work is not dependent upon " Shooting Gibraltars," patchwork tables, mirage, air humidity, or the remainder of the 60 excuses given, though the V-rest and concentric actions and solid pier have been indispensable in this work. E. A. Leopold has clearly stated the difficulty which is responsible for the errors, as follows: " In rifle shooting the trouble is, that the bullet does not start off in the right direction at the muzzle."

The reason why the bullet does not leave the muzzle on the line of fire will be discussed later. Meantime an illustration of the shooting table at the homestead range, the " Shooting Gibraltar," is reproduced for comparison with "any old thing."

The camera was not brought into requisition for purposes of comparing one shooting table with another, but as a matter of curiosity, which might be surmised by any camera fiend with the aid of a magnifying glass. Plate (9) represents the shooting table with a rifle barrel lying in its V-rest, surmounted by a telescope, and bore of rifle pointing to the bull's-eye on the target 100 yards away, and the peculiarity of the picture lies in the fact that the camera caught a picture of the bull's-eye through the rifle bore.

A camerist will readily understand that in order to accomplish such a result it was necessary to line up the optical center, bull's-eye and bore with no image,

PLATE 9.

and it required nearly an hour's work. A rifleman will readily understand that to bring the bore of the rifle in line with the bull's-eye would require several thicknesses of paper; in this case 11 were used, placed between rear ring of barrel and its V-rest, and, as has been explained elsewhere, only the barrel without scope or mountings occupies the V-rest when shooting.

12-inch Barrel Experiments. After the 32-inch, No. 4, full round Pope barrel had been cut into six pieces for short-barrel work, one of 12 inches remained, including chamber, and tests were made with this to discover if blast at the muzzle would deflect the differently shaped bullets more than from a regular 32-inch barrel.

A .32-40 chamber was cut and barrel mounted in concentric rings. In firing from V-rest the experimenter stands by side of the barrel and with this short one the blast is so severe that it causes acute pain in the ears.

Test 61. — On March 7, 1903, with 12-inch barrel, 187-grain cylindrical, 1 to 60 bullet, oleo wad, the size of group was 2.37 inches, no larger than many groups made with full-length barrels. Tests have shown that the best shooting has never been done with this alloyed tin and lead bullet, in any rifle while using black powder.

These minor changes from one test to another do not present a very logical order to the student, but being often made in this haphazard manner, without thought at the time of tabulating, no other method suggests itself of keeping records for easy reference.

Test 62. — March 13. This was like previous one except a lead bullet was used; two 5-shot groups were made, smallest being 1, largest 1.06 inches. Both combined by the pasteboard backing make a 10-shot group of 1.12 inches; not very bad for a 12-inch barrel, and it weakens the muzzle-blast theory, for surely there was blast enough from this barrel, judging from its effects upon one's ears.

(Fig. 47) represents a plot, reduced in size from four to one, of the groups made by this 12-inch barrel in this and the three following tests. The above 10-shot group is represented by black faces and marked "Group 1."

Test 63. — March 7 and 13. In this test the upper edge of bore at muzzle was whittled with a knife, then two groups were made from this mutilated muzzle on the 7th, giving 2.75 and 1.75 inches respectively. Still attempting to obtain a decided deflection from a mutilated muzzle, the cut edge was made deeper and extended one-third around, and seven shots made the 13th, gave a 1.75-inch group with good prints. This last group is shown in cut in heavy-faced circles marked "6," showing their prints about two inches higher than the black-faced normal group.

The muzzle end of barrel was then beveled to an angle of 2½ degrees with a very coarse file, leaving the burrs in bore and the muzzle unfinished. A 5-shot group was made with short side of bore and burr side up; size of group was 1.50 inches, but it was four inches below the normal, shown in cut marked "Group 3." When mutilations were reversed by turning the barrel, the two shots "Group 4" was made, which printed 4 inches above the normal Group 1. The cross in cut shows where a 32-inch barrel makes its normal group.

This deflection of groups up and down resulted from the two mutilations, beveling and extension of burrs into the bore. The group size remained the same as with normal con-

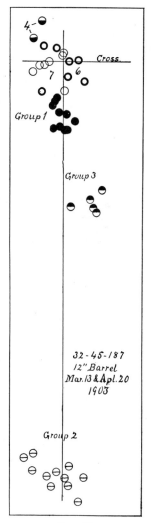

Fig. 47.

ditions, but the next test seems to indicate that burrs in muzzle caused deflection, oblique muzzle having no appreciable effect.

Test 64. — April 20, after squaring up this 12-inch barrel, a center punch was driven into muzzle end of barrel between lands, $\frac{1}{16}$ inch from edge of bore, and muzzle was again squared up. This left in the bore a large bunch of metal which plowed a furrow along the side of bullet as it passed the muzzle, nearly to depth of the grease grooves in bullet.

The barrel was turned in V-rest to bring this muzzle obstruction to its upper side and 12 shots fired from it; size of the group was 2.36 inches horizontal and 1.75 perpendicular, printing $15\frac{1}{2}$ inches directly under normal. No doubt if muzzle obstruction had been perfectly on top as barrel lay in its rest, instead of being approximately placed by the eye, the horizontal group size would have been same as perpendicular. This group is shown at bottom of cut, marked " Group 2," and its size remained the same as normal from a perfect muzzle, but printing in another place.

Test 65. — April 20, after bunch was filed out of bore, a group of seven shots showed that prints returned to nearly their normal position; size of group made was 1.62 inches and is shown in cut in light-faced circles, marked " 7," and is a better one than that made with a perfect muzzle when using alloyed bullets.

The cross shown in this (Fig 47) was drawn $9\frac{7}{8}$ inches below tack in the butt, and only serves to show relative positions of the groups, in this instance indicating positions of one group to another of seven different ones as they printed at 100 yards.

Test 66. — April 24, the above 12-inch barrel was shortened to 11 inches by removing mutilations from muzzle, and with this a 5-shot group made 1.75 inches, the prints being fully as good as grooved bullets usually make, with only a slight tip to one of them.

Test 67. — During the same day as last, a rotation test was made with this 11-inch barrel, giving a group of 2.25 inches, indicating the barrel to be practically the same either side up.

Going back eight years, to the 36-inch, .32-caliber Winchester rifle which gave an average group of 1.56 inches in test 1, while this 11-inch barrel gives groups of only 1.75 inches, what natural deductions might be drawn from such an experience?

For years riflemen have been extremely careful of the bore at muzzle of their rifles, and these various muzzle mutilations would simply demoralize them; yet these tests show that a very serious mutilation does not throw the bullet from the line of bore of the rifle to any greater degree than nearly all first-class barrels (not using the term "first-class" as being "gilt edge"). These tests indicate that the cause of x-error is not due to any permanent condition of extreme muzzle of the rifle bore; here are facts to be placed against speculations which have not been tested.

With few exceptions rifles in common use at different ranges will shoot from four or five to 20 inches out of line of bore at 200 yards, and must be sighted accordingly before ready for use. So any permanent muzzle obstruction or mutilation only needs sighting up to produce as good shooting as before. In other words, a pretty serious permanent mutilation in bore at muzzle simply furnishes a new barrel to be sighted up before fit for target practice. Test 64 shows a group equal to a normal one, while it printed $15\frac{1}{2}$ inches below, no more than some pretty good barrels might do; and in test 63 it will be noticed that the muzzle was badly beveled by a bastard file.

Smooth Bore, .32-Caliber. During August, 1903, Mr. Pope kindly loaned me a .32-caliber, No. 4 barrel, reamed out but not rifled; chambered .32–40. After being mounted in rings and concentric action fitted, it was tested in a thorough manner during the fall of '03, and with all tests paper screens were used at 6, 24, 48, 72, 96, and 150 feet from muzzle. Normal 178-grain bullets, cylindrical and flat end, $\frac{3}{4}$ and $\frac{3}{8}$ inch long, and hollow-base bullets were all tested and records kept — a basketful.

This barrel was utilized to aid in discovering something about the powder-blast force at the muzzle, but it would take a wilder crank than the writer to give cogent reasons for expectations on this line. The basketful of records are amusing at least, and one of the tests, using screens, gave positive records which indicated that some normal bullets, at some part of their flight, turned end over end with a speed exceeding 3300 revolutions a minute, which is faster than speed of a cabinetmaker's buzz saw.

In following the shots 6 and 7, made with 187-grain normal bullets from this smooth bore, No. 6 developed a tip of five degrees at 6 feet from muzzle; at 12 feet, 90 degrees or a true keyhole; at 24 feet, 180 degrees or flat base on; at 47 feet, no hit; at 72 feet, it swung into line at 180-degree tip; at 96 feet, 180 degrees; at 150 feet, 168 degrees, nearly base on.

Number 7 shot developed a tip of 3 degrees at 6 feet; at 12 feet, 30 degrees; at 24 feet, 168 degrees; at 48 feet, 45 degrees; at 72 feet, 90 degrees or true key-hole; at 96 feet, 168 degrees; at 150 feet, 180 degrees, or base on. Both the 6 and 7 shots cut into the same hole at 96 feet, made an inch group at 150 feet; at 48 feet No. 6 missed the screen but came back into the 72-foot screen. The number of tumbles between screens is not known, but No. 6 made 180 degrees tip in the first 24 feet, which computation will show is more than 3000 end-over-end tumbles per minute.

The behavior of cylinder or flat-end bullets, lubricated, when shot from this smooth bore, were very different. None of the nine shots made keyholed at any screen that caught them, though some tipped slightly; most of them made prints like a steel die. Their flight was much farther from line of fire than the pointed ones which developed the buzz-saw motion, and only three reached the 72-foot screen; while all the pointed or normal bullets reached the 150-foot screen, printing near each other through that screen.

Tests were made with normal bullets having flat bases swaged oblique, also hemisphere bases and hollow base with hole running up ⅜ inch, and normal bullets shot butt end first. Screens were set for these at 2, 4, 6, 8, and 12 feet, and to prevent the two-foot screen from being obliterated by muzzle blast, shots were made through a small hole in an inch board. All these shots went through the 12-foot screen at 90 degrees tip except one shot butt end first, and tip was developed gradually through each two feet of their flight. From this test it could not be detected that the muzzle blast acted different upon one style bullet over another, regardless of the great variety of shapes presented by various bases to the blast. Even the normal bullet, shot butt end first, kept that end on without tipping, same as the cylinder one which had the same flat front but very different-shaped base.

In recording these curious tests it was observed that none of the cylinder bullets, shot from different barrels, from 11 to 36 inches long, developed much if any tipping, and this smooth bore did not make them tip. In one test two cylinder bullets were loaded at once into the Ross-Pope, .32-caliber rifle, and screens showed that they separated at 30 yards but made very fair prints about an inch apart at 50 yards, after which they were lost. Perhaps these smooth-bore tests may prove of value some day.

Vented Barrel, Pope. During July, 1903, Mr. Pope loaned me his new muzzle-loading, .32-caliber vented barrel. In this eight rows of ventilating holes, five or more in each row, were bored a little way from muzzle on a spiral and entering bore between the lands, the rifling being done after holes were drilled. It was a very neat job and probably the only barrel made where the rifling was produced after the drilling, or where each hole entered the bore through the grooves. Machine screws were threaded into each vent to permit shooting with open or closed vents, and when mounted with its concentric action it was well tested.

Test 68. — August 3, with normal .32–40 ammunition, bore-diameter bullets, oleo wad, no cleaning, vents closed, three groups were shot; size 1.25, 1.37, and 2.50 inches.

Test 69. — Same day and under same conditions, except vents were open, three groups were made; 2.50, 2.75, and 2.87 inches. There was not enough difference between elevations of the two tests to be detected, and if there had been any difference in velocity, the V-rest would have disclosed it.

Test 70. — August 7, with same barrel, damp, quiet day after rain, giving fine shooting conditions, with 187-grain, cylindrical bullets, 10 consecutive shots were made, five with open and five with closed vents, cleaning between each shot; size of group was 1.62 inches. The vented and unvented shots mixed with each other, as shown in (Fig. 48), with great uniformity. The heavy circles represent shots from vented barrel and light ones from unvented.

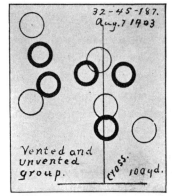

Fig. 48.

Test 71. — During the same day as preceding, 15 consecutive shots were made, using Leopold's oleo wad in breech loading, thorough cleaning between each shot, damp, quiet day; five shots made with vents open gave a group of 1.06 inches; ten shots under same conditions, except bullets were swaged with bases .012 inch oblique, that is, body of the bullets was .012 inch shorter on one side than the other; size of group with vents open was one inch; with vents closed it was 1.37 inches. In (Fig. 49) below, the first group, inclosed with pencil lines, is from vented barrel and square-base bullets; the second group from vented barrel and bases .012 inch oblique; third group from unvented barrel and bases .012 inch oblique. It will be noticed that in measuring Group 3 the upper shot is thrown out thus measuring because, without doubt, that particular oblique-base bullet was entered incorrectly at breech.

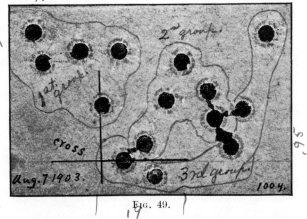

Fig. 49.

This liberty was taken because the original group is clearly illustrated by cut, and this one test is not sufficient to prove that oblique bases are less deflected with vents open than when they are closed. Unfortunately this test could not be repeated, as this fine rifle barrel was destroyed, with thousands of dollars' worth of Mr. Pope's rifles, in the San Francisco disaster.

This oblique-base bullet gave the blast from muzzle a chance to force it from its course if blast had any intention of acting that way. If the barrel was vented, the blast, supposed to be less, would deflect the oblique bullet less; yet the size of 10-shot group with this oblique base, five open and five closed vents, was 1.37 inches, and it printed 1.37 inches from center of a normal group at 4 o'clock, the same place oblique bases always print with a normal or unvented barrel when their short side comes out of muzzle at top.

Test 72. — August 3. In this test a special made screw with blunt point was put into one venthole until point protruded into rifle groove, the top one at muzzle; this plowed a furrow to the bottom of grease grooves in the bullet on its top as it issued; all other vents being closed, five shots were made, giving a group of 1.06 inches. Three other shots were made under same conditions, giving a group of 1.37 inches, and each group printed at 4 o'clock, 3.12 inches from normal, and prints were fine.

This test gave very marked results, showing that bullets uniformly unbalanced during last six inches of flight in bore make as true groups as unmutilated ones, even making truer ones than any except those from a gilt-edged bore. All tests have indicated that a bullet mutilated at base will commence to tip, oscillate, and gyrate immediately upon its exit from muzzle, and this, like other tests of the kind, show that if a bore will uniformly mutilate the bullet and deliver the mutilation from same position at muzzle, no appreciable loss of accuracy is incurred.

Test 73. — Before it was discovered how to obtain any definite conclusions from this vented bore, it was shot 13 times on August 3, using best ammunition, without cleaning, vents open, making a group 5.75 inches horizontal and 4 inches perpendicular. Dirt caked in the bore, commencing at ventholes and extending into the grooves towards muzzle, which was the cause of this bad shooting. It always did this when vents were open unless air was very moist or the bore thoroughly cleaned between shots.

Test 74. — August 9, still hunting for effect of muzzle blast, strawboard screens $\frac{1}{32}$ inch thick, 3 × 6 inches, were placed on a sliding frame in front of muzzle of vented barrel at such a distance that some would be broken by the blast and some would not. Eight were used, both with closed and open vents, and the experiment showed that a larger number of them were broken by muzzle blast when vents were open than when closed. The test did not indicate that the 40 ventholes near muzzle of this barrel had any influence upon muzzle blast.

Utility of Vented Barrels. **Test 75.** — Strips of cardboard were rolled into cylinders three inches long and with several diameters, 3, 2¼ and 1¾ inches, secured by twine, to encircle the barrel at its vented portion; and it was found that they were not affected by any blast from the vents, although the inside surfaces of these cylinders were only from ¼ to ⅜ inch from muzzle of the vents. A still smaller cylinder, however, with inside surface less than ¼ inch from the vents' muzzle, was pushed through and broken at the strings, although where the blast from the separate holes impinged upon the strawboard it was only blackened without being indented or abraded in the least. This same quality and thickness of strawboard, 60 times this distance away but in front of the muzzle blast, is broken by it. (Fig. 50) on opposite page was photographed from these strawboard tubes and shows how they appeared after having been placed over the holes of vented barrel, indicating that the amount of gas which escaped through them had very little force. Only the tube shown on right, the sides of which were only $\frac{3}{16}$ inch from mouths of ventholes, was ruptured.

The upper strip, as seen unrolled, shows markings of smoke from the eight rows of vents. Though the above experiment is not particularly scientific, it demonstrates pretty clearly that the force of escaping gas or powder blast through the holes amounts to very little, not enough to have any influence upon the escaping bullet.

Others have been busy, using different methods for testing the utility of barrel ventilation, and, it seems, have pretty generally decided that holes or slots will not materially act as vents. Powder gases have a forward velocity, following the bullet, and are not turned aside from their course to any extent by ventholes in the bore.

Mr. Pope wrote that a number of years ago a man by name of Spearing invented and made some rifles for army use which were ventilated a few inches from the breech, and that he may have succeeded in venting the barrel in this manner, but it was about all he did do. Such a barrel could not be utilized

for two reasons: first, the blast through the vents would endanger the man who shot it; and second, because the force of powder blast, passing out of the vents, left none for pushing out the bullet.

This matter of ventilating rifle barrels had been pretty well discussed 33 years before this writing, in connection with the old .44-caliber powder-and-ball affair of my boyhood; and during 1899 a machinist in my shop was engaged to drill $\frac{1}{8}$-inch holes in a .32-caliber rifle, five inches from the muzzle, and he also made tools for removing burrs left by the drill. This was done to discover how many holes would be required to materially reduce the muzzle blast, but after several tests with constantly increasing number of holes, no positive information was obtained and the matter was dropped with the expectation of taking it up again later.

About the same time letters from Mr. Pope informed me that he thought it feasible to rifle a barrel after the ventholes had been drilled, thus doing away with all burrs, and would attempt it as soon as the number required could be ascertained.

Four years after this correspondence with Mr. Pope, and after the above experiments with vented barrels had been made, the following communication was sent to Friend Leopold:—

"MILFORD, MASS. Dec. 23, 1903.

"FRIEND LEOPOLD: I undertook in 1899 to determine how many holes were required to vent a barrel in order to have it do good work. I have it. Let x equal number of holes to be drilled in the barrel, then x minus x equals number of holes to be drilled.

"Mr. Pettit, during his stay here, determined where these holes should be drilled in relation to the muzzle. He claims that they should be drilled through the bore as close as possible to and just in front of the muzzle.

"Yours truly,
"F. W. M."

There was a curious coincidence connected with this particular correspondence with Mr. Pope. Though no ventilating system was ever mentioned to me, or known to me at the time, Mr. Perry E. Kent, of Utica, N.Y., commenced his ventilating system, though entirely ignorant of my work and interest in the matter. I understand that since then Mr. Kent has made it his sole business to vent barrels, having a patent upon his method, and claims to be prepared to substantiate his claims for them. In discussing these ventilated barrel experiments with Mr. Leopold, among many letters received the following is quite characteristic:—

"NORRISTOWN, PA., Oct. 24, 1903.

"FRIEND MANN: Your 'notes written on the cars' received. I ventilated my .22-caliber a distance equal to 25 calibers, very thoroughly, by cutting wide slots between the lands, commencing at the muzzle, and found no material difference in the accuracy of the shooting. And we might have known this without trying, if we could think clearly, if we could follow from cause to effect, if we could reason just a little bit; if we were possessed of only a faint suspicion of that quality which we claim puts the genus homo above the other representatives of the animal world. The 'whizzer' proved to us that there is a stream of compressed powder gas and dirt *6 feet long*, issuing from the muzzle at each discharge; and then we ventilate our barrel 6 or 8 inches to get rid of this. If we were possessed of reasoning faculties, we would know that a barrel 3 feet long cannot be ventilated for a distance equal to 6 feet. If the barrel were 8 feet long, with 6 feet of its length well ventilated, then we ought to get rid of the worst part of the blast at muzzle, but not nearly all by any means, as the whizzer only recorded that part of the stream of dirt in which the particles had sufficient velocity and energy to embed themselves in the pasteboard. Such a barrel would be a failure as there would be a great loss of velocity, due to friction of bullet in the long ventilated portion of barrel. The remaining velocity at muzzle, if any, would be so small as to be of no account in any case.

"Nearly two years ago, I suggested in *Shooting and Fishing* respecting Mr. Kent's Venting System, that the improved shooting which he obtained in some cases might be due to lapping burrs and choke-out of the muzzle. And if I were to write for the paper again to-day, I would be compelled to repeat the same statement.

"Yours truly,
"E. A. LEOPOLD."

The Whizzer. That experiments with ventilated barrels might be so complete as to put the question of their utility forever at rest, a test suggested by Mr. Leopold was made by placing before the flying bullet and powder blast a rapidly revolving disk of cardboard; and the method adopted is well represented in (Fig. 51) on the following page.

A stiff piece of cardboard, 36 inches in diameter and circular in form, was mounted 12 inches from muzzle of the barrel so as to admit of rapid revolution; and the illustration shows how it was accomplished. While one of the experimenters is at the wheel causing the cardboard disk to revolve, the other is standing at the side of the "Shooting Gibraltar" with its V-rest, in the act of discharging the barrel. It was a simple arrangement but very effectual.

This cut also serves to indicate the method always used in firing from this rest with concentric actions.

The speed at which this " whizzer " was made to turn was about a mile a minute, or about one-fifteenth the speed of a flying bullet. The bullet in passing through the rapidly revolving disk made a print as if it had tipped.

Without doubt this was due to speed of target, because at the 100-yard butt it made a normal print. The stream of gas, smoke, and partially burned powder left its trail for two feet upon the revolving pasteboard.

As Mr. Leopold says, to get rid of this six feet of powder blast issuing from the muzzle, probably four-fifths of length of a barrel would need to be well ventilated, which would certainly result in lack of sufficient velocity to flying bullet to be effectual.

The thin card & the bullet striking squarely through it except for rotation, would alter the shape of a bullet practically nothing, while a lead bullet striking a hard pc of wood might have its point badly battered & that on one side, it could go almost anywhere. L's reasoning does not hold.

THE BALLISTICS OF SMALL ARMS

In writing up this subject for the *Pacific Coast Magazine* Mr. Leopold says that the moving target taught another valuable lesson. Old hunters have often claimed that their missing game was caused by the interference of a small twig which caused the bullet to glance and leave its true line of flight. This moving target, or "whizzer," deflected the .32-caliber, 200-grain bullet, about one and a half inches at 100 yards, which is a smaller error than good fixed ammunition possesses.

A test that may throw some light upon the subject of the interference of small twigs with the bullet's flight, will be found under the head of "Driving tacks with bullets" on page 326. The slight deflection of the bullet caused by the whizzer where the strawboard was traveling a mile per minute, was remarkable, but may be not more so than the largeness of the deflection as shown in the test just referred to above.

Pope 28-inch, .28-caliber, Barrel and Fixed Ammunition; 1903. During 1902, after much persuasion from Mr. Leopold, the work of devising a practical fixed ammunition load for a .28-caliber rifle was undertaken. Nothing heretofore had been done with fixed ammunition except to allow some other party to use it, but after determining the value of a bore-diameter bullet some interest was aroused, especially in connection with a chamber that would give control over the shell and bullet. Such a chamber had been talked about and discussed in different publications, but no one had seemed enterprising enough to attempt it.

A new barrel was received from Mr. Pope early in 1903, full octagon, No. 3, 28 inches long, with dovetail rib full length for telescope mounts, and a half-round groove in bottom for ramrod.

The special chambering tools received from the Ideal Co. produced a chamber to receive the .28–30 Herrick straight shell after its forward end was reduced .006 inch in diameter. This slight reduction gave control of how the shell should fit the chamber, and how the bullet should fit the shell. Without it nothing would be gained on experimental lines over what had been accomplished by other riflemen.

The chamber was also made $\frac{1}{16}$ inch shorter than the Herrick shell, thus giving control to fit of the shell at forward end. Four special reamers were obtained to fit shells to bullets, and these would leave a shoulder in the shell against which the base of bullet was prevented from slipping on to the high-pressure powder. Swages were made for forming bullets to fit the reamed shells, the 138 and 120 grain being the ones experimented with. Special loading tools were constructed of steel that the several bullets might be entered straight in shell. All these required the work of a year in planning and making, and the perplexities and expense incurred of making alterations by manufacturers from regular factory style was beginning to be realized.

During September, 1903, Mr. Leopold also suggested to drop practice entirely with black powder and do all future experimenting, except muzzle loading, with high-pressure smokeless. This suggestion was utilized as soon as a few tests then in hand were finished, and his long and careful experiments were well in hand for guidance.

Up to this time, however, manufacturers had not taken one step towards perfecting the rifle chambers for smokeless powder with the lead bullet, though this powder was adapted as well as possible by its makers to existing rifles of the black powder class; thus the wonderful improvement inaugurated by smokeless may have been materially handicapped.

Test 76. — August 27, 1903, commenced experimenting with the .28–30 fixed ammunition rifle. It took a Herrick straight shell, $2\frac{7}{8}$ inches long with an outside mouth diameter of .308 inch, 27 grains Hazzard F. G. powder, nitro-primed; ten 5-shot groups were made on this day with .279, .281, and .285 inch diameter bullets, seated in their respective shells. The first made groups of 2.62 and 2.75 inches, averaging 2.69 inches; the second 1.62, 1.62, 2.25, averaging 1.89 inches; and the last, made with .285 bullets, gave 1.62, 2, 2.12 and 3.12, averaging 2.25 inches.

The same day, with bore-diameter bullets seated in front of shell, groups of 1.75, 1.12, averaging 1.44 inches, were made; all the shots of this day making excellent prints though not perfect. Prints made by the .279 bullets, which are

THE BALLISTICS OF SMALL ARMS 119

smaller than bore, were equally good; while front-seated ones made no better than the others.

Test 77. — September 1, six 5-shot groups were made with .279, .280, .281, and .285 bullets, 27 grains powder and 3-grain nitro primers; bullets fitted into their respective shells. The .279 bullet made a group of 2.37 inches; .280 made 2.50 and 2.62, averaging 2.44 inches; .281 made 1.25, 1.87, and 2.12, averaging 1.75 inches; the last, .285, regular factory size, made a 3.37-inch group.

A comparison of groups made with this fixed ammunition with test 19, page 32, which was also made with a .28-caliber rifle and .285 bullet, the one universally taught to be correct for this caliber, is invited. So far as known at this writing no one has been found enterprising enough to test anything smaller or to record such tests. Also compare size of the three-inch group of test 19, the 3.12-inch group of test 76, with this, and it will be observed that in all the .279 printed better than the .285, thus continuing the evidence that an oversize bullet is not required to produce good prints. Is it not a little curious, to put it mildly, that the bullet which has a diameter as large or larger than groove diameter of barrel, the one considered best and usually furnished by manufacturers, should make larger groups than a bore-diameter bullet?

Test 78. — September 22, several groups were made with fixed ammunition, the .279 bullet making 1.37; the .281 making also 1.37 and 2.25, averaging 1.81; the .285 making 1.75 and 1.87, averaging 1.81 inches.

Many of these bullets were caught in oiled sawdust, which has been the practice for a long time in all important experiments, thus giving records of their behavior. This oiled sawdust catches most of them without a scratch. The slightest record made on point with a knife cut or steel die is plainly visible and will indicate to which test the bullet belongs. With this method it was known that the .279-inch bullets take the rifling perfectly. None of them showed any gas-cutting when the Leopold wad was used to retain powder in the shell.

Test 79. — October 6, three 5-shot groups were made with fixed ammunition and the .280 bullets made a 2.12 group; the .281 made same, and the .285 also made the same size group; and after four days' painstaking work on the range, this legend was left written on one of the targets: " Four days' careful tests, with best conditions all around, show about the same groups with .285, .281, .280, and .279 inch diameter bullets, entered in any way into shells, either by fingers or a carefully made loader."

During these four days we performed more vital and varied experiments with this .28-caliber ammunition than could have been executed in four years with ordinary ammunition and conditions. Again, in these tests the chamber and shells were so constructed that we were enabled to shoot all the varied charges and bullets through the same barrel, under the same conditions and hour of the day, instead of changing barrels and other conditions when shifting ammunition. In other words, only one change was made at a time, and this is one of the essentials of a scientific test.

Although W. E. Mann and the writer were unanimous in leaving that legend at that date, it did not influence our enthusiasm or place any obstruction in our path. Evidently all the bullets did so badly that no opportunity was given to individualize. For three years we had talked at and discussed matters on this range of our building and if plenty of wild schemes were not tried it was because Yankee ingenuity could not hatch up something to do next.

Reflections upon Black Powder and Cast Bullets. While in the act of discarding the use of low-pressure or black powder, and the lead or cast bullets, for nitro powders and jacketed bullets, some conclusions drawn from the foregoing tests and experiments are in order. As soft lead bullets are to low-pressure black powder, so are the jacketed ones to the high-pressure or nitro powders, though perhaps in a little less degree. In short, when jacketed bullets were brought out it was suddenly discovered that lead and putty were too near alike for accuracy in shooting.

When the chamber pressure from black or smokeless powder is sufficient to drive the lead bullet over 1450 feet per second, the bullet approximates what has been termed an " express " and accuracy sufficient for target work disappears. This chamber-pressure limit, which acts the same with either powder, seems to be the limit which is admissible for accuracy with the lubricated bullet, as shown by a multitude of tests all over the world where increased speed has been attempted.

It is reasonable to suppose that some phenomena connected with chamber pressure is reponsible for this loss of accuracy; in the past, however, it has been explained in many different ways, but our recorded experiments with short barrels, taken in connection with targets and screens from unbalanced bullets, throw enough light upon the subject to seriously interfere with the conventional reasons which are tossed from one to another, explaining the inaccuracy of old-time " express " bullets, or those having greater velocity than is conducive to common accuracy.

Target practice by thousands of shooters, show occasionally fine groups with lead bullets when thrown with normal lead-bullet speed; but in spite of greatest care days and hours will come when most any form of lead bullets makes large groups. The lead bullet is a very delicate thing to manage before loading, and why should it not be still more delicate at time of discharge, upset, and flight through bore? Tests 1 to 79, already gone over, show that under favorable shooting conditions there is an average of large spreading groups in all but a few fortunate cases; good groups one hour, and with same barrel and ammunition and equal care, poor groups the next. There were very few exceptions where accuracy could be depended upon long enough to ride into another state on an express train after woodchucks, as demonstrated by the 3-inch group shot in New York by a Massachusetts man, as related in test 19, page 32.

What is the cause of the x-error we are searching after? If dirt in bore was the cause, we could clean between shots. If dirty black powder was the cause, we can substitute smokeless, though while this shows a little superiority, it yet remains true that where accuracy is demanded the lead-bullet velocity has been limited to less than 1450 foot-seconds. If a speed of 1800 foot-seconds or over is demanded, we must use a metal-jacket bullet. Cast bullets can be speeded to 1450 feet per second and jacketed ones to 2500 with good accuracy.

122   THE BULLET'S FLIGHT FROM POWDER TO TARGET

A success with any bullet for small arms, so alloyed that it will not upset, has not been recorded. The "putty-plug" theory seemed to indicate that if a bullet was alloyed harder than lead it should have its x-error more or less reduced; but unremitting tests, year after year, demonstrated that a pure lead bullet, bore diameter, with black powder, outshot all others when using any of the seven .32-caliber barrels that were tested. Every attempt that could be imagined was made to overthrow this conclusion, viz. that the bore-diameter outshot every other bullet, but without success. All manner of bullets that could be suggested with any plausible reason, or no reason at all, were tried without success. In all instances, without a single failure, the pure lead, bore-diameter one would show up as the one for accurate shooting.

It is well known that this conclusion, thus forced upon us by these painstaking experiments with black powder, is contrary to all practice and preaching, but because it has been forced experimentally it must be recorded.

Because some riflemen were using an alloyed bullet in preference to the soft lead one, it was surmised that reasons existed for the practice, but the only one elicited was that pure lead had not been given a fair trial. It is a fact that lead is very soft and mutilates easily, hence must be carefully handled; but swaging a bullet to bore diameter eliminated much and made handling easier and safer, and if it could have been known that such a bullet would shoot better than any alloy these same riflemen would undoubtedly have utilized it.

The contention here is not to prove practicability, but to indicate that preceding recorded tests show soft bullets with black powder did outshoot, when care was taken, any alloyed one. Those who doubt had better commence a series of experiments with .32 and .28 caliber barrels and black powder, extend them over a number of years, using bore-diameter bullets against any other that can be devised, and publish results.

This naturally brings up the statement previously made that the x-error is largely due to deformation of the bullet at time of upset and on its journey through rifle bore, making an unbalanced bullet of one that was balanced before being discharged, thus bringing to notice conditions that at present writing cannot be harmonized or explained under the "putty-plug" theory, because this theory seems to indicate that hard alloyed bullets should do better. Why will not the hard alloyed bullets which, it seems, should keep their balance better because less deformed, do better shooting and give better groups?

It is quite a task to convert a running idea from the mind into sufficient form and substance to warrant the expenditure of a hundred or two dollars for physical material, and when machinery is collected, and the last one of nature's laws complied with, and when experiments persist in proving the exact opposite of finespun theory, which so frequently occurs, one can hardly help doubting his own reasoning faculties.

Laflin & Rand High-pressure Sharpshooter Powder. *Test 80.* — October 30, 1903, was the first day's test with high-pressure powder, using Laflin & Rand's "Sharpshooter" brand, and still experimenting with .28–30 Pope, 28-inch barrel and fixed ammunition. Two groups were shot; the 120-grain, 1 to 30 U. M. C. .285-inch bullets, 9 grains powder, oleo wad, gave 1.37 inches; second group made with .280 bullets gave 1.44 inches.

Then 15 shots were made to obtain trajectories, using 9, $10\frac{1}{2}$, 11, 12, 13, 14, and 15 grains powder. The shells seemed to stand these charges for none of the primer pockets expanded. Some of the bullets from any of the loads tipped too much, but those shot from excessive charges printed as well as from the 9-grain charge.

Trajectories obtained in this test were: nine grains gave 2.62; $10\frac{1}{2}$ gave 2.31; 11 gave 1.81; 12 gave 1.75; 13 gave 1.71; 14 gave 1.56; 15 gave 1.44 inches. The reduced Herrick shell, as used, when entirely filled with this powder was found to contain $24\frac{1}{2}$ grains; filled to base of bullet the contents was 20 grains, and from this last measurement the amount of air space left by any charge can readily be ascertained.

By the term "trajectory" is meant the height the ball, at 50 yards, reaches above a straight line drawn from muzzle of rifle to print of the bullet at the 100-yard butt; in other words, when a bullet is shot at a target, 100 yards distant, the sights must be raised so when upon the target the bore of rifle points to a spot some inches above the bull's-eye, the bullet flying in a slightly curved line from muzzle to target owing to force of gravity. The distance between the curved line described by bullet, or the trajectory curve, and the imaginary straight line from muzzle to print of bullet at 100 yards, taken at the 50-yard screen, is termed the 50-yard trajectory for 100 yards. This is the trajectory always referred to in all the writer's experiments.

Test 81. — November 3, a rotation test was made with this barrel and fixed ammunition, using nine grains powder, resulting in a group of four inches, with the penciled cross on the 100-yard butt in its center. This cross had not changed its position for two years, and is $9\frac{7}{8}$ inches below the tack which marks the line of V-rest. It is where the Pope 1902 "Bumblebee" rifle placed 59 out of 65 shots. This rotation test indicated that a nine-grain charge of powder carries the 128-grain bullet with same velocity as the old reliable .32–40 187-grain bullet, front-seated, and shell filled to wad with black powder.

The cross referred to may have been shot away many times, but ends of it are distinct as when drawn, and for two years it was drawn on the face of all original targets before taking down, and the V-rest has not moved since its erection.

Test 82. — November 3, 14 groups were made with different charges of powder; 11 groups being with 8 or 9 grains only. Omitting three of these groups, the smallest was .62, largest 1.50, average 1.30 inches. Again we have some appearance of accuracy.

Test 83. — December 24, three groups were made with fixed ammunition, using 9 grains powder; first was 3, second .75, third 2.50 inches. This was certainly discouraging with fixed ammunition, sending us back to the erratic groups previously made. To be sure one group was very small and would look well in print if the first and third were destroyed, but a double-barrel gun loaded with double B shot, at close range, would murder more woodchucks and offer as much sport as such a rifle.

This closed all fixed ammunition tests for more than a year and this barrel was rechambered for front seating, though never taken into consideration as

THE BALLISTICS OF SMALL ARMS

suitable for woodchuck hunting, and fixed ammunition was absolutely out of question for accuracy with any barrel yet in hand. An examination of tests 76 and 83, however, show the great improvement of high-pressure smokeless powder over the best load of black with nitro priming, and that the .281-inch ball shot better than the .285 or factory size.

Many specimens of powder were examined

Fig. 52.

and tested from time to time, all of more or less interest, and as a matter of curiosity and comparison a photograph was made of four widely different specimens. In (Fig. 52) is represented some grains of Hazzard's F. G. powder and four grains of black cannon powder as used during the late Civil War.

(Fig. 53) exhibits a number of grains of the Laflin & Rand sharpshooter powder, and one piece of modern smokeless powder, as used in some heavy ordnance.

Fig. 53.

Telescope Mounts; an Invention. As early as 1894, while pretty thoroughly absorbed in rifle and ammunition experiments, hunting for the cause of x-error, the necessity for reliable and easily adjustable telescope mountings presented itself. The first special Stevens rifle with its telescope mounts of that year emphasized the necessity still more emphatically and decided that mountings must be obtained which should be accurate and practical. In looking about for an idea it was found that neighbor Jewell, an old-time long-range man, was right in saying that the telescope mounts then on the market were defective in various ways. The one in hand would allow the scope to take two separate positions in the front ring, and the clamping device at rear mount was so well made for security that it would retain the scope tube in a bent position after slight pressure upon it.

As a general rule inventors and manufacturers are not very energetic in producing articles for which there is no demand, particularly is this the case with manufacturers; and after persistent hunting the conviction was forced that, as a rule, riflemen were expecting nothing better. Old customs retain a marked influence, many times unwittingly impeding the progress of even thinking men.

To adjust the scope elevation for a fine rifle by a scale reading 20ths of an inch was an anomaly, and no system of mounts had been devised which, if accidentally or intentionally moved, could be replaced in their old position; the rifleman could determine whether the sights had been moved or not only by the tedious process of going to the range and targeting up the barrel anew, and this must be done on an exceptionally quiet day. If his elevation for 200 yards was approximately found, no amount of persuasion would induce a change to 100 yards; the process was too tedious and troublesome.

It was not necessary to lay awake nights to decide requirements that a proper scope mount should meet and fill, but it took years, rejecting 10 different mounts and accessories, before the eleventh one was made to " fill the bill," and this has been patented just for fun, so any one is at liberty to use any part of it.

A visit of three days was made to the largest collection of firearms in the country, and extended examination indicated that mountings of telescope sights were considered of least importance of anything about the rifle. One had patterned after another for years, and a hundred-dollar rifle was fitted with mountings that " looked like 30 cents."

(Fig. 54) on opposite page gives the first and ninth attempt, the ninth model shown at the left never being completed. With the exception of this, however, all the 11 models were finished and used on various barrels until a better one took their place.

The first attempt was to so construct the mounts that, unless broken, the rifleman could determine whether the scope had been moved from its position or not; the next was to construct so that after an alteration of any kind, by any one, except breakage, it could be reset to its old position with certainty and immediately, although .002-inch variation at mount means $1\frac{1}{2}$ inches at 200 yards with a 19-inch telescope.

These fine fractions of necessity compared with the coarse ones of the rifleman's constant associations account for his shooting day after day at a fixed and unalterable distance, and when this becomes tiresome he stays at home.

There was not much difficulty in meeting most of the requirements of a proper mount, and the eleventh model, shown in (Fig. 55) on following page, was the one finally patented. Its successful attachment to the barrel, however, for a long time seemed impossible, though the different models made and thrown aside were mainly to make these mounts more artistic in mechanical construction and more rigid. In the left corner may be seen the dovetail block marked b' in the patent drawing (page 129), which being driven on to the longitudinal dovetail, provided on the barrel, becomes immovable in the correct position for receiving the removable rear mount shown as the central figure of the cut. The block at the right is for the removable front mount which contains the windage, but this mount, after being once set and pinned is only correct for its particular

Fig. 55.

barrel. (Fig. 56) represents a block made to be firmly soldered to the barrel when no dovetail has been cut in it by the manufacturer, simply taking the place of the longitudinal dovetail, particularly when a round barrel is to be utilized. This is drawn and included in the patent.

The drawings given on Plate 10, of the eleventh mount, are a reproduction from the U. S. patent, issued Sept. 11, 1806. The writer has built jigs and tools, and manufactured 12 for his own convenience, and four years of constant use have not suggested any alterations. They are built of cast steel, solid and heavy, so solid and heavy that most riflemen to whom they have been shown prefer the good old-time mountings when a telescope rifle should have been kept in a glass case where it would not be exposed to the wind.

Fig. 56.

The drawings also show the utility of the dovetail rib running full length on top of the barrel for receiving these mounts, which has been repeatedly re-

PLATE 10.

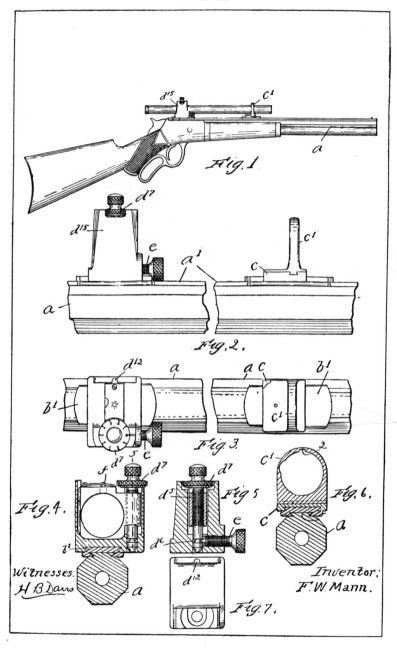

ferred to in connection with various experimental barrels. The novel features of this mount, in a nutshell, are as follows: —

Excessive rigidity, and by its certainty of position is not susceptible to jars or shocks; easy removal of the telescope from mount, and also the replacement of the same to its correct position with certainty; the transference of the mount from one rifle to another without tools, and in every instance the mount will find positively the same position on the respective barrels as before, and because the mount has but one lateral position and cannot be replaced in any other than this position, no sighting up is required for either rifle after sighting up of the front mount has been once performed; the mounts can be attached at any position, or changed to any position on the rifle barrel between the muzzle and the breech without disfigurement of the barrel, because there are no slots, screw holes, solder, or rings which determine its position, thus allowing the use of a telescope of any desired length; a micrometer elevating screw with a powerful clamping device for holding said screw in whatever position it may be set, and a large graduated head on the elevating screw, and means for determining the position of said head on the elevating screw whereby manipulation of the screw is convenient and its reading distinct.

The longitudinal dovetail running the entire length of the barrel has solved the hitherto insurmountable difficulty of constructing a firm and reliable telescope mount. After the rifle is accurately sighted this micrometer mount will produce a range finder with the rifle. An unknown distance is found by reading the elevation after it has, by test, been set for the unknown distance, by having the bullet hit the object aimed at.

At the end of four years the old timer, Nathaniel Jewell, of Milford, who first pointed out the defects of commercial mounts, had educated the clubmen at the Milford range in the use of telescopes and how to remedy their defects, so none of them would appear any more at this range with open or peep sights.

Quite a correspondence was exchanged between E. A. Leopold and the writer relating to telescope mounts, and the adaptation of target rifles to field work, which is fairly well set forth in a single one of the many letters passed, as follows: —

"MILFORD, MASS., Jan. 9, 1904.

"FRIEND LEOPOLD: Your letters are in. Thanks for the required charge of F. G. for .28-caliber, etc. That matter is settled. Your account of 'Recreation, Pleasure of Sport as some call it,' was to the point. I have never read anything from your pen of this nature before, and I have been thinking since. If you will bear with me, read how my thoughts run.

"With scope mountings that cannot be quickly and satisfactorily adjusted, there is good reason as you say, for smiling at a person out for sport who would try to adjust such mounts when he did not know where to put them, even if they were easily adjusted. With mounts that are non-adjustable like yours, one would naturally shoot his rifle aiming over for distance, for he could do no other way. Three guesses are then required, one for distance, one for how much the ball would drop for said distance, and one for how much to hold over after he had guessed how much the drop is. This form of fun for a scientific rifle crank should be out of date. It seems unscientific.

"Again, in the above form of shooting I cannot see why fixed ammunition, or a repeating rifle, would not be as good as any, and a $17 one as good as the $47 rifle. Also with your .22–15 which throws the bullet wild, I cannot see the need of a high-power scope, or of micrometer mounts. All you can do is to blaze away, or rather snap away. Here is where you are. You do not need much accuracy in the mounts because your bullet flies wild. And because you cannot adjust your mounts it makes but little difference whether your bullet flies wild or not. So why should you not think that fixed ammunition is adapted to your sport? You are mistaken. You have adapted your sport to fixed ammunition, and not your ammunition to your sport. Neither have you adapted your rifle or your mounts to your sport. You demand much of your range rifle and then discard it when you are out in the field. I have often wondered just what riflemen think the range rifle is for. I have never had any idea that it was for range work alone. Range rifles and field rifles should be interchangeable, not by government rules, but by the good sense of certain individual men who claim to think for themselves.

"Yours truly,

"F. W. M."

Accurate Fixed Ammunition Difficult.

It may be well to explain that the experiments made during 1903 with this .28-caliber, fixed ammunition rifle, were made by one who lacked confidence in this form of work, and the variety of tests made may have lacked the one which might have been worked into an accurate load; also that in one respect the painstaking instruction given by Mr. Leopold was not obeyed. This was not intentional, because it was expected this work would be continued, the many tools provided for it being carefully preserved.

The labor and care necessary for keeping the shell in order for gilt-edged fixed ammunition, and for producing it, is all out of proportion to the very simple matter of front seating when the chamber is constructed for this latter form of loading. All kind of difficulties arise with fixed ammunition in connection with making tools for it, while with front seating they are far less. As to accuracy of the two forms, the various tests recorded are self-explanatory.

Respecting speed in loading while hunting in the field, during which front seating has been condemned, there are two sides. The hunter who is after the largest possible amount of game represents one, and he who proposes to kill with a single shot represents the other. He who is after the greatest amount of game should have along an understudy armed with a double-barrel gun, charged with double B shot, with which to finish the business attempted by the rifle. The other will be satisfied with a little game if every bullet furnishes one specimen. So far as speed in loading is concerned, one shot an hour may be the average when after woodchucks; one in two days is likely to be sufficient when after deer, except when they are hunted in story around the campfire, or after the first shot, due to hasty loading, has made a miss, or after the home-coming.

.28-30 Pope 1904 Rifle.

This barrel came from Mr. Pope early in 1904, and a special chambering reamer was obtained for the purpose of making an experimental chamber so it might take a reduced Herrick .28–30 straight shell, $2\frac{1}{2}$ inches long, being $\frac{1}{4}$ inch shorter than factory made. One inch of front of the chamber was same as bottom of the grooves, viz. .285 inch, and the shell was reduced to match.

With due care these chambers at their front end can be made very central with the bore; with a sharp reamer and properly made pilot it is surprising how perfectly the bottom of grooves leave their mark in the new chamber, and these marks all disappear when the chamber is lapped out a little with fine emery.

The solid base of the Herrick shell is $\frac{3}{16}$ inch deep, and inside depth of this new shell is $2\frac{1}{16}$ inches. Filled with sharpshooter powder, leaving air space for oleo wad, it contains 20 grains, the same volume as the 1903 fixed ammunition shell after the 120-grain bullet was seated.

This chamber permits of front seating with a bore-diameter bullet without a seater; the loaded shell, acting as a plunger, carries the bullet to its place without jamming or bruising, as no force is required. By shortening the shell equal to length of the body of bullet, a .285-diameter one can be seated in chamber, but not in the rifling without mutilation or force. This system is very similar to fixed ammunition when using a new steel shell most perfectly made, as has been done by some. The bullet is discharged into the rifling from a .285-diameter chamber perfectly central to bore, in one case, and from a .285 shell in the other.

Test 84. — April 19, 1904, a trajectory test was made with this barrel; front-seated, Leopold wad, 128-grain swaged bullet. The trajectory with $10\frac{1}{2}$ grains sharpshooter powder and $7\frac{1}{2}$ U. M. C. primer was 2.90; with $11\frac{1}{2}$ grains 2.35; with $12\frac{1}{4}$ grains 2.15; with 13 grains 2; with 14 grains 1.80 inches. Six shots making these trajectories formed a group at 100 yards of 3.50 inches perpendicular and .50 inch horizontal; every one of the shots would have cut a ramrod if placed perpendicular at the target. A 28-grain charge of black powder was tried at same time, giving a trajectory of 3.31 inches.

Trouble with Smokeless and Rifle Bores.

Test 85. — April 23, 1904, four 5-shot groups were made with 11 grains smokeless, oleo wad, 138-grain, bore-diameter, front-seated bullet; smallest group was .87, largest 1.63, average 1.30 inches; trajectory 2.88 inches. Two 5-shot groups were then made with black powder under same conditions, resulting in 1.62 and 1.50 inches; trajectory 3.31 inches. Fifteen shots were then made, with slight alterations in bullets, but the groups were wild and some bullets tipped badly.

After cleaning the next day a ring was discovered $\frac{1}{16}$ inch in front of chamber, irregular and extending three-fourths around bore, and a similar one was found $\frac{3}{8}$ inch from muzzle. Mr. Pope cut off both ends of this barrel, re-chambered it with same tools first sent him, and reported that bore was badly pitted for two inches up from new chamber.

Both the 1902 and 1904 barrels had been cleaned with extra care, using the highly recommended powder solvents as sold by the proprietor of the formula, and some others as highly recommended; but it was evident smokeless powder and $7\frac{1}{2}$ primer were causing trouble with the bore, and no successful method of prevention was found. Though both barrels were cleaned with three times as many swabs as after black powder, and immediately after the day's work, the ring appeared at the muzzle in one night after leaving a swab saturated with the powder solvent at that point.

The swab was rusted to the bore when found, and repeated experiments showed that it did not result from the solvent alone, but by some primer acid left in the dirty swab. There could be little question but the breech ring was a rust ring and not due to excessive pressure, and at this time it did not seem possible to keep these .28-caliber soft steel barrels from deteriorating rapidly. Grease and swabs had no effect, as moss would grow in the barrel nearly every night. Highly magnified, this moss was shown to be small straight cylinders of a black substance which seemed to be forced out of small holes in the steel.

THE BALLISTICS OF SMALL ARMS 135

Ruined Rifle Bores *vs.* **Smokeless** *vs.* **Primers.** The question of what was ruining so many valuable rifle barrels in the use of this high-pressure powder was becoming a serious one, and we determined, if possible, to discover by a series of tests whether it was the powder, the primer, or both combined. Many of the tests had little bearing upon this particular question, but were interesting and helped to decide some other things, as will be seen.

By supporting a steel plate in position over a gas jet and laying upon one edge of it a $\tfrac{3}{4}$-inch square bar of steel, to protect the operator, one primer after another may be exploded by the heated plate when placed upon it behind the bar, and the jet of flame rising from behind may be watched. It was found that one or two out of every five of the $7\tfrac{1}{2}$ primers failed to explode, only sizzling a little like damp powder, the anvil in these remaining undisturbed, while the anvil in those that do explode is expelled forcibly enough to often remain stuck in the boards at top of room.

Under the same conditions all the $2\tfrac{1}{2}$ primers did explode, but one or two out of five have a defective explosion, and the operator cannot detect that the flame extends any higher from the $7\tfrac{1}{2}$ than the $2\tfrac{1}{2}$ primers, or that the detonations are any louder or more violent.

Upon application the U. M. C. Co. replied that one primer was no stronger than the other, but the mixture being different one would properly ignite smokeless powder, while the $2\tfrac{1}{2}$ was more suitable for black, and the failure of the $7\tfrac{1}{2}$ to uniformly explode by the application of heat was known to them.

In another test an empty shell in the .32-40 rifle was utilized in which primers were exploded in the normal way. Reducing strength of hammer spring until some would fail to explode, then returning it back until all primers from same box would, it was found that about one in five sizzle into the shell while others gave the proper detonation. This does not apply to any particular make of primers, but shows very clearly a matter of importance. It indicates that a firing pin that makes no misfires may not strike hard enough to properly explode the primer when primer is not at fault, and this differs from the way chemists would expect a fulminate to act.

In any ammunition the firing pin should not be driven entirely through the primer, yet it should be struck with sufficient force or hangfires are liable to occur even with good primers. The construction and length of firing pin should be adapted to its purpose, especially when using high-pressure powders, kept in order and free from rust so its action may be uniform. The fact that the charge is ignited does not prove the firing pin is acting properly. Reducing the hammer spring showed that primers of different makes exploded with much lighter strokes than others. The $9\frac{1}{2}$ U. M. C. would not explode with full power of the spring, and information was obtained from one factory that these primers should only be used in a bolt action.

Still another experiment was made by placing a bullet in an empty shell, .32-40, with three pounds or less pressure, and the explosion of a $7\frac{1}{2}$ primer moved the bullet forward a little until it was stopped by the rifling, while if the bullet was entered into the shell with over three pounds pressure it would remain unmoved by the explosion. Computation indicates that the $7\frac{1}{2}$ primer in this empty shell gives a pressure of at least 36 pounds to square inch, or about double the normal atmosphere.

Filling the shell with some inert substance to represent the powder bulk brought in complications which made the test useless. The primer residue thrown into the barrel with these and other like tests was large in quantity and very acrid in nature. Smokeless powder should not be accused of destroying barrels; the manufacturers of primers admit it is the primer that does the mischief. Although the same primer exploded in a full charge of black powder will act but feebly, if any, upon the barrel, it is quite evident that this powder not only neutralizes the primer acid, which the nitro powder does not, but its larger bulk of hot gases dilutes the primer residue and drives more of it out at muzzle.

.28–30 Pope Continued. As previously mentioned, this barrel was rechambered to remove breech ring made by acidity of charges, and half an inch was also cut from muzzle, then subjected to further tests.

Test 86. — May 17, 1904, five groups were made, front seating, 11 grains powder; smallest 1.25, largest 2.50, average 1.75 inches.

Test 87. — Four groups were shot with black powder, nitro priming; smallest 1.12, largest 2.50, average 1.75 inches. May 21, five more groups were made with same powder; smallest 1.12, largest 2, average 1.75 inches. This corroded barrel did not seem to be shooting small groups.

The constant tabulation of these tests and groups is important if positive shooting information is to be obtained. It is a rifle study to show that the causes assigned in past for shooting errors are incorrect, and to determine, if possible by a system of exclusion, the real cause or causes and the laws through which they are acting. There has been enough of theory and too little genuine experimental knowledge for real advancement in ascertaining facts.

These records do not compare with those published from crack rifle ranges because they are not picked groups or competitive shooting, yet may aid more towards the discovery of real difficulties and methods of correcting them.

With this kind of shooting there is no buckle to the rifle or sights to get out of line; no mirage or varying of sights between shots. The operator does not have a speck of dust fly before the shooting eye to account for a poor shot. Each charge of powder is weighed by fine chemical balances so trajectories are not changed by varying charges, and the same shell is used for all shots in the group so the fit is uniform.

A careful experimenter, keeping alone at this work for years, ought to discover that there is some definite trouble causing this irregular shooting, not only in this slightly corroded barrel, but in nearly every other which is not yet excluded.

Test 88. — This was made June 14, to test result of using different air spaces between powder and bullet in the shell. Ten and a half grains of sharpshooter powder gives one-inch air space with bullet seated in the rifling, and represents the first of seven shots made, as shown in (Fig. 57) on following page. No. 2 shot was entered $\frac{1}{4}$ inch beyond front of chamber and an increasing air space of $\frac{1}{4}$ inch was given after each succeeding shot, hence the 7th shot had $2\frac{1}{2}$ inches air space. The 7-shot group gave 4 inches perpendicular and

.75 inch horizontal. As indicated by cut the sharpshooter powder explodes with fair uniformity with excessive air space, but gives higher trajectories as the air space is increased.

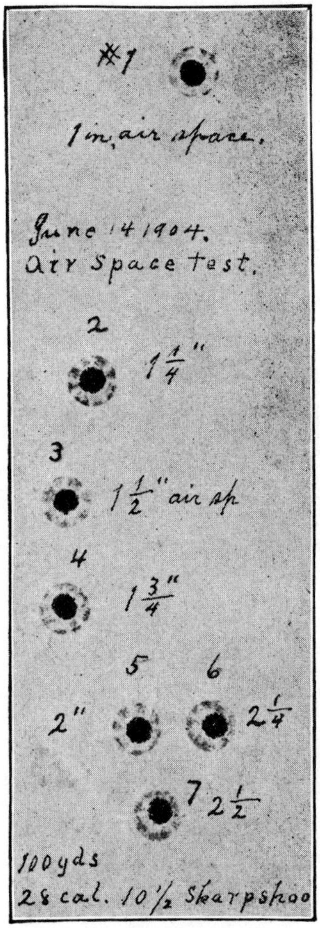

Test 89. — During same day as preceding three 5-shot groups were made with bullets .280 body diameter and a base band of .290; 10½ grains powder, bullet seated with some force in front of shell but not in rifling, as the shell had been cut off, thus placing body in the bore and leaving base in chamber, ¾-inch air space; smallest group 1.12, largest 1.25, average 1.19 inches.

Test 90. — This experiment was made under same conditions except with one inch air space and with bore-diameter bullet loosely seated in rifling, making groups of from 3 to 4 inches. A comparison of these two last tests indicates that smokeless powder, when large air space is present, is useless unless there is some initial resistance while it is being ignited, there being practically no frictional resistance afforded by the loosely fitting bore-diameter bullets, while the large base band of the two-cylinder bullets used in previous test furnished the obstruction needed.

The peculiar property of nitro powders, under different degrees of compression, or with varying air spaces, is commented upon in Mr. Pettit's letters which have been quoted.

FIG. 57.

Test 91. — On June 18, 1894, to compare the reduction of air space in .28-caliber barrel with the 88th test, seven 5-shot groups were made. One shell was shortened ¼ inch; another ⅜ inch; one had ¼-inch brass plug soldered into its base with ⅜ inch cut from its muzzle. The groups proved to be higher and lower as air space was reduced or increased; smallest group made was 1, largest 1.75, average 1.50 inches. The inch group happened to be one shot from least air space, but all groups were good. Some bullets in all groups printed badly, but neither the 1903 nor 1904 .28-caliber barrels made fine prints with any ammunition.

Test 92. — On July 4, the day in which everybody was licensed to do any kind of wild shooting if it only made a noise, the sixth group shot in this test was up to the scratch. It was larger than the target (8 × 11 inches); the three shots which did hit it made a 4.75-inch group. Two dry, well-fitting swabs were pushed through bore between shots of this group; otherwise than this the best ammunition was used.

Fig. 58.

W. E. Mann, in commenting on this particular group, says that it looks as if the excessive friction caused by a peculiar surface left in barrel after passing a tight-fitting dry swab through it caused a higher form of explosion, thus badly deforming the soft lead bullet so it must go wild.

Many trials with this sharpshooter powder demonstrated that the bore must either be left untouched between shots or very thoroughly cleansed. Six other groups, five before and one after this, are given in next test.

Test 93. — To confirm test 92 the details of it are here recorded. Six 5-shot groups were made, without cleaning, using nine grains powder, 120-grain 1 to 30 alloy .285 bullet, seated in chamber, a very close imitation of fixed ammunition. Smallest group made was .87, largest 1.62, average 1.19 inches, and the small groups of this test begin to show that something is needed, and later tests indicate it is an alloyed bullet. The sixth group of this test as before mentioned was larger than the 8 × 11 target.

The fourth group (Fig. 58) is printed above to assist the reader to fix in mind the real size of a 1.62-inch group.

The absence of detail in these various tests may be a mistake, making them uninteresting to a casual reader, but an excuse for omission lies in the fact that no amount of detail or explanation would do justice to the steady day by day experimental labor performed on this range. It may be taken for granted that the details were well discussed, planned, and investigated by W. E. Mann, as well as writer. Both the 1903 and 1904 barrels were a great trial, due largely to the action of smokeless powder and primers upon barrels not adapted, this last being ruined during the middle of first season.

A .28-9 Barrel, 1904. The annual woodchuck hunt at "Medicus" preserve was fast approaching and a rifle must be obtained that had some promise of accuracy. As the bore in the .28-caliber, 1903 rifle was in fair shape, its fixed ammunition barrel was rechambered for Herrick shell, one inch shorter than factory make, and $\frac{3}{4}$ inch shorter than shells tested with the previous barrel, the .28–30, since air-space tests with this indicated that less was desirable. Front end of chamber remained the same in diameter, viz. .285 inch, and dies were made for forming shells. A fancied necessity must have helped this barrel along.

Test 93 a. — July 26, with bore-diameter, alloyed 1 to 16 and 1 to 20 bullets, with base bands, pushed into place by shell, nine grains powder, leaving $\frac{7}{16}$-inch air space, five groups were shot, shown in cuts (Fig. 59) on opposite page. Smallest is .75, largest 1, average .87 inch. The first and second groups were made with 120-grain alloyed 1 to 16 bullets; the third, fourth, and fifth were made with 138-grain 1 to 20 alloyed bullets. Place a nickel on the groups.

Test 94. — On July 30, .28-9 barrel, nine grains powder, oleo wad, 120 and 138 grain bullets, front-seated, were used, and air space was made to vary from $\frac{7}{16}$ to $\frac{1}{4}$ inch by brass soldered into base of shells; no cleaning, bullets all entered or pushed to their seat by shell; eight groups were made, smallest .75, largest 1.25, average 1.06 inches.

Test 95. — On August 6, this experiment was made similar to the last, with nine grains powder and also with 11 and 12 grains "lightning" powder, oleo

THE BALLISTICS OF SMALL ARMS 141

 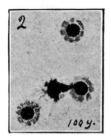 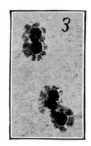 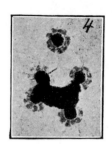

FIG. 59.

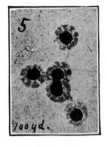

wad, air space varying from $\tfrac{7}{16}$ to $\tfrac{1}{8}$ inch, 120 and 138 grain bullets, all dropped into chamber and seated with loaded shell, no cleaning; nine groups were made, smallest .87, largest 1.50, average 1.03 inches.

This steady shooting for two days with front-seated bullets might well be compared to the fixed ammunition with same barrel, made in test 76 and following, which were so discouraging.

The woodchuck hunt came in September, and this barrel, with a remodeled Stevens action, Sidle scope, went to Dr. Skinner's range at Hoosick Falls. This rifle outshot, at 100 yards, Dr. Skinner's old reliable .38-55 which had, in its day, made many wonderful scores, but after a few shots during this summer of 1905 this barrel was entirely ruined. In spite of care given, with less than 100 shots, it became a nutmeg grater, the steel being corroded away, supposedly by the $7\tfrac{1}{2}$ primers, as happened to the .28-30 barrel of 1904.

Discarding Two .28-8, 1905 Barrels. After the success with .28-9 barrel, as to accuracy, and its failure regarding durability, twin barrels, .28-caliber, were received from Mr. Pope, hoping for greater durability, and by his advice they were made of smokeless or high-pressure steel. For fear the primer acid might ruin one before the ammunition could be properly worked up and rifle properly sighted, the twin was obtained to fall back upon.

The recoil from nine grains sharpshooter powder was so slight, and previous barrels so heavy for chuck hunting, these barrels were reduced to pipestem proportions, of which the military rifle is a sample.

To test a theoretical idea the rifling was made only .0015 inch deep, half the usual depth, and shell was reduced still further by $\frac{1}{16}$ inch, making shell only $1\frac{3}{8}$ inches long; filled with sharpshooter powder to oleo wad it was found to contain $11\frac{1}{2}$ grains. This reduction of air space was made, hoping to use the $2\frac{1}{2}$ U. M. C. black powder primer in place of the $7\frac{1}{2}$ smokeless that so quickly destroyed two barrels.

All these changes were enough to destroy the value of any barrel, and so they did, and they made it necessary to reduce the charge from nine grains, with its $\frac{1}{4}$-inch space, to eight grains with $\frac{3}{8}$-inch space, altogether against previous plans. The nine grains opened nearly all the primer pockets, and the pipestem barrel jumped so much when discharged that all hopes of accuracy, when used at muzzle and shoulder rest, were abandoned.

That the nine-grain charge would show its power if given a chance was soon demonstrated in following manner: one of the best Ballard actions attainable was annealed and reduced in weight for this barrel, the machinist working three weeks drilling and filing and finishing. Every possible part was reduced where metal could be removed, and there was some of it. Unknowingly, he cut the metal away where it ought not to have been, and the fifth nine-grain charge from it set the breech block back into the receiver, making of this three weeks' labor and highly prized action an attic treasure.

Later, a $\frac{7}{8}$-inch steel tube was slipped over this pipestem barrel and filled with melted tin, and after cutting a thread on the barrel at its muzzle a nut was screwed tightly down, holding tin and steel tube firmly in place. This barrel was fitted to another Ballard action, sight blocks, and telescope mounts, after which it shot good groups; having been mislaid, however, none of these tests can be found.

After finally removing false tube and repairing the spoiled Ballard action with a well-fitted steel band, this made the finest show of any rifle in the rack. Its twin barrel is still wrapped in tissue paper, as received from Mr. Pope, chamberless and hammerless, and with it stands a new .30-caliber Krag, smooth-bore pipestem barrel, also finely finished though still unchambered. Very few riflemen have the opportunity of looking through such a bore as this unrifled one presents. When bringing it to the eye great things can be seen as well as imagined, like the telescope to the astronomer's eye.

This year, 1905, closed all lead-bullet tests and experiments, and all rifles, 24 different barrels, ammunition, chambering tools, and some 40 odd swages were laid away.

Remodeling a Rifle. Experimental Shells. During 1905 an attempt was made to lighten up the .28–30 1904 barrel and, in a remodeled Stevens action, make a handsome and light rifle for chuck hunting to please a friend. The full octagon, dovetailed rib barrel was turned down as much as was thought advisable, leaving two inches of original dovetail at front mounting but removing remainder to within three inches of breech. A forestock was fitted to within two inches of the muzzle to receive a ramrod. The chamber was still further reduced to take the reduced Herrick .28–30 straight shell, made $1\frac{5}{16}$ inches long, inside measurement being $1\frac{3}{16}$ inches, while regular factory shell is $2\frac{15}{16}$ inside.

The bore of this barrel had not been entirely ruined, for rust rings had been cut from both muzzle and breech, and this was third chamber cut. When finished, however, it proved too heavy to suit the one for whom it was made and would not make steady groups even with front seating, — a failure in every way except in its fine appearance.

While closing experiments with the cast-lead bullets, a word or two in regard to various shells used in connection therewith, and an illustration of them, as shown on following page (Fig. 60), may be of some interest. They show different manipulations of the Herrick shell.

Figure 1 is the Herrick .28-caliber straight shell, 2.50 inches long, for fixed ammunition.

Figure 2 is the 120-grain cast bullet.

Figure 3, the same shell slightly shortened and reduced to improve fixed ammunition.

Figure 4, the shell shortened again and still further reduced to bring outside diameter of its neck to groove diameter of .28-caliber barrel.

Figure 5, the shell shortened to 1.50 inches and reduced to bring neck diameter to groove diameter or .286 inch.

Figure 6, a Pope 138-grain, .28-caliber, cast bullet for muzzle loading which being swaged to bore diameter was used in nearly all .28-caliber barrel work.

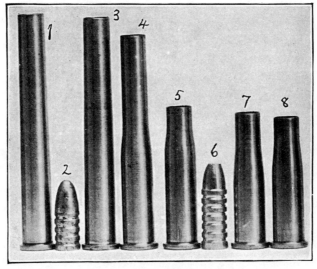

FIG. 60.

Figure 7, the shell further shortened and shoulder carried further down, to reduce air space, leaving the neck, groove diameter as before.

Figure 8, the shell shortened again, still further reducing air space.

At one stage of bullet experimenting, three new bullet molds were produced in an unsuccessful attempt to cast a 138-grain, bore-diameter bullet composed of one part tin to 20 of lead, and with slight base band so as to avoid sizing down and swaging up a grooved bullet. The swaging of a grooved bullet is apt to close up grooves more on one side than the other, thus unbalancing it. None of the three molds cast satisfactory two-cylinder bullets.

THE BALLISTICS OF SMALL ARMS 145

.30-caliber, 21 & 8 twist, 1904 Barrel. Jacketed Bullets.

While making the usual annual visit to Mr. Pope, he was informed that a rifle was wanted that would stop a woodchuck, stated with some emphasis. After considering a moment he replied, "Why not try a .30-caliber smokeless?" meaning by this a .30–30 Winchester which carries a jacketed bullet.

Though this was an entirely new idea, of using metal-cased bullets for accuracy, a plan instantly presented itself for testing them under a new system; one similar to bore-diameter lead-bullet tests, though Mr. Pope did not suggest this for accuracy, but for its killing power.

After watching the magazines for years, it was not learned that metal-jacket bullets would do "gilt-edged" target shooting, and so far as could be discovered all riflemen claim that jacketed bullets did not upset when discharged; this was the very thing desired, notwithstanding test 13, June 25, 1898, indicated that a bore-diameter bullet with an enlarged base band that did not upset was valueless. At this time, however, it was implicitly believed by me that the metal-jacket, two-cylinder bullet would upset.

Two of these .30-caliber barrels were received, one with regular 8-inch twist and the other 21-inch, the slowest possible for Pope to cut without making a new rifling head. By 8-inch twist, or pitch, is meant that grooves in a rifle make a complete turn in its bore every eight inches.

One barrel was chambered regular for the .30–30 shell, but never so used; and soon as new chambering tools arrived one barrel was cut off and both were chambered, leaving front end exact diameter of bottom of grooves, that was .308 inch.

Shell reducers were made and steel swages for working over the 170-grain metal jacket, and after several months and persistent efforts a few bullets, nearly correct in shape for this caliber, were completed. The difficulties encountered in obtaining swages, extending over these months, will hardly be of interest in this connection, but the time spent prepared one to admire the extreme accuracy with which the factory turns them out. After sawing many of the metal jackets open, freeing from lead, and measuring all parts of jacket with micrometers, their perfect uniformity was apparent. With all this accuracy and uniformity, to put this work of art into an ordinary shell and commercial chamber is a monstrosity; that their shooting made large groups was to be expected; that these large groups result from their great velocity, is an idea descended from old lead-bullet days when accuracy was not expected from "express" bullets. A new Crank's Corner should be instituted at Walnut Hill.

Test 96. — September 2, 1905, a rotation test was made with the 8-inch twist barrel, resulting in a hollow group of 6.25 inches, and the 21-inch twist made a group of 10 inches.

These were both pipestem barrels, refuse ones, because turned too small for the .30–40 Krag, and were experimental, rifled but never finished outside, and, as will be observed, both shot a larger rotation group than any since the Stevens .22–15 was tested by a stranger to the V-rest. The barrels were so light that they recoiled badly in the rest.

Test 97. — December 20, 1904, with the 8-inch twist barrel (cut for 220-grain government bullet), bore-diameter, 170-grain, soft-point U. M. C. bullet with a base band $\frac{1}{16}$ inch long and diameter of .308; the base band fitted the chamber while body of bullet .303 lay lightly in bore; 18 grains lightning powder, no wad, bullet dropped into chamber and seated with loaded shell. This chamber and these bullets were designed after the exact pattern of all of my smokeless powder chambers and ammunition using lead or cast bullets and several of my old black powder rifles dating back to test 13, June 25, 1898. After I had discovered that metal jackets would upset, and showed Mr. Leopold the following groups and sent him the bullets, he christened them "Two-Cylinder Bullets," a very appropriate name.

This was a test to discover if bullets would upset, and one does not go 16 miles alone to a range, on a New England December day, unless there is something doing. Three groups were made, as illustrated by (Fig. 61) on opposite page, measuring 1.44, .87, and 1.25 inches repectively, and were the first metal bullets shot on this range. They were caught in sawdust and found to have upset to bottom of their grooves, whole length of their bodies, and every print was perfect.

Test 98. — January 2, 1905, with same 8-inch pitch rifle and conditions similar to last except bullets with $\frac{3}{64}$, $\frac{1}{8}$, and $\frac{1}{4}$ inch bands were used, all seated by loaded shell; body of bore-diameter bullet fitted the lands easily, bore not being throated in front of chamber as in some high-pressure rifles, three groups were made. The $\frac{1}{4}$-inch base band group was 1.62, $\frac{1}{8}$-inch band 1.75, and the $\frac{3}{64}$-inch band group 1.25 inches.

Test 99. — January 20, test was the same as the two preceding; two groups were made, the $\frac{1}{16}$-inch base band group being 1.50 and no base band group was 1.25 inches. One of the bullets with base band tipped slightly, but it was the first shot. The five with no base bands made perfect prints

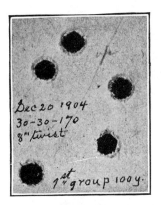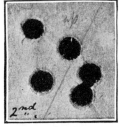

Fig. 61.

and formed a group $\frac{3}{4}$ inch higher than the other. Since no wad was used there was gas-cutting with no base-band bullets in corner of each groove. They all upset, however, and filled to bottom of grooves except in extreme corner into which even a lead one will hardly upset.

Test 100. — August 19, 1905, with 21-inch twist barrel, an experimental one; 18 grains lightning powder, no wad, 170 grains U. M. C. swaged, two-cylinder with $\frac{1}{16}$-inch base band, entered by loaded shell; first group made was 4.75 inches, and all bullets tipping five degrees. Another 5-shot group was made and only one bullet reached a 12×14 inch target.

In respect to the degree to which a bullet may tip before it tumbles, Mr. Leopold determined by his experiments that a tip of six degrees seems to be the limit.

Test 101. — This was made same day as the last, using 23 grains lightning powder which is the factory load, no wad; size of group 2.62 inches, all bullets making fine prints and all but one perfect.

It is very likely that the V-rest and concentric action could not prevent or overcome the jumping or springing of this crooked pipestem barrel; the rotation test, which gave 10 inches, would indicate how much the barrel was bent, and this is the first time since the V-rest and concentric action was invented that any doubts arose regarding their ability to manage the barrel, but this barrel was slender and crooked and the charge heavy.

Another factor encountered here was the fact that the high-velocity bullet must make a curved flight in this particular barrel, or the barrel must alter its position to allow ball to fly straight. To test this point two groups were made, using a 15-pound weight on the front ring and 10 pounds pressure of the hand on rear ring of barrel, and with 23 grains powder size of groups made were 1.75 and 1 inch. All the prints were perfect.

In connection with this, the preceding test should be compared, as they disclose the reasons why this barrel was fitted with its unusual twist. With 18-grain charge the group made was larger than the 12 × 14 target. With 23 grains the bullets made perfect prints and fine groups, in spite of crooked barrel.

After about 70 shots from this 21-inch twist barrel the 170-grain bullet began to tip and make larger groups. Increasing the powder charges from 23 to 25 grains righted the tipping bullets and brought them back into the regular size group. In cutting a 21-inch twist with a rifling head made for a much steeper one, the land is all cut away except its leading side, thus leaving in this experimental barrel a land which has no shape comparable to that in normal rifling.

Test 102. — September 2, after putting steel tubes, one inch in diameter, on both these .30-caliber barrels, filling with melted lead, and putting $\frac{5}{8}$-inch hexagon nuts upon muzzle of each to retain the tubes, making weight of tubes and barrels seven pounds, a rotation test was made with the 21-inch twist, printing a hollow group of 14.75 inches, with a center $5\frac{1}{8}$ inches below the tack, and $\frac{5}{16}$ inch to left of a perpendicular line drawn at the tack. Riflemen will readily understand that the center of a complete rotation test group of a bent barrel determines the position of the group if same barrel was straight, and that the 5½ inches representing the center of this group below the tack indicates clearly the exact fall of the bullet in 100 yards.

At the time this rotation test was made a 5-shot group was made with barrel right side up, giving only .61 inch. The same day the 8-inch twist made one group, to test the value of other work, also another group was made with the 21-inch twist for same purpose, both with two-cylinder bullets having $\frac{1}{16}$-inch base bands. The 21-inch made a group of 1.06 and the 8-inch one of 1.12 inches.

Reflections; Pipestem Rifles and Jacketed Bullets. These two .30-caliber barrels decided all and much more than was expected of them. They have proved that with 18 grains lightning, high-pressure powder the 170-grain bullet is upset to bottom of the grooves through whole length of its body, and its soft nose settles back and takes grooves as well. With 23 grains powder behind they would certainly do this. If the 170-grain bullet will upset in front of only 18 grains powder, certainly the 220-grain government bullet will upset before 34 grains of powder.

The upset of a jacketed bullet must occur, and why so many riflemen contend that it does not is unaccountable. By actual test 600 pounds' weight will settle the metal-covered bullet down into a swage and make it impinge on sides of the die. Why should not 17 tons' pressure per square inch, which it receives in rifle bore, make it upset?

Again, these .30-caliber barrels gave better groups and lower trajectories with cylindrical, bore-diameter bullets with no base bands, than any other form of this bullet. It recalls the Chase-patch system where paper was used as a cover to lead bullet; so used it produced a bore-diameter bullet, though this was not the reason advanced among these who were using that system.

To the surprise of some riflemen acquaintances, increasing the velocity at muzzle, in the 21-inch twist, caused the bullet to make perfect prints and 1.06-inch groups with 23 grains powder, while they tipped badly and would not reach the target from the 18-grain charge, a charge that did almost perfect work in an 8-inch twist duplicate barrel.

The Winchester, .30–30 barrel which is rifled for this bullet has a 12-inch twist, the Remington a nine-inch, yet this 21-inch makes perfect prints when the bullet is balanced in the bore and not shot from a shell into a bore much smaller than the steel-covered ball; no wonder a sharp pitch is demanded in the factory barrel when bullets are to be so treated.

The small-sized groups made from V-rest with this 21-inch, though the barrel was strained out of line by an iron nut against a steel tube filled with lead, causing a rotation group of 14.75 inches, were remarkable.

Of more importance than the above conclusions were facts unexpectedly determined from these two differently pitched riflings, in proving or greatly strengthening an important statement made by Mr. Leopold in 1904. The tests respecting this statement formed the major part of all experiments with these lead-filled steel-tube barrels, and will be discussed farther on.

It will be noticed that the shell and bullet for these tests, with these barrels, were theoretically designed, and remained unchanged throughout all shooting with the normal 8-inch twist, a barrel the accuracy of which exceeded all lead-bullet rifles except the gilt-edged one of 1902. This ammunition was devised in a few moments, stimulated by Mr. Pope's remark, " Why do you not try the .30-caliber for chucks? "

Though these barrels were cleaned with less care than the two .28-calibers which were destroyed so rapidly, they are yet in good shape. Probably the large powder charge used in these diluted the primer acid, thus carrying more of it out at muzzle, or perhaps the harder steel barrel made some change for the better. The primer acid did affect these barrels, however, in about the proportion of this charge to the old lead-bullet charge, and also in proportion to the number of different days on which the several barrels were tested.

THE BALLISTICS OF SMALL ARMS 151

Ammonia vs. Primer Acid. Mr. Pope informed the writer in 1905 that he was employing concentrated ammonia to clean his .30-40 long-range rifle, and thought it was doing good work, and it was immediately put to test with the two .30-caliber barrels in hand. It worked well, entirely counteracting the primer acid, the swabbing on the day after shooting brought away no signs of any continued corrosion following swabbing of the day before, so it was voted, by a majority of one, a godsend to riflemen.

When the utility of ammonia is fully understood in this connection riflemen need have no fears regarding corrosion; it counteracts the primer acid and dissolves away the copper and nickel that may adhere to the steel. A few simple directions are all that are necessary to combine with one's own ability for observation.

The ammonia should be what is known by druggists as "stronger" or concentrated, which is about 28 per cent, having a specific gravity of .89; and this must be kept in well-stoppered bottles, else its strength deteriorates very rapidly. Our experience teaches that this stronger ammonia may be freely used in the steel barrel, and when the copper and nickel have been fully dissolved away the swab will lose the blue and green color which these metals impart to it; also that swabs well saturated with the ammonia will quickly remove the results of an ordinary day's shooting from the bore. If there is any question of copper or nickel being left in bore, it may be filled with the ammonia for a few minutes, but if left an hour or more some samples have been found to attack the steel.

The rifleman's judgment must be used in this matter, but keep all oil or other solvents out of the barrel until the ammonia has done its work, then the barrel may be treated the same as after the use of black powder. The addition of other ingredients to this stronger ammonia, as recommended by English authorities, has not been found necessary by the writer.

It is an interesting experiment to test the action of ammonia upon a brass shell or a jacketed bullet, noting the time taken to generate the characteristic colors from each. The green and blue colors produced by the contact of ammonia with brass and nickel are explained in text-books upon chemistry.

NOTE. — The term "primer acid" used in these pages is an expressive term rather than a chemical one. By it is meant the intensely acrid primer residue.

Accurate Ammunition Difficulties. 7 mm. Caliber Rifles.

Three smokeless barrels were received from Mr. Pope on January 1, 1906, one of them, a .28-caliber intended for lead bullets, never received a chamber and was never tested; the other two, one 24 and the other 30 inches long, had the same bore though termed 7 mm., each with 16-inch twist and lands $\frac{1}{32}$ inch wide. Longitudinal dovetail sight blocks were soldered upon these full round barrels, before finishing or browning, and in position for front and rear telescope mounts, thus obviating the necessity of slotting or cutting, leaving the barrels as large as the smokeless steel blanks would finish out, amounting to little flush of an inch in diameter. The sight blocks being two inches long, and encircling a third of barrel's circumference, are held firmly by solder, and are the outcome of the patented dovetail rib which, in its original form, extended the entire length of top flat of all octagon barrels. This block containing the longitudinal dovetail is covered by and contained in the patent described in detail on page 129.

These jacketed bullets materially changed the aspect of rifle experimenting in many respects; they seemed to open the way for fixed ammunition and another attempt was made towards producing it. Chambering reamers and reamers for shell mouths were gathered; bullet swages for the soft-nose bullets and jigs for reaming the mouth of shell central with its exterior circumference were made, and shell reducers.

The 7 mm. factory shell has no flange that works with ordinary actions, so the .30–40 government shell was shortened from $2\frac{1}{4}$ to $2\frac{1}{16}$ inches and the inside mouth diameter sized down by a die to .004 inch smaller than bottom of groove of the 7 mm. When the shell was reamed to take the bullet an internal shoulder remained for the bullet to rest upon, thus preventing it from reducing the proper air space in loaded cartridge. The completed shell when filled to the base of bullet would contain 37 grains Laflin & Rand lightning powder, and as 28 grains became the standard charge an air space of 25 per cent was left between powder and ball.

The 170-grain, soft-point, U. M. C. 7 mm. bullet was cut off and a new point formed, swaged cylindrical and bore diameter, leaving $\frac{1}{16}$-inch base band and

THE BALLISTICS OF SMALL ARMS 153

this was left .001 inch larger than rifle grooves; weight of bullet was then 130 grains.

When a thing is done it does not follow that the doing was an easy matter, and making of swages, dies, and reamers presented innumerable difficulties which may be of some interest. Perhaps it is not generally understood that no chambering reamers can be had of any rifle factory, all the great factories being very timid about sending out such things, besides having a reputation for throwing all manner of obstructions in the path of experimenters.

A new steel reamer will not make a hole the size of the reamer, and this is an experience that must be allowed for. The shell mouth must be reamed while it is in the sizing die, as no reamer will cut the inside of a thin brass shell with certainty unless the outside is supported; besides, when the shell is removed from the die, it expands by its own elasticity and the reamed hole expands with it, while it is also larger than the reamer before removal from the die, and the shell will not go into the chamber because the sizing die was made with the chambering reamer. This necessitates the making of new dies, and no reamer to make them with.

To avoid trouble, which is difficult of explanation, three dies were made by as many special reamers, makeshifts which odd jobs demand, a die for sizing the body of shell, one to reduce neck, and another to shorten the shoulder so it will not jam in the chamber. The shell neck die must be small enough so when the brass expands after leaving it, the neck will be .001 inch smaller than the chamber, which size makes a close fit.

After these operations the reamed hole or shell mouth has changed its size three times, and is now .001 inch smaller than base of the bullet; in seating the bullet, therefore, in shell it expands it nearly .002 inch so the latter will not enter chamber of the rifle. As the shell mouth reamer cannot be enlarged, the next best thing must be done, and in this case the chamber of the rifle was carefully lapped out with emery until the shell would go in. These were only the beginning of difficulties, leaving enough besides to the free imagination of any experimental investigator.

To give a better idea of the work imposed, without trying to represent all

154   THE BULLET'S FLIGHT FROM POWDER TO TARGET

difficulties, 18 of the swages made for experimental purposes with .28-caliber rifle and black powder were photographed as exhibited in Plate (11).

Any mechanic can gather from the illustration something of the amount of labor involved, and, although not every experimental rifle called for so many, all did require from six to 15 each. In many instances a variety of bullet

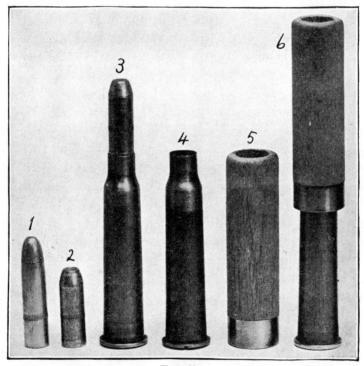

FIG. 62.

forms were made in same swage by using different plungers at point of ball and at its base. This form of swage is easy to make and repair, and any variety of bullet points can be made in same swage by shaping the plunger with different cherries.

(Fig. 62) above represents the perfected fixed ammunition for 7 mm. rifle, the difficulties attending its making having just been touched upon.

PLATE 11.

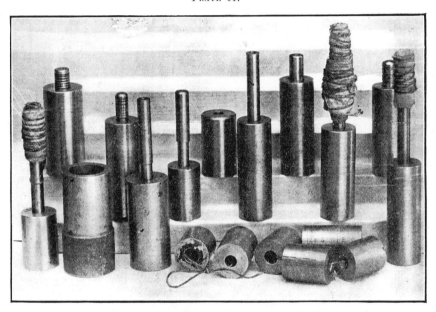

Figure 1, in (Fig. 62), is the 175-grain, soft-nose, 7 mm. factory bullet.

Figure 2, same bullet cut to 130 grains, and swaged to bore diameter, leaving $\frac{1}{16}$-inch base band, .001 inch larger than grooves of rifle.

Figure 3 gives a completed cartridge, fixed ammunition.

Figure 4, the .30–40 government shell shortened and reduced at neck to 7 mm., having a shoulder upon which base of bullet impinges after being entered $\frac{1}{16}$ inch.

Figure 5, a soft pine cylinder, reamed to fit body of shell, to protect the complete cartridge.

Figure 6, method of protecting the cartridge for carrying.

The next cut (Fig. 63) exhibits the .25, 7 mm., and .30-caliber metal-jacketed bullets, caught in oiled sawdust, and their mutilated soft-lead points.

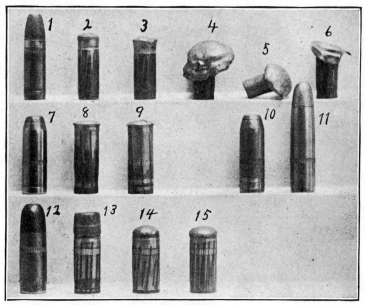

Fig. 63.

A careful inspection of the various mutilations in connection with means used to produce them will be instructive and interesting.

Figure 1 is the unshot 25-caliber jacketed bullet, experimented with during 1906.

Figures 2 and 3 were shot with 22 grains lightning powder.

Figures 4, 5, and 6 were shot with 24 grains powder.

Figure 7 is an unshot 130-grain, 7 mm. bullet, cut down from the 170 which is shown by figure 11.

Figures 8 and 9 shot with 28 grains powder.

Figure 10, unshot bullet similar to 7.

Figure 11, unshot 170-grain bullet from which 7 was cut.

Figure 13, the 170-grain U. M. C., .30-caliber, shot with reduced load so soft nose was not lost.

Figures 14 and 15, the .30-caliber bullets shot with 18 grains powder.

It will be observed that in figures 2 and 3 the soft noses were not driven back into the jackets, but have the appearance of being wiped away as putty might be by the fingers; and it appears that the sharp edges of jackets sheared off the soft lead as it came over their sides.

Figures 4, 5, and 6 mushroomed in the oiled sawdust finely, and all similar bullets fired with this charge did the same.

Figures 8 and 9, fired with 28-grain charge, indicate that metal jackets at forward end are expanded a little, and the soft lead has been wiped away as in figures 2 and 3. These last were the bullets used in woodchuck hunting during spring of 1906.

Even with its reduced charge, figure 13 indicates that its soft point settled back and filled rifle grooves for half its length. All these three varieties of bullets were bore diameter and their metal jackets upset to bottom of the grooves for nearly the whole length of their bodies.

In spite of all difficulties and labor involved, as illustrated by swages and shells, and more which followed, by February of this year ammunition and barrel were sufficiently advanced for trial, and with slight show of impatience a start was made for range 16 miles away.

Test 103. — February 26, 1906, with 7 mm. 30-inch barrel, 130-grain bullet having body diameter of .27975 inch, and base diameter of .286; 31 grains lightning fixed ammunition, bullet entered in shell $\frac{1}{16}$ inch; size of group made was 5 inches.

Other shots were made with 35 and 37 grains powder, to test trajectory, but went so wild as to be useless. The bullets caught in oiled sawdust, however, showed satisfactory upset. The bore immediately in front of chamber was enlarged by these five shots so that other bullets fired on this date were loose before upset. This barrel was cut without choke.

Test 104. — March 2, experiment was identical to test 103 except the swage was changed and body of bullet was made .281-inch diameter to fit the enlarged throat; fixed ammunition, with 31 grains powder; one 5-shot group made .81 inch, and a 4-shot group made .56 inch, as shown by (Fig. 64), the center one

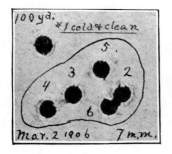 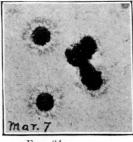 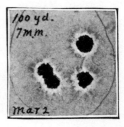

Fig. 64.

being made in next test of March 7, two groups only being shot on March 2d. In the .81-inch group it will be observed that the first shot is marked out.

Bullets caught in sawdust indicated no gas escaping. A trajectory group was also made at this time, the 31-grain charge showing .87 inch, and 37-grain charge showing .75 inch, same trajectory as the Lee navy, .236-caliber.

Test 105. — March 7, bullets from last test were submitted to Mr. Leopold, who thought there was quite an escape of gas all around them, the metal being blackened in a peculiar manner their whole length.

More ammunition and a day's time was given to this test, which was practically same as the last, using, however, a 34-grain charge which gave trajectory of .75 inch. One 5-shot group only was made, resulting in .75 inch, as illustrated under test 104. The last four of the bullets were recovered from sawdust and found blackened their whole length, except in places where metal jacket did not touch the barrel. After searching nearly an hour for the first bullet, it was obtained and found clean as before shooting. This bullet was fired through a perfectly clean bore, which indicated that the other bullets were blackened by a dirty bore and not by gas-cutting.

Having obtained this barrel for woodchuck purposes for the use of brother on the homestead farm, shooting on the range was limited to 24 shots in order to preserve its gilt-edged quality. It was fitted to a new Winchester single shot, double set trigger, with Sidle nine-power $\frac{7}{8}$-inch telescope tube and Mann's patent micrometer mountings. The cut on page 154 shows the ammunition first used, and now in use except the body diameter of bullet was increased to make it bore diameter after first 5 shots. Brother did not care for weight if the rifle would only shoot, and in testing it at different distances up to 400 yards, he pronounced it a success.

The homestead farm, all around the house, has been triangled with surveying instruments and all distances are known in the various directions. There is a 100, 200, and 300 yard range from window of his upstairs den, and this den has been our club room while shooting, the factory in Milford containing tools and machinery for manufacturing.

24-inch 7 mm. Barrel. With this 24-inch barrel, which was like the last one tested except in length, five shots were fired into the iron turnings in basement of machine shop, which enlarged its throat .001 inch, thus obviating the difficulty experienced from the first shots of previous rifle.

Test 106. — March 12, with new 24-inch, 7 mm. barrel, 130-grain bullets swaged to bore diameter, leaving base band of .286 inch, fixed ammunition, no wad, overcoat and mittens, seven shots were made; the 5-shot group made with a 28-grain charge gave .50 inch; the 2-shot group made with a 32-grain charge gave .37 inch $1\frac{3}{8}$ inches above the .50-inch group, as (Fig. 65) shows on following page. The trajectory of 28-grain charge was 1 inch; of the 32-grain charge ⅞ inch. Only the seven shots illustrated by cut were made from V-rest, when barrel was assembled, same as previous one for W. E. Mann, and taken with writer to Dr. Skinner's woodchuck preserve at Hoosick Falls for practice.

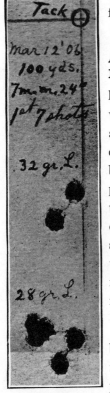

Fig. 65.

A Shooting Match. Ammunition for the two 7 mm. barrels was designed from knowledge gained while manipulating lead bullets and pressure tests upon jacketed ones, wholly theoretical, but it was not found necessary to change after the first shot except as before noted where bore became enlarged. Quite a number of shots were made with lead bullets and reduced charges of powder, in an old Milford pasture, with this 7 mm. barrel, Winchester action, and 19-inch telescope sight, to obtain the windage of front mount and correct elevation, lead bullets being used to avoid unnecessary shots with jacketed ones. Several days later a shooting match was indulged in between W. E. Mann and the writer, really for the purpose of obtaining elevation with metal-cased bullets for "chucks," as follows: —

Test 107. — May 2, 1906, using woodchuck rest invented by Dr. Baker, near the homestead range, even shots were made by W. E. M. and the odd ones by F. W. M. Plate (12) presents the whole 13 shots, the three upper groups being upon original targets and the others made up from pasteboard backing which is always used at the 100-yard butt.

Omitting the first, which is usually wild from a freshly oiled bore, numbers 2 and 3 made a .25-inch group 2½ inches above black paster; changing elevation of sights, numbers 4 and 5 made a .56-inch group, one inch too low. Shots 4 and 5 showed a mistake in computation through carelessness, the micrometer mount not being at fault. Again changing elevation and No. 6 hit the ⅞-inch paster with center of print ⅛ inch from its center, 8 o'clock position, and No. 7

PLATE 12.

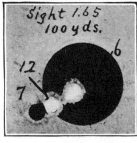

Targeting 7 m.m. 24" Barrell.
Even shots W.E.M. Odd shots F.W.M.

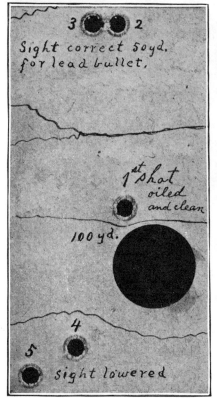
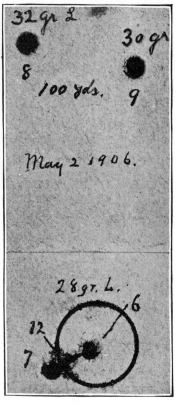

print nicked the paster at about same position; No. 8, using 32-grain charge instead of 28, elevation of sight the same, printed $3\frac{1}{4}$ inches above left-hand edge of paster; No. 9, with 30-grain charge, printed $2\frac{7}{8}$ inches above right-hand edge of paster; No. 10, using 28-grain charge at 125 yards, same elevation, printed $\frac{7}{8}$ inch below center of paster and $\frac{1}{8}$ inch to its left; No. 11, with same charge, at 125 yards, printed one inch directly below center of paster; No. 12, at 100 yards, printed in the paster $\frac{7}{16}$ inch from center, at 8 o'clock; No. 13, with same charge and elevation, at 50 yards, printed $\frac{3}{16}$ inch from center, 12 o'clock.

This was alternate shooting between two individuals with same rifle, lying on ground and from Baker's woodchuck rest, neither having shot this rifle before. The first group of two shots at 100 yards made .25 inch; second made .56 inch; third of three shots after elevation was correct, .44 inch, all at 8 o'clock on paster; fourth, at 125 yards, two shots made .25 inch; fifth, with 30 and 32 grain charges, made 1.18 inches, and the 50-yard shot was .19 inch from center.

The reasons for above test are not far to seek. It has been an axiom with the writer that the rifle, in fine rest work, is to blame instead of its holder for inaccurate shooting, and has silently differed from the alleged causes offered by most riflemen, or all riflemen so far as known. Even at double rest and high-power telescope sight, they attribute off shots to the holding when there is reason to believe that the cross hairs of their scope were not off center $\frac{1}{4}$ inch. From the abundance of tests already given it is evident that the general spreading of groups is due to the rifle and ammunition, and this shooting-match test caps the climax of proof.

It was previous to the suggestion of Mr. Pope's to try the .30-caliber on woodchucks that Mr. Leopold sent special directions to give up at once and for all the old black powder, and after a long correspondence with the latter gentleman he advised a trial of the 7 mm. rifle. Still continuing correspondence, a .25-caliber was suggested for woodchucks, so three special barrels of this caliber were ordered and received during October, 1906. It has been found to simplify matters somewhat, in experimental progress, by taking orders from the right parties and hitching them to the line of least resistance.

The woodchuck rest, first brought into notice by Dr. H. A. Baker, and constantly employed by the writer while in the field away from regular rifle ranges, is a most convenient and simple article as (Fig. 66) shows. This hollow iron

FIG. 66.

tube, with its sharp point, carries within it the swabbing stick, can be easily entered into the ground, while the sliding rest is quickly moved to any desired height and fastened by set screw. In tramping the fields or woods, it makes a

FIG. 67.

good walking stick by sliding rest to the top. It can be used in place of tripod for small cameras, and is no mean weapon at close quarters, stiff and strong but quite light in weight.

Dr. Baker recommends it for opening a generous space between the barbed wires of an obstructing fence; by setting its sharp point into one wire, catching the other upon sliding rest, moving it up and carrying wire along, then fastening by set screw, space for passing through without catching the clothing is quickly afforded. The usual position taken while shooting from this rest is well represented by the photograph on preceding page (Fig. 67).

.25–36 Marlin Factory Barrel. A 26-inch factory barrel was utilized while the three specials were being made, in order to determine size of bore, pitch, and style of rifling so a set of tools might be made up. The bore diameter was found to be .251, and groove diameter .257 inch, pitch nine inches. Several swages were made for the 117 and 85 grain, .25-caliber, U. M. C. soft-nose bullets. Chamber reamers were made for the .25–36 Marlin shell and the .30–40 government, besides all other reamers, dies, and reducers that might be needed; eighteen reamers were required for this ammunition alone, to be used with this and three special barrels received later.

Test 108. — On July 16, 1906, with factory 26-inch barrel, factory ammunition, factory chamber, two groups were made, 1.87 and 2.18 inches respectively. See (Fig. 68), page 165, for the 2.18-inch group.

Test 109. — July 28, with swaged, jacketed, 117-grain, two-cylinder bullets, having $\frac{3}{32}$-inch base bands, pushed far as possible into rifling and shell entered with no wad, leaving base band of bullet in the chamber, the first group gave 2; second 1.30; third, with factory ammunition, 2.25; fourth, the same as first, .81 inch. The trajectory with 24-grain charge factory ammunition was 1.44; same charge and two-cylinder bullet, 1.06 inches.

Test 110. — August 2, practically same as the last, except bore was throated for $\frac{1}{8}$ inch to take the .285-inch base band and not leave it in chamber; 117-grain, front-seated bullet, 24-grain charge; 30 consecutive shots were made, six groups; smallest .62, largest 1.37, average .93 inch.

Test 111. — August 4, with 24-grain charge, bore-diameter bullet lying free in bore without any base band; two shots were made which flew wild and were lost, supposed not to have taken the grooves. These same bullets with $\frac{3}{32}$-inch base bands, to give charge a slight initial pressure, upset the whole length into grooves, and soft point of old style U. M. C., 117-grain bullet follows into grooves for nearly half its length.

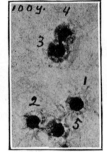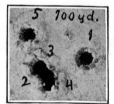

Fig. 68.

Here again we have bullets acting very similar to lead ones in test 90, page 191; evidently in this test the limit was reached where a bore-diameter, jacketed bullet, with extra air space and no base band, would upset; that is, the caliber was so small and bullet so light they would not act as sufficient obstruction to the charge behind, needing the slight base band to hold long enough to insure proper ignition.

Test 112. — August 4, same day as the last, bore being throated out as in test 110, 20 consecutive shots were made; smallest group .56, largest 1.25, average 1 inch; trajectory 1.12 inches. Two of these groups are shown above in (Fig. 68), smallest measuring .70 and the other 1 inch, which happens to be the average of whole 20 shots. It will be noticed that the smallest group made, .56 inch, is quite a little smaller than the .70-inch one in cut.

Test 113. — August 10, five inches having been cut from muzzle of this 26-inch barrel, leaving its length 21 inches, 15 consecutive shots were made with all front-seated bullets; smallest group .56, largest 1.25, average .93 inches; trajectory 1.12 inches. This Marlin barrel was robbed of five inches to make three short barrels for other tests.

Test 114. — September 1, with this 21-inch barrel, rechambered for .30–40 government shell, and throated $\frac{1}{8}$ inch to take the $\frac{3}{32}$-inch base band of bullet, twenty consecutive shots were made; smallest group .87, largest 1.06, average .93 inch; trajectory with 24 grains lightning, 1.30, 26 grains 1.12, 28 grains 1.06 inches.

The .30–40 government shell, filled to base of bullet, which sets into the shell $\frac{3}{32}$ inch, contains 39 grains lightning powder. The 28-grain charge, which gives 1.06-inch trajectory, same as 24 grains in the .25–36 Marlin shell, has 28 per cent air space. The .25–36 shell, filled to base of bullet, which is set $\frac{3}{32}$ inch into the shell, contains 27 grains powder, and the air space left by a 24-grain charge is 11 per cent, giving a trajectory of 1.06 inches. Thirty-two-grain charge in .30–40 shell, using 117-grain bullet, gives a trajectory of .75 inch, about 2500 foot-seconds velocity, same as Lee navy .236 caliber, and, so far as tried, this latter shell with this charge works well. In a rifle fitted up for Mr. Leopold, a special .25 Marlin, the .25–36 shell does not appear to stand the 24-grain charge.

Metal Jackets, Short Barrels, .25-caliber. Experiments conducted with short barrels during 1902 were very complete so far as black powder and cast-lead bullets are concerned; but new conditions were presented by the advent of smokeless, high-pressure powder, and bullets with metal jackets, so a series of experiments were undertaken to test these new conditions, commencing in August, 1906, with .25-caliber jacketed bullets, and short barrels were rifled and chambered for high-pressure work.

Five inches was cut from muzzle end of the .25-caliber Marlin, and this was threaded on outside, concentric to its bore, then cut into three pieces of unequal lengths, the chamber end being reënforced by a steel ring two inches long, which was screwed and soldered in place.

(Fig. 69) illustrates the three short barrels and their chamber ready for use; *a*, the chamber, and *b, c*, the nuts with their ventilating holes joining the barrels.

Fig. 69.

By unscrewing nut *b* the front barrel, 1.37 inches long, is removed, leaving one of 1.32 inches beyond the chamber; unscrewing nut *c* takes away another barrel of one inch and leaves one of only .32 inch attached to the chamber; whole length of combined three barrels being 2.72 inches outside the chamber.

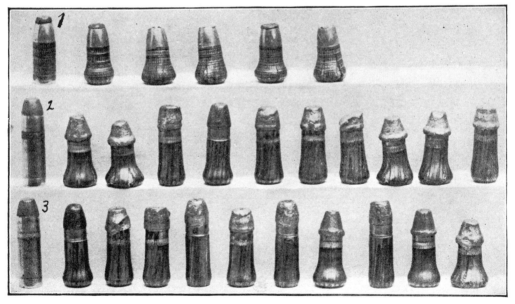

Fig. 70.

It will be observed, by referring to (Fig. 70), there can be little question about upsetting of bullets even when jacketed with metal, and of bore diameter.

Figures 1, 2, and 3 represent the unshot bullets, 1 being the 85-grain, 2 and 3 the 117-grain bullets, all others having been shot from the .32, 1.32, or 2.72 inch barrels, through a pasteboard target into a box of oiled sawdust, which was placed 21 inches from their several muzzles.

Fixed ammunition with full 24-grain charge was used, and all but three of these illustrated bullets were bore diameter except the .06-inch base band, and this enlarged base band was entered loosely into the shell .06 inch. Details of the several tests made with this ammunition need not be tabulated because they so closely follow those of four years before, made with lead bullets, but an experimenter on these lines would not find it difficult to select from the preceding cut such bullets as were discharged from the respective .32, 1.32, and 2.72 inch barrels.

The manner and extent to which these soft-nose, metal-jacketed bullets upset is better shown by the magnified cut (Fig. 71).

The first figure shows an unshot factory bullet, having a diameter of .258 inch and length of .916 inch; second figure shows conditions after shooting from the 2.72-inch barrel and recovered from oiled sawdust, with a length of .800 inch, diameter at base of .325, and soft point enlarged to .375 inch. Even its metal jacket, near junction with soft point, was expanded to groove diameter or .257 inch, and before its exit from this very short barrel the soft point had settled back and filled grooves half its length. Clearly, if the soft nose would settle back in this barrel, it must do so in a longer one.

The third figure shows condition of bullet after discharging from the 1.32-inch barrel and presents still greater deformity, with length reduced to .680 and base spread to .430 inch, while its soft point, even in this very short barrel, settled back into grooves before exit, and was spread to .326 inch after exit. These bullets, which were so perfectly recovered from oiled sawdust, in practically same condition as they left the short barrel bores, make a very interesting study by throwing no little light upon action of bullets before the powerful blast of powder behind.

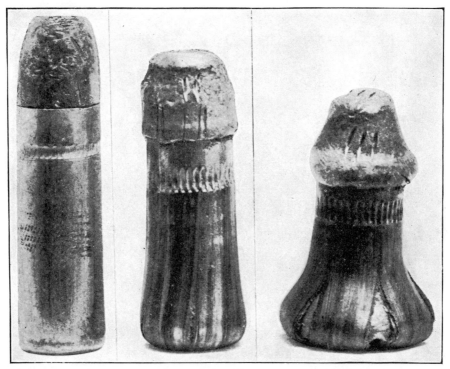

Fig. 71.

In the next cut (Fig. 72) a number of .25-caliber jacketed bullets are exhibited, magnified to be more easily examined.

Figure 1 shows the unshot 85-grain, and 6 the unshot 117-grain bullets; figures 4 and 5 the condition of 85-grain bullets after discharging from 1.37-inch barrel; figure 7 condition of the 117-grain bullet after shooting from .32-inch barrel; figure 8 from 2.75-inch barrel. The factory ammunition from which figure 7 was shot allows the bullet to extend into shell .50 inch, the regular .25-36. As the barrel that fired this shot was only .32 inch long, the bullet traveled .81 inch before exit, and its soft nose did not take the grooves because it stood out from muzzle at time of firing. Figure 8 being fired from the 2.75 barrel, allowed its soft nose to take the grooves for half its length.

170   THE BULLET'S FLIGHT FROM POWDER TO TARGET

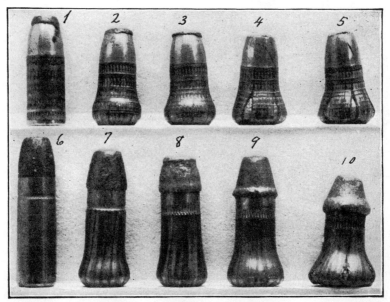

FIG. 72.

Short Barrels and Full-mantled Bullets, .30-caliber.

During 1907, a year later than the experiments with .25-caliber short barrels and soft-nose bullets, some tests were made with .30-caliber short barrels, to determine action of the 220-grain, full-mantled bullet before the blast of a 35-grain W. A. powder charge.

Procuring a barrel 11 inches long from the Winchester Repeating Arms Co., it was cut into several pieces, which, joined together with vented steel nuts and fitted to a high-pressure concentric action, presented the appearance shown by cut.

FIG. 73.

The length of shortest barrel, outside the chamber, was half an inch; with next joint added made one and a half inches; the next three and a half, and the last one, adding four inches more, made a barrel seven and a half inches long. Since in the .30–40 fixed ammunition, furnished by the Winchester people, the bullet extends into the shell half an inch, the four barrels were practically 1, 2, 4, and 8 inches respectively in length, or joined as represented in cut, eight inches instead of seven and a half.

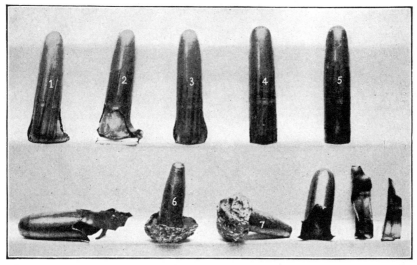

Fig. 74.

(Fig. 74) above exhibits the results, remembering that all bullets were caught in oiled sawdust without mutilation except that received at muzzle of the several barrels.

Figures 1, 2, and 3 were discharged from the four-inch barrel.

Figures 4 and 5 from an 8-inch one.

Figures 6 and 7 from an inch barrel.

The bases represented in figures 1, 2, and 3 were upset by the muzzle blast and jackets broken on one side, figure 2 presenting the broken side. Figures 4 and 5 were not deformed, owing to fact that they were discharged from a barrel of sufficient length to take care of them. The bases of figures 6 and 7, bullets discharged from an inch barrel, were rolled over upon themselves and their jackets blown to pieces. Three of these pieces are shown to the right of figure 7, and one to the left of figure 6, figures 6 and 7 being the lead cores remaining after the jackets were removed by powder blast.

(Fig. 75) exhibits, in figures 7, 8, 9, and 10, the condition of the same full-mantled bullets after discharges from the two-inch barrel. It will be noticed

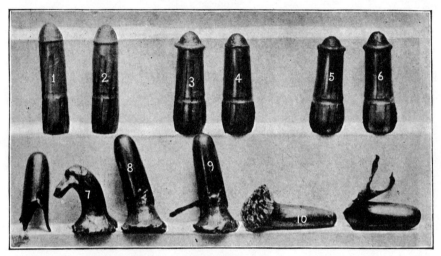

Fig. 75.

that the base of the jackets of each were blown to pieces and lost, but a part of the jacket of figure 7 is shown at its left and of figure 10 to its right, figures 7 and 10 being only the lead cores.

Following the experiments with full-mantled bullets other experiments were made with soft-point, metal-cased ones, and some of the results are illustrated by figures 1, 2, 3, 4, 5, and 6 in the above cut. Figures 1 and 2 show results when discharged from a four-inch barrel, with their bases enlarged from .308 to .332 inch, and soft-nose bases from .280 to .308 inch, showing, however, no marks of rifling upon the enlargements of soft nose. Evidently the point received its enlargement, same as its base, after exit from muzzle.

Figures 3 and 4 were discharged from the two-inch barrel and exhibit bases enlarged, after leaving the muzzle, to .375 inch, and soft point of figure 3 to .345 inch. Figures 5 and 6 were discharged from an inch barrel and show their bases enlarged .080 inch above normal size, and their points .055 inch. With this base enlargement no rupture of the jackets took place in soft-nose bullets, since they have a continuous metal covering over their base, while the full-mantled bullet covering is continuous over the point but not over the base.

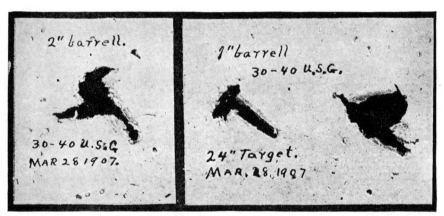

Fig. 76.

The prints made by two of the mantled bullets, or what is known as the government bullet, when fired from the two-inch and one-inch barrels respectively, are quite interesting and (Fig. 76) gives their actual size and appearance. Although the shot fired from the two-inch barrel makes but one print, undoubtedly the jacket was entirely separated from its core when the target was penetrated. The shot from the one-inch barrel sent the jacket through in one place and the core in another, as cut shows.

Special .25-36 Marlin, 14-inch Pitch. By an error this barrel was only 21 inches long instead of 24 as ordered. It was $1\frac{5}{16}$ inch full round, fitted with dovetail sight mount blocks, properly soldered before barrel was finished or browned, and contained no chamber. Bore diameter was .251 inch, from end to end without choke, and lands were left only half standard width. The usual pitch of this style is nine inches, but this was ordered with 14-inch twist with expectation that it would be too slow for the 117-grain bullet.

After being fitted to special Stevens action it was chambered for the .30–40 government shell reduced to .25-caliber, concentric rings and action were added, and it was taken to the range for trial.

Test 115. — With a bore-diameter bullet, $\frac{3}{32}$-inch base band, .001 inch larger than groove diameter, fixed ammunition, 24-grain charge lightning powder, leaving 40 per cent air space, size of group was 1.25 inches and bullets tipped badly; with 26-grain charge, size of group 1.81 inches, and bullets made fine prints; with 28-grain charge, 27 per cent air space, size of group 1 inch, and fine prints; trajectory with 28-grain charge, 1 inch. The straightening up of tipping bullets by increasing their speed with the larger charge, by only two grains, will be noticed, as was also the case in tests 100 and 101, page 147, with the .30-caliber.

Test 116. — October 12, with same barrel, government shell, fixed ammunition, no cleaning, 117-grain bullet cut down to 100 and properly swaged to bore diameter, 28-grain charge, trajectory of fifth shot was 1.06 inches, size of group .81 inch; with the 117-grain, U. M. C. short, soft nose, present factory make, swaged to bore diameter, 28-grain charge, trajectory of second shot was 1.12 inches, size of group 3.75 inches; again, with the 100-grain bullet, swaged bore diameter, and 28-grain charge, size of group 1.25 inches; with 20 grains of sharpshooter powder, 100-grain bullet, fourth shot trajectory was 1.25 inches, size of group 1.44 inches.

Test 117. — October 16, with fixed ammunition, no cleaning, 117-grain bullet, 22 grains sharpshooter powder, air space 37 per cent, trajectory 1.06 inches, size of group was 3 inches and fine prints; with 24 grains of same powder, trajectory .85 inch, size of group 2.25 inches; with 26 grains same powder, trajectory .75 inch, size of group 1.87 inches and fine prints; with 30 grains lightning powder and air space of 23 per cent, trajectory .80 inch, size of group 1.19 inches and fine prints; with 32 grains same powder, air space 18 per cent, trajectory .75 inch, size of group 1.50 inches and fine prints; with 24 grains sharpshooter powder and 100-grain bullet, air space 31 per cent, trajectory .75 inch, size of group 1.25 inches and fine prints.

In all these tests a single shell was used for consecutive shots until it stuck some in the chamber, and the primers or primer pockets gave no trouble. At the .75-inch trajectory the shells would stick some at third shot with the sharpshooter powder, and fourth or fifth with lightning powder, while the 1.06-inch trajectory does not make shell stick during first five shots. No Frankfort Arsenal shells were tested.

Test 118. — November 6, under same conditions as the previous with 22 grains sharpshooter powder, 117-grain bullet, size of group was 2.50 inches, tipping badly; with 24 grains same powder and bullet, size of group 2.12 inches and fine prints; with 26 grains same powder and bullets, trajectory .75 inch, size of group 1.50 inches and fine prints; with 30 grains lightning powder and same bullet, size of group 2.87 inches, though all the bullets went into the same hole except one, which printed 2.87 inches out. Here again it is found that speeding up the bullets, as in the .30-caliber, 21-inch barrel, makes fine prints.

Test 119. — November 10, same conditions continued, 30 grains lightning powder, 117-grain bullet, size of group was 2.62 inches; another group of 1.62 inches showed that any single group may be a lucky one and is not a reliable test of the shooting qualities of a rifle. With 32 grains same powder, size of group was 1.62 inches; with 22 grains sharpshooter powder and 100-grain bullet, size of group was .75 inch, and another group made with the same was .62 inch; with 32 grains lightning powder and 117-grain bullet the group was 1.19 inches.

Test 120. — This was made under same conditions as the last, except the bore had been lapped out 14 inches from chamber, making it about .00075 inch larger than remaining seven inches of muzzle end. Six 5-shot groups were made with 86 and 100 grain bullets, and 20, 24, and 26 grain charges sharpshooter powder and 30 and 32 grains lightning. All shots made were consecutive and without cleaning; size of groups were 1.06, .87, 1.12, 1.25, and 1.25 inches.

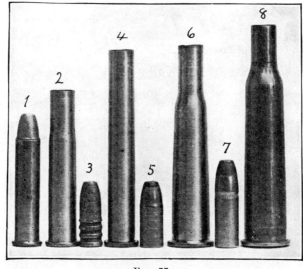

Fig. 77.

What caused spreading of all the groups made with this 14-inch twist Marlin barrel? It was not wind. It was not the sights. It was not irregular powder charges or irregular ignition, because groups spread horizontally as much as perpendicular. If the 14-inch twist was at fault, it was not because the bullets had the appearance of having tipped, for they made fine prints, perfect as far as could be seen, though with high-speed metal bullets it cannot be absolutely determined that prints were perfect.

The average size group with the 100 and 88 grain bullets all through the tests, with any charge of powder up to a trajectory of .75 inch (which equals the .236-caliber Lee navy) was .93 inch, while with 117-grain bullet the groups averaged 1.87 inches, taking only the groups where bullets did not tip. No doubt the 14-inch rifling was better fitted to the 86-grain bullet than the 117-grain, but this does not give any answer to the question, Why? Should the spreading of 117-grain bullet group be attributed to tipping when there was no appearance of any; to the wind; to mirage; to holding, or to some speck flying before eye of the shooter as he attempted to make certain shots in these groups?

Experiments with this special Marlin barrel may be more readily appreciated by reference to (Fig. 77) on opposite page, which shows development of some of its ammunition.

Figure 1 is regulation rim-fire cartridge.
Figure 2, the .25–20 shell.
Figure 3, a .25–21 cast bullet.
Figure 4, the .25–21 straight center-fire shell.
Figure 5, a .25-caliber, 85-grain U. M. C., soft-nose, metal-jacketed bullet.
Figure 6, the .25–36 Marlin shell.
Figure 7, a .25-caliber, 117-grain U. M. C. bullet.
Figure 8, the .30–40 U. S. government shell, sized at neck to take the .25-caliber ammunition used and described in tests 115 to 120. Experience shows that the .30–40 government shell can be reduced in neck to .25-caliber at one operation when dies are in proper condition, using Leopold's bullet lubricant, a lubricant which fills all requirements in this particular, far ahead of any other that has been tested; and without the Mann bullet press, or something similar to take its place, many of the tests recorded would never have been attempted.

Mirage vs. Telescope. That a mirage does exist under certain peculiar conditions of the atmosphere is not questioned, but because it has frequently been held responsible for errors in rifle shooting when using telescope sights, pretty conclusive tests were instituted during September, 1904, to ascertain its influence.

The experiments were conducted in an open field of several acres, slightly elevated over surrounding land to the east and south, and some 200 feet or more above tide water. There was a grove of pines 300 yards to the east and a similar growth covered a part of the horizon to the south, north and west being open.

During the September days of experimenting, varying conditions of atmosphere prevailed, from hot noon sun, to cool morning and evening, air surcharged with moisture and rain, a stiff dry breeze under the hot sun, and a similar breeze with cold and clouds.

Observations were made at three points of the compass, from two steel standards or beams, set firmly into the ground to the depth of 30 inches, on top of which were firmly fixed heavy iron V-rests to receive telescope; these rests were accurately machined so that the two brass rings in which the scope was mounted lay accurately upon the V-surfaces, thus obviating the slightest lateral movement.

One of the rests was set due north and south and the other $4\frac{1}{2}$ degrees south of east with one of its ends pointing $4\frac{1}{2}$ degrees north of west.

One hundred and nine yards south of the first rest a paper disk, two inches in diameter, was attached to a post $2\frac{1}{2}$ feet above ground, the range passed over being comparatively level; 145 yards north another similar disk was fixed to a telegraph pole, 30 feet above ground. Two hundred yards east of the second rest a like disk was attached to a plank five feet above ground, and all three disks were adjusted so when the telescope was laid in V-rests the cross hairs came nearly to center of the disk towards which it was pointing.

From the east standard, as shown, (Fig. 78) on opposite page, the line passed over a deep ravine for first 80 yards, the next 40 gradually rising to within two feet of the line, still gradually rising until at 110 yards the line passed but a few inches above barren, sandy soil, affording an ideal spot for mirage.

The telescope used was a 16-power Sidle, one-inch objective, made particularly for V-rest work, mounted in brass rings so adjusted that while rotated in V-rests no movement of cross hairs upon respective disks were observable. Five or six, or more, observations were taken on each of the following days, Sept. 8,

19, 21, and 26, and all verified by three different persons, each of whom was very much interested in detecting mirage. The manner of taking these observations is well illustrated by the first day's work, September 8, and by observing cut of the east and west standard, as shown, with telescope lying in its V-rest.

Fig. 78.

The first observation was taken at 1 P.M. during hot, bright, sultry sunshine; during middle of afternoon a breeze and heavy dark clouds changed conditions very materially while the next observation was taken; at 4 P.M.

the scope was again turned upon the disks from the several directions, under a bright sun, and still again a few minutes later with sun under heavy black clouds. A little later still rain began to fall, during which an observation was taken, and still others during the evening, the last being taken at 10 P.M. by reflecting artificial light upon the white disks; these several observations being all taken to the north, south, and east, and varied in every possible manner, yet the cross hairs, throughout every observation, kept their exact place on the several disks, not varying a hair.

These observations were repeated on three other days with same results, being so uniformly the same as to make their repetition tedious, but results were carefully tabulated and preserved.

They all demonstrated that no mirage could be detected, in slightest degree, to affect telescope lenses or change the position of cross hairs on target up to 200 yards.

Space covered by Cross Hairs of Telescope. The amount of space covered on the target, at 100 and 200 yards, by the cross hairs of a telescope has been considerably discussed, and the question was interesting enough to provoke an accurate measurement of the same, using the 16-power Sidle scope.

Placing a white disk of paper, two inches in diameter, at 200 yards and another at 100 yards, a strip of black paper was cut and laid partly across the white surface central with horizontal hair of the scope. By trimming this black strip, little by little, a point was reached when the strip was identical in width with the cross hair as it appeared when magnified upon the white disk. It took two to play this game, one with eye to scope and the other with scissors at the target, and from the scope the 32d of an inch strip, cut from the black, could be detected.

Besides deciding the amount of surface on target covered by the hairs, it was found there was no deception in noting distances, movements, or size of objects as seen upon the target through the scope. The two cuts, on opposite page, one exactly representing the disk at 200 yards and the other at 100, the position of cross hairs and amount of space covered from sight upon each, will fix in mind this matter better than elaborate description.

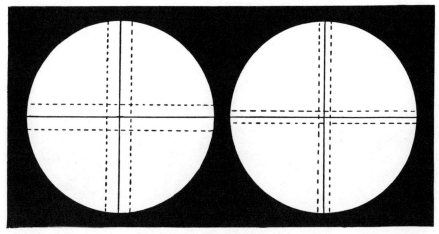

Fig. 79.

Figure 1 represents the two-inch disk as seen at 200 yards, and figure 2 at 100 yards. The black lines represent the center of cross hairs and dotted lines the space on the two-inch disk, by exact measurement, which cross hairs cancel from view; in other words, the amount of magnification of the hairs in this 16-power scope upon the respective disks, 100 and 200 yards away, is too little to be of any material consequence in rifle practice.

Distance Measuring with Scope. That short distances on a target can be as accurately estimated through a good rifle telescope, 200 yards away, as if immediately under the eye of good judges, was demonstrated in an interesting test made for that purpose.

Placing a two-inch black disk at a distance of 200 yards, an inch black paster was moved to and fro until the observer at scope, by raising his hand, indicated that its center was two inches from center of the black disk. This test was experimented with and varied a number of times, and in each instance

182 THE BULLET'S FLIGHT FROM POWDER TO TARGET

the observer's judgment, when exercised through the scope as to distance, proved accurate as if the disk and paster were immediately before him upon the table, and his eye measurement of distances is unusually accurate.

This proved a very satisfactory test in regard to the alleged deception of distances on a target, when viewed through a telescope. No difficulty was found from mirage, aberrations, or reflections through the scope lenses, though contrary to some theories that have been occasionally advanced in different rifle journals. Absolute demonstration by competent experimentation is worth more than theoretical conclusions which are derived from imaginary circumstance.

FIG. 80.

Conveniences. The wide-awake rifleman is always looking for conveniences that will simplify detail work at range or in the field, and (Fig. 80) of various receptacles for accurately weighed smokeless powder presents suggestions.

It shows three forms and sizes of vials which proved indispensable, not only during the experimental stages with dense nitro or high-pressure powders, but also in field work after woodchucks.

It is admitted that, for close uniform trajectory, charges of nitro powders must be accurately weighed and not measured. Many hours spent in testing the accuracy of several commercial nitro powder measures, demonstrated that the charges thrown by them vary too much, though skillful handling of them may reduce somewhat their irregular working.

Actual test on the ranges showed that one grain in weight increased the powder charge and diminished air space, thus bringing in an error, in some instances equal to all others combined.

The first vial in the cut, containing nine grains sharpshooter powder, proved very convenient for the .28-caliber bullet, while second and third vials, containing 23 grains lightning powder for the .30–30 metal-jacket ammunition, was equally convenient. The fifth, with its 28 grains for the 7 mm. special 130-grain metal-jacket bullets, and the sixth, containing 32 grains for .25-caliber, 117-grain, and 85-grain soft-nose bullets, were ready at hand when wanted. These different charges of nitro powders, as many as will be needed on range or in field, can be accurately weighed by balances at home, and vials loaded and corked ready for use in shell when needed.

A difficulty was encountered at first when attempting to pour nitro powder from commercial vials into either of the three shells used with above-named ammunition, but tips of vials 1, 4, 5, and 6 fit either the .30, .28, 7 mm., or .25 caliber shells perfectly so not a grain of powder will fail to find its way into them. The expense of these special vials is very little, and the excuse for showing them lies in the fact that as the charge increases the straight model was found impractical in length.

Superiority of Bore-diameter Bullet Discussed. Through all the tests and experiments made at the homestead range the bore-diameter bullet as before mentioned has shown its superiority; with high-pressure and black powder, using cast or metal-jacket bullets, in all rifle barrels, this was the most accurate one tested. The Chase-patch bullet, which had done such accurate work on many ranges, is made into bore diameter by its patch of one thickness of paper. The Pope system uses one which is bore diameter or less

184   THE BULLET'S FLIGHT FROM POWDER TO TARGET

with exception of the enlarged base band, which goes to bottom of the grooves and pushes down from muzzle to breech any large collection of dirt that may be in the bore. The constant use and mention of bore-diameter bullet throughout all our rifle work is not so much to indicate its practicability or to advise its use, but rather to throw light upon our search for the x-error. What may we discover from the fact of its superiority in accuracy?

Looking at conditions of any of these bore-diameter bullets at time of discharge, and when receiving their heavy pressure from powder gas, it will be observed that base of the bullet can make a forward movement and gain some velocity before the gas pressure is obliged to start the point, at which time the bullet is fully upset, has filled the grooves, and is ready for flight through the bore without further mutilation.

In this case the bullet is not obliged to be reduced in diameter, which elongates it just at the time when the heavy breech pressure is in operation, and the necessity of having the point start faster than base is obviated, this last condition being invariably present where the bullet is driven into a bore smaller than itself. This not only lengthens the bullet but produces extra friction and, no doubt, increases the chamber pressure above that required for uniform ignition of powder charge. With a groove diameter or oversize bullet the point is made to travel faster than base while the base is receiving its greatest pressure, often amounting to over 17 tons to square inch.

This lengthening of the bullet while under this great pressure from the rear reminds of lead-bullet times, where the bullet simulated a putty plug in a popgun and is thrown out of balance, sometimes on one side and sometimes on the other, without any dependable uniformity.

An important characteristic of the bore-diameter jacketed bullet is exhibited when it makes lower trajectories with same charge of powder than the average groove diameter one. With the .25-caliber Marlin and factory ammunition, the trajectory of its oversize bullet was 1.62 inches, while with same charge the bore-diameter one was only 1.06 inches, as per test 109, page 164, which also gave smaller group.

It was found by experiment that with .30-calibers, and larger, no base band is

THE BALLISTICS OF SMALL ARMS 185

required for accuracy and the metal bullet may be made bore diameter throughout its entire length. The leakage of gas did not cause inaccuracy, as indicated by test 99, page 147, where it also made the best groups and lower trajectories.

If an oversize or groove-diameter bullet will make a two-inch group and marksmen are satisfied with this amount of accuracy, it will be quite difficult to give satisfactory reasons for changing their methods. But when velocity, powder charge, chamber pressure, or abrasion of rifling is concerned the bore-diameter bullet takes precedence over any other. That gas-cutting and erosion would be more pronounced with this remains to be proved.

Experiments made on homestead range have dealt with accuracy of flight, regular powder ignition and escape of gas around the groove-diameter base band, but more particularly with the question of accuracy for which all marksmen are searching. Perhaps the demand for gilt-edge accuracy in a smokeless barrel for 15,000 shots cannot be satisfied, but there are doubtless some who would pay the price of a smokeless barrel if its accuracy would win a single trophy although at the end of the match its gilt edge was dulled.

Flight of Bullets; Screen Shooting. As it became important, in course of these experiments, to determine the tip or action of rifle bullets in their flight from muzzle to target, a series of thin paper screens placed at intervals along the range promised the best medium for our purpose. As early as 1901 E. A. Leopold recommended this method as a convenient one for deciding several problems that were in question.

Setting paper screens through which bullets must pass in their flight from muzzle to target, at right angles and at intervals along the range, to include frequent spaces from inches to feet or yards at will, introduced some very interesting developments which were utilized more or less in connection with all experimental barrels after 1901, and there was no indication that a few screens exerted any influence towards obstructing the bullets' flight; that is the bullets flew as well with as without screen obstructions.

During some of the first experiments screens were placed promiscuously along the range considerable distances apart, and some general results were obtained. It was good practice to prepare for some finer work, and a little later intervals were made four feet, then two; even then it was found that movements of the bullet could not be read from the several screens with any degree of certainty, and the intervals were reduced to six inches; then the flight of the bullet began to be intelligently recorded, though the time required to read results of a single shot was often extended to two and three hours.

Inaccuracy to some degree was usually present in setting the screens along the rough, outdoor range, and air conditions were not always the most desirable. It was necessary that the screens be so placed that the crossing lines on each should not vary a line's breadth from a straight line between muzzle and target, through the whole 100 yards, and this was not always a perfect success.

When telescope lay in V-rest, placing a thickness or two of paper under its rear ring changed the position of its cross hairs at 100 yards from line of fire to line of sight. This was not, however, the line of sight which ordinarily obtains with a rifle where the sighting eye is placed an inch or more above the bore and meets the bull's-eye at target. The above explanations should be fully comprehended in order to appreciate the illustrations of screen shooting which follow.

It is to be regretted that slight errors will constantly creep in, but all this screen shooting was first started for recreation, though later, as results began to be obtained, it was suggested that records, in spite of errors, would be of some interest to riflemen. To eliminate all errors had become impossible because this gilt-edged, 1902, Pope barrel, had lost its accuracy and another has not been found to replace it, hence these particular tests could not be corrected by repetition.

Test 121. — Plate (13) photographically illustrates the experiments of this test, and from it something may be learned.

With Pope, 1902, muzzle-loading, .32–47 rifle and 200-grain bullet a normal group of five shots was first made at 100 yards, with result shown within the penciled lines of largest target on plate. The same target also gives prints of two mutilated bullets shot through the screens, one to extreme right and other in left lower corner.

PLATE 13.

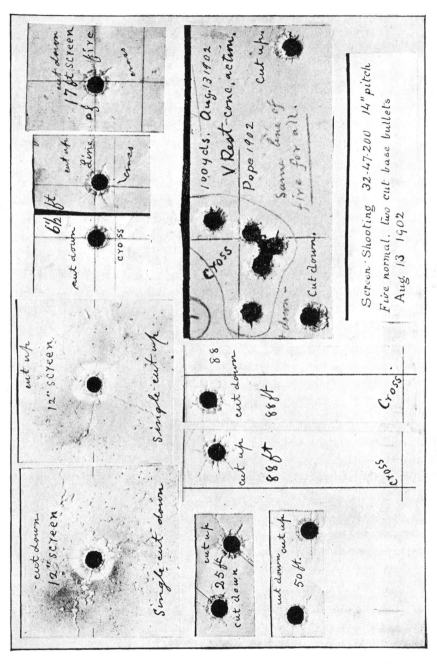

The mutilated bullets were made by cutting one grain, by actual weight, with knife from edge of the bullet's base, extending cut into base end and up its side nearly the length of first base band. All the screens within 24 inches of rifle's muzzle were protected from its blast by placing a board four inches away, with a $\frac{3}{8}$-inch hole through which bullets were shot, and the unblackened ring so plainly seen on the two 12-inch screens on plate, around prints of the bullets, may be explained by one who studies the phenomenon to suit himself.

It will be observed that the two mutilated bullets were so entered in rifle's chamber that one left its muzzle with cut edge up, and the other with it down, and tips of these two retain the same relative position throughout their whole flight of 100 yards, as indicated by screens through which they passed, showing that tip of bullets was governed by cut in edge of their respective bases, and positions of prints on the butt bear a relation to the phase, up or down, at which the cut edges issued from muzzle. This tip is not discernible at screen placed 12 inches from muzzle, but the $6\frac{1}{2}$-foot and other screens show tipping of both bullets very plainly.

At 17-foot screen the bullet, with cut in its base down, has moved .02 inch to left of line of sight; dividing 100 yards, or 300 feet, by 17 and multiplying by .02 gives .36 inch; but print of bullet is .08 inch to left at 100 yards, which indicates that the screen is .03 inch in error or bullet did not fly straight from muzzle to butt.

At 88-foot screen the bullet with its cut base down is .26 inch to left; dividing and multiplying as before, and it should have printed at target .84 inch to left, but it printed .80 inch, showing error in work of .01 inch, or bullet did not fly straight. The two errors indicated above and in Plate (13), were taken at random with no thought of selection, and they indicate that bullet did not vary from a perpendicular plane, at the screens, over .04 inch. Hundreds of other screens used in these experiments indicated that a small error existed in placing them, or in the travel of the mutilated bullets.

With the other, or cut base up, however, the showing is somewhat better, as will be noticed by careful measurements on the plate. In $6\frac{1}{2}$-foot screen the print is .04 inch to right; dividing and multiplying as before gives 1.30 inches at 100 yards, while the actual print is 2 inches to the right, showing error at $6\frac{1}{2}$ feet of only .004 inch.

All these errors are so slight that the experimenter could calculate very closely where any of these bullets will print at butt by measuring their position in relation to the cross on screen properly placed only 17 feet from muzzle.

In Plate (14) the rise of bullet above the cross, as may be seen, is quite interesting, the cross representing a straight line from muzzle to center of a normal group. The plate shows 88-foot screen lying upon the 50-yard one, with cross the same for both, thus enabling all screens to be shown on one plate.

Test 122. — Plate (14) illustrates this test which was conducted on same day as the last, August 13, 1902. A 5-shot normal group is seen at the top of plate inclosed with a pencil mark, and prints of two mutilated bullets at right and left; bullet with its notched edge issuing from muzzle down being at the left, and the other with its notch up at the right lower corner, the cut up one 4 o'clock and cut down one 10 o'clock, the latter being about half the distance from normal group as former.

By referring to Plate (13) it will be noticed that the bullet with cut base down printed about half the distance from perpendicular line of cross on the left that the one with cut base up did on the right, though nearer 7 and 4 o'clock respectively, due without doubt to some error in loading whereby the notch did not come out in a strictly down position at muzzle, while in most other tests the prints were uniformly at 4 and 10 o'clock, the 10 o'clock being half the distance from the normal group of the 4 o'clock.

The distance between prints on the 88-foot screen in Plate (14) is 1.05 inches; dividing 306 feet by 88 and multiplying by 1.05 gives 3.66, while the prints at butt, or 306 feet, are 3.90 inches. At 50 yards they are separated 1.80 inches; dividing 306 by 150 feet and multiplying by 1.80 gives 3.67, but prints are 3.93 at butt, showing that without doubt the bullet left a straight line in its flight by at least .04 inch in 50 yards.

Similar computations are easily made from other screens, noting, however, that what has already been termed the 100-yard butt is really 102 yards or 306

PLATE 14.

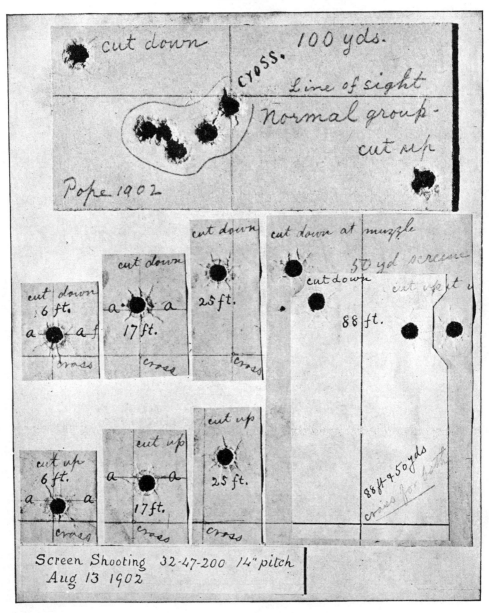

feet. It should also be observed that lines $a\,a$ represent line of fire on the 6 and 17 foot screens, that is the line from center of muzzle to tack on butt, which is 9 inches above the normal group. It is not certain that lines $a\,a$ are absolutely accurate, though carefully made before the shot, simply illustrating a fact often stated by physicists that absolute dependence cannot be placed upon any one mechanical test unless it can be verified. For instance the shot is .06 inch below the line of fire at 6 feet, while a normal one would have passed only .004 below; multiplying the difference by 306 divided by 6 gives 2.63 inches, while at the butt it is only .88 inch low.

If it was positively known that line $a\,a$ on the 6-foot screen, cut up shot, was correct, we should begin to know what was going on between the muzzle and six feet away where the bullet gets its first full tip and half its first oscillation. It is doubtful, however, about this line being correct. Both Plates (13) and (14) indicate clearly that after 17 feet, or even $6\frac{1}{2}$ feet, the bullet travels, so far as shown, in a perpendicular plane to the butt, or to within .06 to .12 inch of it.

Test 123. — The next day after preceding, or August 14, the test illustrated by (Fig. 81) on following page was made.

It shows the path of one bullet, with one and one-half grains cut from edge of its base and loaded so as to issue with its cut edge up from muzzle, through eight screens to the butt; the arrow connected with cross indicates direction of its print, 3.60 inches away, 4 o'clock; same cross serving for the 88-foot screen.

Print at the butt is 3.25 inches to right and 1.60 below line of sight. At 88-foot screen it is .97 inch to the right; multiplying this by 306 divided by 88 gives 3.40, but print at butt is 3.25, which shows that print at the 88-foot screen was only out of line .04 inch.

A magnifying glass placed upon the 12-inch screen indicates that the bullet had possibly a slight tip but so slight as to be unsafe to state in what direction. At 24 inches, $6\frac{1}{2}$ feet, 10, 17, 25, and 50 feet the bullet shows equal tips, but illustrations shown further on indicate that if there had been a 12-foot screen it would have shown no tip.

192   THE BULLET'S FLIGHT FROM POWDER TO TARGET

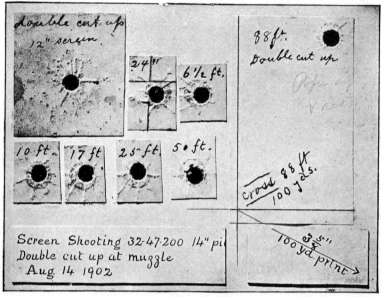

Fig. 81.

Test 124.—This was made on the same day and illustrated by (Fig. 82) on opposite page. In this experiment the notch cut in bullet's base was placed to bring it out down, and the line of sight is shown upon all but the 12-inch screen; on the 25 and 88 foot screens and 100-yard butt it is made to coincide for sake of compactness.

The print at butt is 3.93 inches, 10 o'clock, in its usual position; and this particular bullet shows slight tip on the 12-inch screen, 10:15 o'clock, and on the 24-inch, 8:45 o'clock, being fully developed on the $6\frac{1}{2}$-foot, 5 o'clock, and $16\frac{1}{2}$-foot, 3 o'clock.

If a screen had been placed at $11\frac{1}{2}$ feet, it would have shown no tip, as indicated by slight tip showing on the 10-foot screen; in other words, the screens indicate that the bullet made one complete oscillation, or one and a sixth circle on the clock, between the $6\frac{1}{2}$ and $16\frac{1}{2}$ foot screens. This one-sixth of the circle described beyond the complete oscillation of bullet exhibits its gyratory movement, indicating six oscillations to every gyration.

The cross on the 25-foot screen is marked " correct " which signifies that it was placed before the shot, verified after and found correct. The print on this screen is .26 inch to left, which multiplied by 306, divided by 25 gives 3.18, but print at butt is 3.75 to left of perpendicular, showing clearly that the shot was

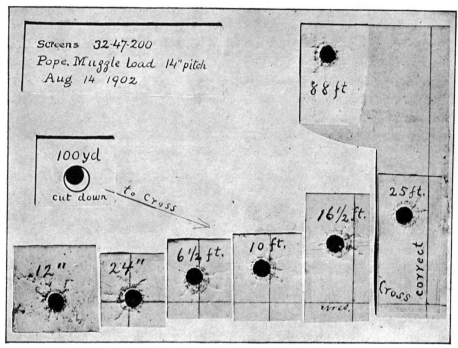

Fig. 82.

.04 inch out of line at the 25-foot screen. At the 88-foot screen the print is 1.11 inches out which multiplied by 306 divided by 88 gives 3.84, while the print is 3.75, showing print at this screen to be .04 inch out of line of flight.

Undoubtedly the 24-inch screen, as represented in cut, was improperly placed as it could not be focused with the scope and was placed by sighting through bore of the rifle. The print in 10-foot screen is .10 inch to left, which multiplied by 306 divided by 10 gives 3.06, showing that the print which was 3.75 at butt was only about .03 out of line at 10 feet.

If deductions be carefully drawn from the cuts, as shown in Plates (13) and (14) and two following cuts, and compared with the usual excuses given by riflemen for their off or wild shots, some important information may be gained. In one of the shooting houses in Pennsylvania there are tabulated on the boards 60 and odd reasons, given by different marksmen, for their off shots, possibly not one of which contains the truth so clearly indicated by the original screens as shown in these and the following illustrations.

Test 125. — This was made on same day as two preceding ones and its results are shown on Plate (15). Two bullets were used in this test, No. 1 being a normal 200-grain, 32-caliber, and No. 2 was mutilated by removing one grain of lead from the edge of its base by a single cut. The crosses on all screens represent the line of sight or the line joining center of the muzzle to center of a normal group at the butt. The prints of 1 and 2 at the butt are those of original target, the cross being same for both.

Very few normal shots were made with screens, and this was a fortunate one. Its course can be traced to the right in 25-foot screen, to the left of line at 88-foot, showing the bullet made a curve at the 25-foot or this screen was improperly placed. It was not verified, but the 50-foot one for No. 2 shot was.

The print of No. 2, at 50 feet, shows that at butt it should be .97 inch to right while it was 1.50 inches, indicating that at 50 feet it was .09 inch out of line, while the normal or No. 1 shot was .04 inch out at 25 feet if the screen was correctly placed. At the butt No. 1 print is nearly perfect, the break in paster being an exaggeration; but none of the screens through which this shot was passed showed fine prints.

Very naturally it was desired that screens should all have been verified, but there is quite a difference between using a fine steel scale and magnifying glass at the desk and setting up paper screens in woods over a rough range. These plates and cuts give photographic illustrations of original screens, prints, and pencil lines as made in the woods, and no doubt errors have crept in.

PLATE 15.

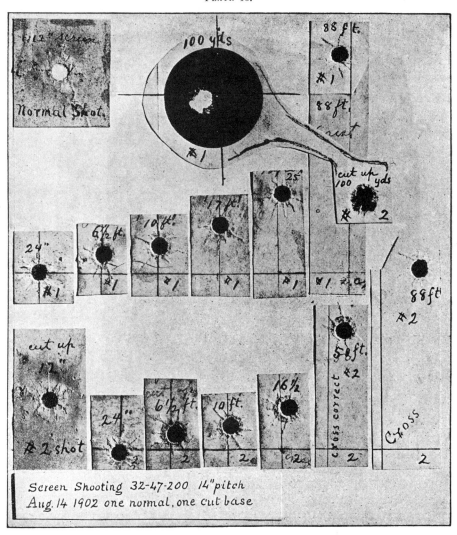

Test 126.— This was made August 15, 1902, and shown in (Fig. 83) following. It particularly illustrates the tip of the 200-grain bullet with one and one-half grains cut from edge of its base and entered so that the cut was down when emerging from muzzle. It was passed through screens two feet apart up to 30 feet, then four feet apart up to 50 feet. It printed on the butt 3.80 inches from

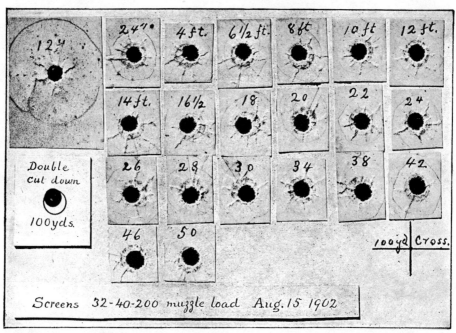

Fig. 83.

cross, as indicated by the cut. If we indicate the tip by direction of the base in bullet's print, as has been the uniform custom of the writer, and the print of base shows at 3 o'clock, then the point stands at 9 o'clock.

The cut of screens shot through in this test gives the maximum tip at 6½ feet and no tip at 12, while at 16½ and 18 feet the tip was large, and again no tip at 22; maximum tip again at 28 feet and at 34 no tip. Undoubtedly the base of this tipping bullet changed its position .03 inch to either side of its center, thus making a movement of .06 inch every 10 feet of its flight. If this movement of the base was in a circle .06 inch in diameter, its annular velocity was 25 inches per second.

After many tests which did not determine the conditions or motions of a tipping bullet, the screens were placed nearer each other, as in this case, which solved the cause of the difficulty.

Test 127.—This was made August 23, and is illustrated by Plate (16). Three Pope muzzle-loading, 200-grain bullets were used, No. 1 and 3 being mutilated by cutting one grain of lead from the edge of their bases, while No. 2 was a normal unmutilated one. The print made near the cross was by the normal or No. 2, that of No. 1 is seen at the upper left-hand corner, and No. 3 is connected to cross by dotted line from the right.

If these bullets did not leave a straight line, their prints at the butt would be 3.10 times farther apart than on the 99-foot screen, where the whole three are shown, and 4.10 times farther apart than where printed on 75-foot screen. No. 3 works out very closely, to within .01 inch, but No. 1 is so far off that some error is indicated in noting its flight after leaving the 99-foot screen. It did not print on target at the butt, but was taken from the board. The distance between Nos. 1 and 2 at 99 feet is one-third greater than at 75 feet, as it should be, indicating that No. 1 was in exact line of its flight at 75 feet and 99 feet, and should not have printed on the butt as noted.

In Plate (16) screens are also shown for No. 1 shot at 79, 83, 87, 91, 95, 103, 107, and 111 feet, and it will be noticed this shot made a perfect print at 100 yards, a large tip at 79, 91, 99 and 103 foot screens, and less tip at others. It became a non-tipper at every 10 feet, and the 75-foot screen was the only one which caught it at its no-tip position.

PLATE 16.

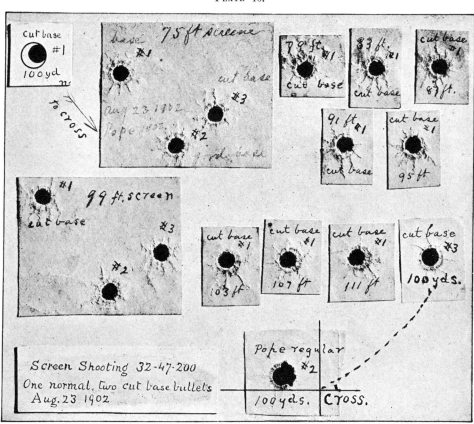

Test 128. — On September 7, 1902, with three mutilated base bullets, substituting the 187 for 200 grain ones, 26 screens were placed for each shot, six inches apart, commencing $19\frac{1}{2}$ feet from muzzle and ending at 32 feet. Figures on screens in Plate (17) give their several distances from muzzle in feet; first shot printing on upper row of screens, second on middle, and third on lower row of both segments of the plate. The line drawn from each print was laid out accurately with an eyeglass before prints became mutilated, and indicates direction of the bullet's tip without regard to the amount.

By studying these tipping bullets from Plate (17), it will be observed that No. 1 tips at 21 and $28\frac{1}{2}$ feet in same direction, No. 2 at $19\frac{1}{2}$ and 27 feet, and No. 3 at 20 and $27\frac{1}{2}$ feet in same direction, showing that these bullets made one oscillation in $7\frac{1}{2}$ feet, while the cut on page 196 indicates that the 200-grain bullet required about 10 feet for a complete oscillation.

By closely following screen prints on Plate (17), it will be noticed that No. 1 loses its tip at $21\frac{1}{2}$ feet, 4 o'clock, and does not lose it again until some time after 12:30 o'clock, when it passes the last screen. No. 2 loses its tip at $22\frac{1}{2}$ feet, 2 o'clock, and does not lose it again until after 11 o'clock. No. 3 loses it at $21\frac{1}{2}$ feet, 4:30 o'clock, and does not lose it again until after 3 o'clock, when its record is lost. By referring again to cut on page 196, the same peculiarity will be noticed.

All the prints of oscillating bullets indicate that the maximum tip on screen is about halfway between the screens which show no tip. Taking this conclusion from our observations, we are warranted in stating that position of maximum tip of the bullet travels left-handed a certain number of degrees for each oscillation, and if screens had been continued for 60 feet or more, the left-hand motion of this maximum tip could have been determined.

A rough computation shows, taken from cut on page 196, that this maximum movement of base makes one complete circle in about 60 feet. On Plate (17), No. 1 bullet shows its last tip at $21\frac{1}{2}$ feet, and if at 32 feet the tip was about to disappear, as it looks, the angle thus formed is 93 degrees, indicating that position of maximum tip would pass around in a circle in left-hand direction in 42 feet. This discrepancy between 60 and 42 feet is of no moment, since the above

PLATE 17.

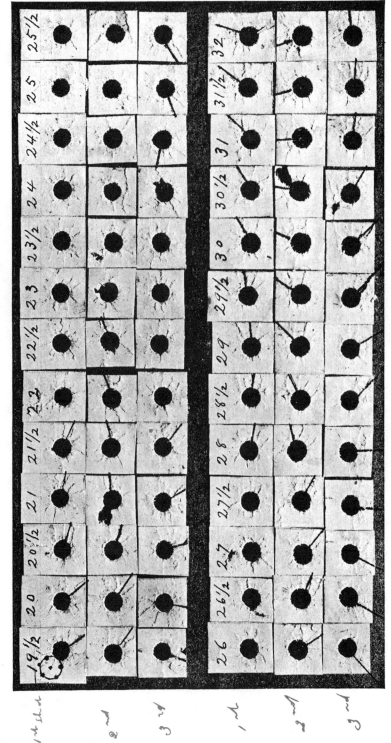

is not an accurate determination of distances, but an attempt to show the different movements of the bullets as indicated by these illustrations.

Another interesting incident is worth noticing in connection with Plate (17), before passing, shown by dotted ring on $19\frac{1}{2}$-foot screen. Before the rings on the telescope were fairly adjusted its rotation test was marked by the five dots in ring, the center of which represents the line of fire of V-rest. As a bullet drops .036 inch in first $19\frac{1}{2}$ feet of flight it would have printed, if rifle bore was straight and bullet a perfect one, .036 inch below center of the circle, which is .19 inch above where it did print. Multiplying .19 by 306 divided by $19\frac{1}{2}$ gives the distance the bullet should have printed at 306 feet out of the normal group, about 3 inches. Since the rifle bore was known to be straight, this calculation must be pretty accurate under the supposition then entertained that the bullet flies in a straight line after leaving the muzzle.

There were no butt prints of the screen shooting illustrated by Plate (17), but the reliability of this shooting must be unquestioned because of the immovable V-rest, concentric action, and Pope's 1902 rifle with its straight bore, all of which were essential for carrying out screen shooting with mutilated bullets if information was to be obtained. There was no flip or buckle to rifle; line of fire and line of sight each projected from same point and were absolutely the same from shot to shot and from day to day.

The average size of first 11 groups made with this rifle on as many different days was .63 inch, named the " Bumblebee " rifle, and the only straight bore out of 23 thus far used on the homestead range.

Comparative 100 *vs.* 200 Yard Butts. Test 128 a. — Riflemen as a body disdain to honor the 100-yard range. Perhaps they have good reasons for doing so, and perhaps they are as numerous as those quoted from the boards of one of the shooting houses in Pennsylvania, where 60 or more were found for off shots; but any one who is seeking the real reason or reasons for off shots can find them as well on the shorter as the longer range, and the latter is more convenient. It was not practical to make a 200-yard range at the homestead even if desired, and it was found to be unnecessary.

To make these tests undisputable, however, by digging a ditch and raising target eight feet above ground, a 175-yard range was obtained from the V-rest, and five groups were made at that distance, three of them through a 100-yard screen for comparison with normal 100-yard shooting, shown on Plate (18). Skeptical riflemen, or those particularly interested at 200-yard range alone, should measure up these shots with a fine scale, and a little assistance will be afforded in following explanations.

The three 175-yard papers which formed original targets are cut in exact proportion to their respective 100-yard papers which were used as screens at the latter distance, being shot through in order to reach the further target. The three 100-yard screens are represented at top of plate, the three 175-yard targets being below and numbered to correspond. By close comparison it will be found that the identical bullet which made a good shot at 100 yards made equally as good at 175, size of group being proportional to distance.

By studying plate it will be observed that width of No. 1 group at 100 yards is .75 inch; at 175 yards it should be 1.30 inches, and a careful measurement of group shows it to be exactly that. The width of No. 2 group at shorter range is 1 inch, and at longer it should be 1.75 inches; measurement shows it to be only .06 inch larger. The width of No. 3 group at 100 yards is .28 inch, and at 175 it should be .50 inch, while actual group made is .41 inch, indicating that in this group bullets varied .05 inch laterally from uniform curve at 100 yards.

The up and down or perpendicular errors in these groups should not be confused with side to side or horizontal, as the 6th bullet in group 3, and 9th in group 1, are defective in speed, and it is well known that errors of this character show greater at long range, representing altogether a different question than that discussed in these tests.

This variation of No. 3 group is the maximum of the three, and these were the only three 175-yard groups made with screen at this range, and are not selected cases. This 175 and 100 yard screen work clearly indicates that at 175, 100, or 50 yards, 88 or $16\frac{1}{2}$ feet, the great trouble in rifle shooting lies in the fact that the bullets do not start in the line of fire at the muzzle, and riflemen who wish to do some shooting had better fight it out on that line.

PLATE 18.

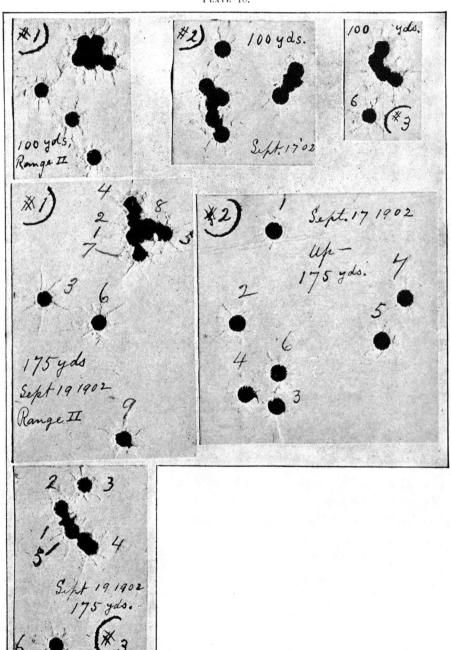

Plank and Screens. Some years ago H. M. Pope discovered that a bullet flying very close to any flat surface was deflected so as to print at target several inches out, but towards the side occupied by the flat surface, at 200 yards printing quite wide of the target. He was correct, and it was soon discovered, by experimenting, that the longer the flat surface presented and the nearer the bullet passed to it without grazing the greater the deflection. This phenomenon being contrary to snap shot reasoning, it was determined to test the matter with greater care, hoping and rather expecting it might disclose some other secrets regarding bullets' flight and, awaiting a better designation, we term it "plank shooting."

During 1901 four days were given to the sport and it proved very interesting as well as instructive. The cut on page 206 and Plate (19) illustrate prints from this plank shooting very accurately, being drawn to exact scale of one inch to four from original targets; that is, one-quarter inch represents a full inch of the targets which received the shots.

Three four-foot, and a four-inch, pieces of planks were obtained which readily afforded four different lengths of surface for experimenting, a four-inch, 4, 8, and 12 feet, and a frame was prepared for placing them, commencing nine feet from rifle's muzzle when in Pope machine rest. As planks were set on edge, shooting could be along either side of surface desired, and all shots represented on the cut or plate were made from same line of sight. The black-faced ones about the cross were normal groups, made without interference from plank. The paper targets represented by the cuts were two feet four inches long, reaching across the 100-yard butt, but did not prove long enough to catch all the shots.

Very early in the shooting it was found necessary to use paper screens to obtain any positive knowledge of bullet's path along the plank, so channels were cut across its surface 12 inches apart and deep enough to catch and hold the edges of screens and present them at right angles to line of fire.

With the aid of W. E. Mann every possible experiment that imagination could invent was indulged in on this range, and riflemen were invited to suggest experiments, of which this plank shooting was a sample. All was for a purpose, however, even if they resulted in hundreds of useless ones not worth recording.

206 THE BULLET'S FLIGHT FROM POWDER TO TARGET

Test 129. — This was made July 17, 1902, and illustrated by (Fig. 84) following showing prints of six normal and 18 plank shots, line of sight being the same for all. Eight feet of plank was utilized, extending from a point nine feet from rifle's muzzle towards the target with its surface to the right and at varying distances from line of fire — .32-caliber barrel and 187-grain, muzzle-loaded bullet, left-hand twist.

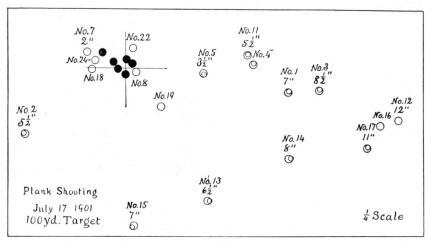

Fig. 84.

The normal shot group is indicated by black faces; plain circles indicate that the bullets printed fine, and small concentric circles within indicate roughly the amount of tip and direction of point as it cut target. Shots are numbered in the cut as made, and the distance of each from cross, or center of a normal group, is given in inches. For instance, No. 7 shot passed $\tfrac{3}{4}$ inch from the face of plank and, as will be noticed, made a perfect print two inches from the cross.

Numbers 8, 18, and 24 passed $\tfrac{5}{8}$ inch from surface of plank; 5, 11, and 19 passed $\tfrac{3}{8}$ inch; No. 4, $\tfrac{1}{4}$ inch; and 1, 3, 12, 14, 16, and 17 were very close to plank, but not having screens during this day's shooting exact distance is not known; their closeness could only be judged from the fact that shots 13 and 15 grazed the plank and printed as cut shows. No. 2 also grazed the plank for 24 inches, and a curious phenomenon in connection with 15 was shown by its grazing the plank sufficiently to exactly counterbalance the normal plank deflection, making it print in line but 4.75 inches below normal.

Shot 22 passed a piece of plank only four inches long, and close as possible without grazing, and it shows no deflection. These prints have quite a degree of interest although shots were made without screens; after this, however, screens were used, so the following plates give more positive information.

208   THE BULLET'S FLIGHT FROM POWDER TO TARGET

Test 129 a. — Extended over July 18 and 19, shots illustrated by Plates (19) and (20) were made. Plate (19) is a plot from two targets, separate ones being used each day, and shows in black face a group of six normal shots, using same rifle as in previous test, 187-grain, muzzle-loaded bullet, left twist, machine rest, line of fire same in all shots. Ten shots were made along the 12-foot plank to the right and five with plank to left of flying bullets, and prints are designated as on Plate (19). Plate (20) is from a photograph of the several screens placed along the plank, through which 12 of the 15 hits shown on Plate (19) were passed; also showing prints of three cylinder bullets which missed the butt, therefore not seen on Plate (19). The numbers above each column of Plate (20) refer to corresponding numbers over prints on Plate (19), and the figures immediately above each screen print represent the number of feet it was placed from muzzle end of plank, which added to nine, the distance of end of plank from muzzle, gives distance of screen from muzzle of the barrel.

Shots 1, 2, and 3 have five screens each, while others have only two. The line to right or left of each screen print was drawn after shot was made before removing screen from plank which held it, surface of plank acting as guide for pencil, so the line very accurately represents surface of plank along which bullets flew and the distance of shots away from it. The bullets to the left of plank were caused to tip at about 11 o'clock; those to the right, about 4 : 45 o'clock.

In all prints made by the 187-grain bullets, by other tests, where tipping was caused by cut or plugged bases, direction of the tip made a complete circle every 7 1–2 feet, while these plank tippers did not change position of their tip from 11 o'clock during whole space traveled along the 12-foot plank. Undoubtedly the planks were not set exactly parallel to line of fire, being set by the eye, near enough, however, to indicate that the bullets did not change their course while passing. The large deflection came after the plank was passed.

No. 6 bullet passed $\frac{7}{16}$ inch from plank and was deflected 3 inches at butt, while No. 5 passed $\frac{2}{16}$ inch farther away and was not deflected. No. 4 passed $\frac{3}{4}$ inch from plank, was deflected $2\frac{1}{2}$ inches at butt. No. 1 grazed the plank for 18 inches at the 10-foot screen, thus throwing it off from plank and causing it to print as seen on Plate (19); while No. 2 did not graze the plank, but passed

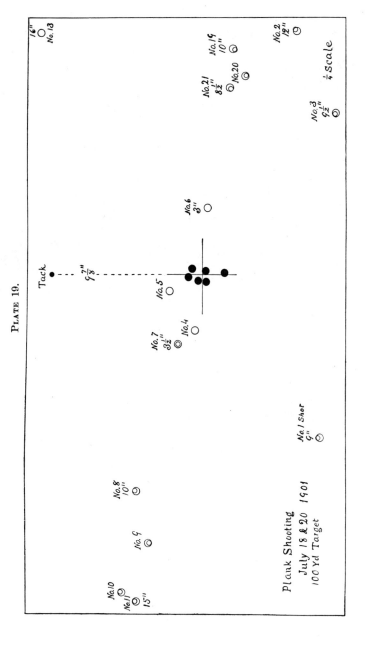

Plate 19.

only a thickness of paper from its surface, causing it to print, as shown, to the extreme right and lower corner of target. No. 3 flew so close to plank that it pinched the 10-foot screen between itself and plank, as Plate (20) shows.

Thus the test of three days' plank shooting, besides other things, demonstrated that a bullet shot nearer than half an inch to a hard surface and for a number of feet, was deflected towards the line of the surface after leaving it instead of away from it; and when passed along the surface more than half an inch away it was deflected from the line of surface; when shot half an inch from surface, the bullet is not deflected either way.

I shot in Hartford 1/2" left of 12 ft plank 125 ft long Pope 32/200 bullet, it was deflected about 12" Right at 200 yds

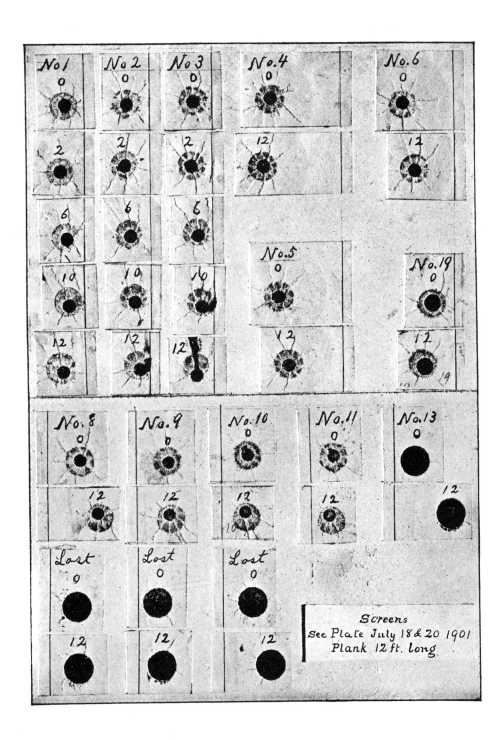

Unbalanced or Mutilated Bullets. Given, a "gilt-edged" rifle barrel, V-rest, concentric action, an ideal shooting day, and normal bullets may be so mutilated by the experimenter before shooting, each bullet different from the others, line of fire remaining the same for all shots, that a five or ten shot group made by these bullets will duplicate, shot for shot, a group made by any expert marksman, thus making two targets identical as to appearance. This may appear like a wild statement, but is susceptive of proof, as will be shown by the six following tests with their photographic illustrations.

Test 130. — Aug. 11, 1903, 5 : 30 P.M., dead calm affording perfect weather conditions, with Pope 1902 rifle, 187-grain, soft-lead bullets, some with cut points and others with plugged holes, but all swaged bore diameter after plugs were inserted; Plate (21), showing one original 12-shot group and one 3-shot, is quite important and interesting because of perfect weather conditions, with gilt-edged rifle and V-rest. As will be observed by careful reading of Plate (21), first five shots were made with unmutilated bullets, forming a group above the cross of .61 inch; the other six were mutilated in different ways and made a group of 6.50 inches, No. 8 bullet not printing on target. The 3-shot group, made with mutilated bullets, is also shown. On lower part of Plate (21) both original targets have been plotted into one, showing prints of five normal bullets by light rings and mutilated ones by heavy ones, except the lost No. 8. On the plot is also shown manner in which some of the bullets were mutilated before shooting, and in this test all mutilated bullets were so loaded as to cause the mutilations to issue at muzzle up, or at top of bore.

One bullet, with hole .10 inch in diameter near base, carried to center and removing 3.2 grains lead, was deflected 3.50 inches, 4 o'clock; others with .125-inch holes near front end, carried to center, and removing 6 grains lead, were only slightly deflected in two shots, as shown by 10 and 2 *a* prints, and showed no deflection in two other shots, indicated by 6 and 1 *a* prints.

Bullets heavily mutilated, as shown at lower corner of Plate (21), by cutting their points, and emerging from muzzle with their light side up, printed at 12, 9, and 3 *a* at 10 o'clock instead of 4, where they would have printed if lightened

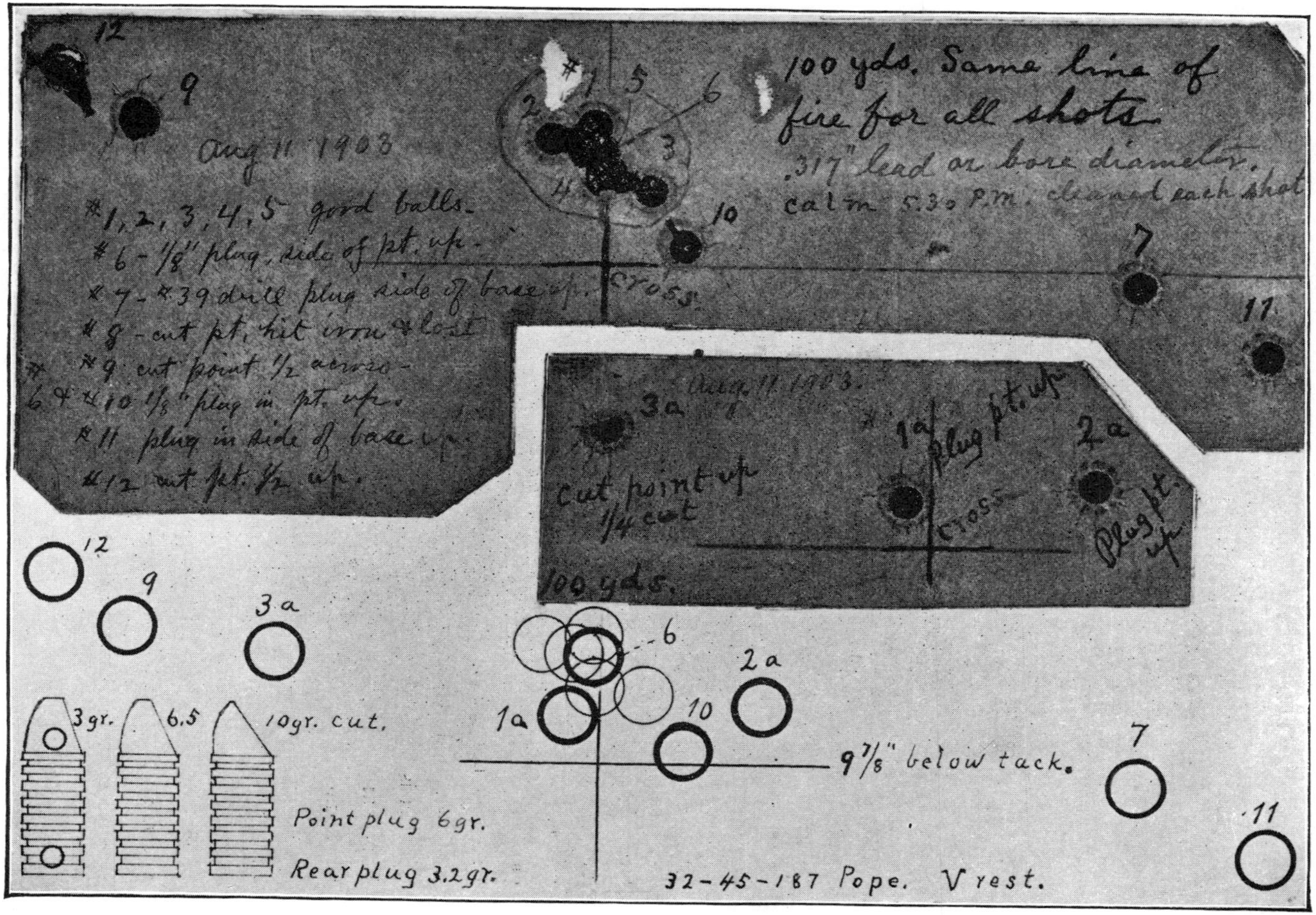
100 yds. Same line of fire for all shots.
.317" lead or bore diameter.
calm 5.30 P.M. cleaned each shot
Aug 11 1903
#1, 2, 3, 4, 5 good balls.
#6 - 1/8" plug, side of pt. up.
#7 - #39 drill plug side of base up.
#8 - cut pt. hit iron & lost
#9 cut point 1/2 across.
#10 1/8" plug in pt. up.
#11 plug in side of base up.
#12 cut pt. 1/2 up.
cross
Aug. 11 1903.
cut point up 1/4 cut
1a plug pt. up
2a Plug pt. up
100 yds.
3gr.
6.5
10gr. cut.
Point plug 6gr.
Rear plug 3.2gr.
9 7/8" below tack.
32-45-187 Pope. V rest.

at bases instead of points. This marked peculiarity of opposite effects produced between point and base mutilations can only be explained at this stage by experimental screen shooting; theories do not count.

Test 131. — On same day as preceding and under the same favorable conditions four groups were made, illustrated by Plate (22), one normal with dirty shooting, of .94 inch, and another with careful cleaning between shots of .75 inch. It will be observed that the latter exhibited lower trajectory than the former, reaching the target .75 inch higher. If these two groups, and the .61-inch group shown on Plate (21), be compared with that marked " Group 2 " on Plate (22), a marked difference is indicated, yet this latter 2.75-inch group was produced by one slight change in the bullets.

In this " group 2 " bullets were swaged with slightly oblique bases, similar to experiments while snow shooting in 1901, where 120 out of 122 bullets were made oblique at time of upset or before reaching muzzle of rifle. In this group shots 1, 2, 3, and 4 had their bases swaged .012 inch oblique, about equal to two thicknesses of ordinary book paper, and shots 5, 6, 7, and 8 were swaged .006 inch oblique, or half the amount of first four, and in each case the obliquity was entered to emerge on the quarters at muzzle, as written on plate. Only one shot of the eight is contained in the circle which would have inclosed the whole 15 normal shots presented on Plates (21) and (22), 10 of which were shot with rifle clean and five dirty.

This is what Mr. Leopold calls a hollow group, made with a new gilt-edged Pope barrel, but not with normal bullets, that is, not such bullets as made the above-mentioned normal groups; instead, the bullets were swaged intentionally oblique. Yet this " group 2 " bears a striking resemblance to the usual 100-yard groups made with a good rifle, double rest, and first-class fixed ammunition. It is so similar as to size and grouping of shots to common everyday double-rest groups that it would not be recognized as a hollow one, but the two normal groups shown on Plate (22) plainly indicate that it is; its similarity to an ordinary group being too marked to be funny.

Observing the conditions under which this group was made, its size was not

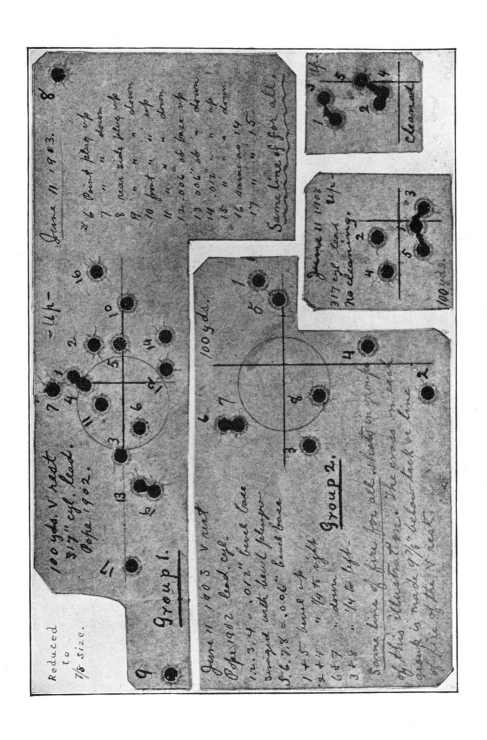

216   THE BULLET'S FLIGHT FROM POWDER TO TARGET

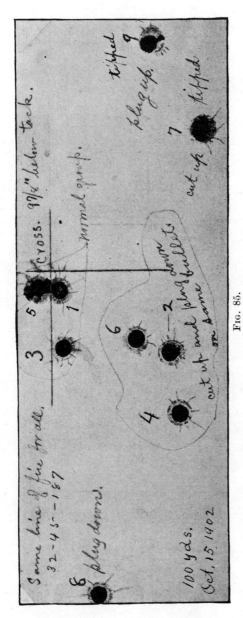

Fig. 85.

due to wind, to mirage, or errors of sighting, to rusty or crooked barrel, irregular flip or powder charge, or any other 60 odd causes for bad shooting, but exclusively by obliquely swaged bullets, obliquity being made to correspond in amount to the measured ones recovered from snow in 1901. From above description the first group may be easily understood by reading table written upon the plate.

Test 132.—October 15, 1902, one 9-shot group was made, as shown by (Fig. 85), under same conditions as previous ones, except mutilation of bullets was varied, though prints 1, 3, and 5 were made by normal or unmutilated ones.

The object of this test was to determine whether the blast at muzzle deflects a defective bullet or whether its deflection is due solely to the fact of its being unbalanced; that is, having its center of weight out of its center of form, and so out of the center of rifle's bore.

In mutilating the other six bullets of this group the knife cut removed one grain of lead from the edge of base, and the plug removed 2.10 grains by boring a hole into the base about halfway between its edge and center, thus unbalancing it the same amount as the one grain at the edge of base. The hole was plugged.

Shot 7 was cut but not plugged, entered to issue from muzzle with cut up. Shot 9 was plugged but not cut, and issued plug up. Shot 8 was plugged and issued with plug down. Shots 2, 4, and 6 were both cut and plugged and all entered to issue with cut up and plug down.

It will be observed that the plugged bullet, without a cut, printed at 9, while the cut one without a plug printed at 7, near each other; with cut on one side of edge and plug on the other of the same bullet, nearly balancing each other, the bullets are made to print at 2, 4, and 6, halfway between 7 and 8, and on line between them. It is easy theorizing how the muzzle blast could glance from the base of a bullet with a cut edge, but not so easy from a plugged base.

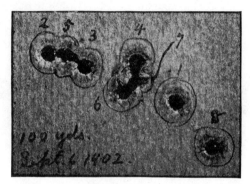

FIG. 86.

In this test, however, it seems to make little difference in results, rather giving a chill to our muzzle-blast theories.

Test 133.—September 6, 1902, one 8-shot group was made to show the effect of slight mutilations upon otherwise fine-shooting bullets, and illustrated by (Fig. 86). The first shot had .83 grain of lead cut from its side, extending through the last two base bands, and the cut was entered to come out up at muzzle. No. 2 was mutilated same as 1, but the cut issued down at muzzle. No. 3 was a normal, unmutilated one. No. 4 had .12 grain cut from the last base band but not through the corner of its base. No. 5 had 1.50 grains drilled from the base of its point and entered to come out cut down. No. 6 and 7 had .50 grain drilled from their several points and made to issue down. No. 8 was mutilated same as 5, but entered to come out up.

This gives a very unique group from an almost perfect rifle, one very similar in shape to an occasional normal one from an average rifle, its unique feature being in knowing the particular part where the bullet was unbalanced and the amount of mutilation or unbalancing of each shot.

Test 134. — This was an 8-shot group made May 20, 1903, after the Pope 1902 gilt-edged barrel had lost its accuracy and was making twice the size groups of the year before. In this group, as shown by (Fig. 87) below, shots 1 and 2 had a small hardwood plug fitted into a hole in one side of their points; 3 and 4 were perfect bullets at time of entering in front of loaded shell; 5 and 6 had a large plug in the side of point, and 7 and 8 a large plug in their side at first front band. All plugged mutilations were entered to issue up at the muzzle.

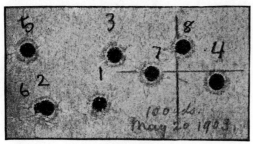

Fig. 87.

The shape of this group can be better comprehended when the four previous tests, with their illustrations, are carefully compared. With this group it is fair to question why the two normal shots, 3 and 4, are separated by 1.20 inches, or why they did not stay together like the 5-shot groups as seen on Plates (21) and (22), or all the shots during one summer up to November 3, 1902. It will be noticed, however, that after that November day this gilt-edged barrel has not made one fine group. It is not a question here of what happened to the rifle bore, but what did the rifle bore do to the bullets in shots 3 and 4? There can be little doubt that it unbalanced these shots just as surely as the other six in this test were intentionally unbalanced by plugging.

Test 135. — This was made August 30, 1902, to determine if possible the effect of a very slight mutilation on the base of an otherwise perfect bullet, thrown from an almost perfect barrel, as this gilt edge then was. Such small groups had been so uniformly made that W. E. Mann suggested this test as a practical though severe one. (Fig. 86) shows a group as made at 100 yards with bore-diameter, breech-loaded bullets, shots 2, 4, and 6 being normal ones, while 3 and 5 were prepared by boring into their bases near the corner with a slim point of a knife-blade. The point was made to enter about .06 inch, and the amount of lead removed would not be noticed except placed under a glass. These mutilations were entered to come out up at muzzle, and the cut indicates, as in all other tests, that when mutilations of base, however slight, are made to issue up at the muzzle, the shot is thrown towards the right from normal, and at 4 o'clock. This and preceding tests were not selected, being the only ones made with such slight mutilations.

Many experiments, similar to and including tests already tabulated, indicate that any form of normal group from a common rifle, with fixed ammunition or front seating, used in target work, can be imitated, shot for shot, at will from the V-rest and a perfect rifle, under fine weather conditions, without changing line of fire from shot to shot. With such a rifle

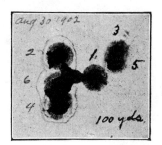

FIG. 88.

and immovable rest, we can imitate any one shot of another rifleman's group, by using an oblique-base bullet, having the correct obliquity and so loading it that its short side will emerge at muzzle on the correct quarter; and this mutilation is probably similar to the very thing which produced the shot we are about to imitate. Any shot in a six-inch group at 100 yards, or in a smaller group, could thus be easily duplicated and by care the whole group could be reproduced. This statement was made at the head of this article, and the reader will find proof at hand, experimental and theoretical.

Test 136.—This was made September 6, 1902, and the cut (Fig. 89) shows its results. The 5-shot normal group is shown at the left within pencil lines. No. 1 shot, as marked in the cut, had one grain lead removed from the corner of its base. No. 2 and 3 had one-half grain removed in same manner. No. 4 was mutilated by twisting a thin knife-blade .06 inch into its base near the corner. No. 5 same as 4, but point was entered .12 inch into base.

220   THE BULLET'S FLIGHT FROM POWDER TO TARGET

All these five mutilated bullets were entered to issue from the muzzle with mutilations up, and the group shows that very small mutilations are plainly indicated by prints when all other errors are excluded.

FIG. 89.

Test 137.—This was made September 5, 1902, to determine the effect produced by monkeying with the lubricant, and (Fig. 90) below exhibits the six shots made. The bullets for shots 1, 4, and 6 had the lubricant entirely removed from one side of the four grooves nearest their base, one quarter the distance around, while 2, 3, and 5 were normal shots, their normal lubricant not being disturbed. The test seems to indicate that irregular lubrication produces no appreciable effect, though a single test of this character is hardly conclusive and properly lubricating a bullet is not likely to injure it.

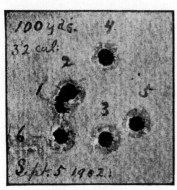

FIG. 90.

In passing this subject, however, it is well to note that by other tests it was shown that by removing the lubricant from a cylinder bullet it was thrown entirely off the butt.

Gyration and Oscillation.
Gyration is produced by a constant force acting at right angles to its motion, and oscillation can be produced by a momentary force applied at right angles to either end of the axis of a spinning body. The reasons why are not generally understood and we shall make no attempt at an explanation. But the visible motions of gyration and oscillation should be recognized by all students, so some illustrations are introduced as an aid.

In the cuts presented, (Fig. 91) shows a top so rapidly spinning upon its axis as to appear motionless, and (Fig. 92) was an attempt to photograph the

Fig. 91.

Fig. 92.

same top after it had been made to gyrate. Its gyrating spindle will be readily recognized, though the body of the top only gave up its bright spots to the camera.

In this case gyration is a constant motion, with no visible oscillation, at right angles to the force of gravity, and a little observation will convince a novice that this gyratory motion is not what causes a top to hum, nor can it be termed a wobble. The axis of gyration in this case is a perpendicular line drawn through the point of the gyrating top. Every gyrating body has an axis of gyration which the reader can more easily understand without than with a definition.

The hum of a top is produced by the rapid movement of its point upon its support, due to the center of form endeavoring to describe a circle around the center of gravity. This motion is identical with the movement which causes the center of form of an unbalanced bullet to make a small spiral in the air.

To illustrate oscillation a gyroscope, counterbalanced by a sliding weight on a rigid arm, is selected, though the photograph here reproduced would have been much improved if the rigid arm upon which the weight slides had the appearance of being in line with axis of the spindle of revolving disk. The photograph was taken with the disk in rapid motion and the whole system supported upon the hardened steel point of the standard; that part of the arm protruding from the weight to the right was painted white so the desired image could be photographically produced.

Fig. 93.

When the disk is not revolving the slightest touch to the painted end of arm will cause the whole system to move with perfect freedom around and upon point of the standard, but not so, however, when the disk is rapidly revolving between its pinions; then the arm becomes rigid and requires considerable force to move it, and will remain stationary at any position placed or to which it may be pushed.

This counterbalanced, spinning gyroscope represents very closely, in some particulars, the rifle bullet when making a horizontal flight; one, however, is supported at its center of gravity by a stationary standard, and the other is not. If some slender object like a lead pencil is pressed gently against one side of projecting arm of gyroscope, as represented in cut, when disk is rapidly revolving, the system will not move around nor will the arm apparently move in the direction it is being pushed. Instead of this it moves at right angles to the pressure and downwards, a motion not indicated in the cut. If the arm is struck a sufficiently sharp blow from the right, it does not move directly downward but downward and to the left and returns to its starting point, thus forming a circle as shown in cut, and it will continue this circular motion over and over. Of course the far end of the spindle of the revolving disk describes this same motion. This is the movement of oscillation without gyration, and the rigid arm combined with spindle of the revolving disk actually executes the motion of oscillation as exhibited by a flying bullet.

It is difficult to realize this interesting motion, and doubtful if it can be fully comprehended without the aid of the apparatus illustrated.

A bullet unbalanced at its base is spinning with its axis of form in line of fire as it leaves the muzzle, and the front or balanced part of the bullet would follow this line. When its unbalanced base is released, however, it leaves the line of fire and follows the tangent of bore spiral, thus imparting a sudden motion to the rear end of rigid axis of spin and at right angles to it. This produces a motion of the axis of a spinning bullet identical with that shown in the cut at one end of the arm of the gyroscope.

Every time the bullet completes an oscillation its axis has returned to the same line at which it received the impulse identical again to the illustration. At this point the axis of bullet is again in line of fire where, if the bullet printed, it would show no tip.

Our experiments indicate that oscillating, tipping bullets also gyrate, and the manner in which these motions combine is extremely interesting, if it can be understood, and two or three simple diagrams are introduced to illustrate this, photography having failed us here.

Let then the lines which form the circles of the three diagrams in (Fig. 94) represent the movement of the center point of base of an oscillating, gyrating bullet; the first shows three oscillations to each gyration where this center passes through its circles as numbers indicate, commencing at its maximum tip at 1 and passing successively 2, 3, 4, 5, 6, 7, 8, 9, 5, 9, and so on, going through its center at 5, or no-tip position, three times, for each gyration, meanwhile describing the three circles as represented. The center figure represents the path of this center where bullet makes five oscillations to one gyration, and the third where seven oscillations occur.

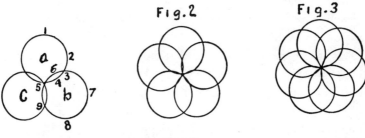

FIG. 94.

While the center of the base is describing one side of either of the three figures, the point also describes the other side of a similar figure, and the bullet is flying point on when the center of its base or point occupies the center of its respective figure.

A Spinning Bullet. Hoping to discover something experimentally regarding the movements of an elongated body spinning upon its long axis, with aid from our shop superintendent, P. H. Kimball, the bullet-spinning machine, as illustrated, was designed and built. After contemplating the finished product it appeared a rather elaborate and expensive affair, but putting it to the tests required proved that it was none too extravagant or complete for delivering the necessary speed to a spinning bullet for the desired experiments, and with it a bullet was finally made to spin like a top. Many a laugh was indulged in over the pranks played by the little bullet before sufficient speed was obtained to give it stability upon its point. When necessary speed was obtained, the spinning bullet could be carried about the room in its little cup like a toy top. It either spun good or not at all.

This taught, though not fully comprehended at the time, that there is a definite pitch of rifling demanded for each form of cartridge used in modern rifles, below which no factory presumes to venture.

THE BALLISTICS OF SMALL ARMS 225

Fig. 95.

(Fig. 95) is from a photograph of the complete spinning machine, simple but practical, as results show, and as a whole needs no description. (Fig. 96) gives in detail one end of the machine, the brass tube in the head of machine having a small pulley attached to its upper end with a string belt about it. This tube was reamed to a diameter of .315 inch, into which the bullet is placed, and a slight smear of oil on the bullet prevents its dropping out of tube by its own weight. When sufficient speed is obtained the bullet is pushed downward through the tube into a glass or porcelain cup by a quick but steady motion of the pointed brass wire, seen lying across lower part of framework supporting the tube.

Fig. 96.

The first obstacle met was in attempting to use a lead bullet. When sufficient speed was obtained for spinning, the bullet was found enough unbalanced to make it hug one side of the tube with such tenacity that it could not be pushed downward by a force short of stopping the machine.

Fine tool-steel bullets were required and accurately made before they could be properly ejected from the tube. No method at hand was reliable for deter-

Fig. 97.

mining speed of the revolving tube or spinning bullet except by observing its musical pitch; it had no comparison, however, to the spinning speed of a modern bullet from an ordinary rifle.

(Fig. 97) above is reproduced from full-size photographs of four different spins which were made by the machine, and the spins would continue in good form, in some cases, for one or two minutes.

It will be noticed that these bullets are made in the form of the .32-caliber grooved and ungrooved, with same diameter, though their points were necessarily made sharp. Their base was made concave in order to receive the pointed wire which expelled them from the revolving tube.

Many curious observations upon the behavior of these spinning bullets might be recorded, because their motions seemed innumerable, and one in particular was interesting. A small camel's hair brush pressed against one side of the base as it stood upon its point would push it all over the concave cup, and even out of it, without causing it to tip or gyrate to an observable degree. Quite another motion was very pronounced if conditions were right, that is, if the point of spinning bullet met with no obstructions upon the surface of the glass; when its gyrating movement began it moved about a point midway of its long axis; in other words, the point described the same circle on the glass plate that the base did in the air, as seen in the cut on following page.

(Fig. 98) is reproduced from five photographs taken at various times, while trying to catch an image capable of showing the gyrating, spinning bullet, and the second figure in cut was the successful trial, where bullet and its reflection from the glass plate are shown to be gyrating about their centers of gravity. While the spin is slowing down the gyrations grow larger and more rapid until nearly invisible, and the bullet nearly over on its side though still supported by its point. Instantly a change occurs, the bullet losing all tendency to stand on its point, and immediately spins about its short axis with great speed, as seen in the third figure of cut, the entire energy of its spin about its long axis being converted into an energy of spin about its short axis.

The first figure of the cut is this same bullet, with its reflection by glass plate upon which it spins before it began to gyrate. The spinning bullet properly delivered from the machine always goes through these forms of spin, as seen in the three upper figures of the cut, when unconfined on a flat surface. The two lower figures only show two of the many forms of tumbling the bullet undergoes after falling below its proper spinning speed, if on a concave surface.

Extended observations on homestead range, including hundreds of tests

228 THE BULLET'S FLIGHT FROM POWDER TO TARGET

and measurements, had practically decided that the bullet in tipping, oscillating, and gyrating, did so about its center of gravity and not its point or base as a turning point, and these actions of a steel spinning bullet upon its point show experimentally, to the unaided eye, that when it either oscillates or gyrates it does so upon its center of gravity.

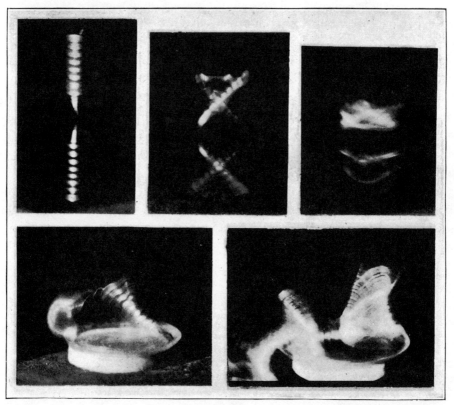

Fig. 98.

There are no principles involved in the spinning bullet respecting its center of gravity that are not embodied in a toy top; but the shape of the former is so much changed from the latter that the bullet, experimentally, makes its various movements about its center of gravity as a turning point, while the top is usually prevented from doing this by excessive obstruction which its point finds upon the surface on which it is supported.

If a bullet gyrates upon its center of gravity in its flight through the air, it must cut a picture in the air similar to (Fig. 99), which was produced mechanically by mounting a normal lead bullet at its center of gravity upon a ball and socket joint, with means provided which caused it to make an accurate gyration. The left figure was photographed while the bullet was in continuous gyration, but the other resulted from four separate exposures upon the same plate with bullet occupying four different stationary positions in its gyration.

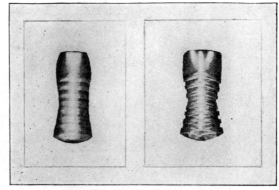

Fig. 99.

To make still plainer, if possible, the figure a tipping, gyrating bullet cuts in its flight, two photographs of the Zischang .32-caliber bullet are introduced, enlarged two times, on following page (Fig. 100). The first was mechanically moved three degrees upon its center of gravity between two exposures, and the second six degrees, representing respectively a three and six degree tip of a flying bullet, the center of gravity being approximately on the black dot in the center of first figure.

Tipping Bullets Deceptive. One becomes interested in the details of a subject as soon as experiments begin, and the unexpected is always occurring. Most riflemen are apt to be satisfied that they know which bullets tip and the amount of tip by examining a target at 100 or 200 yards. They are quite liable to be satisfied that the amount of tip is known when the screens, if any, are examined through which the bullets have been passed on their way to the final target, but a photographic illustration of an original target will serve

230   THE BULLET'S FLIGHT FROM POWDER TO TARGET

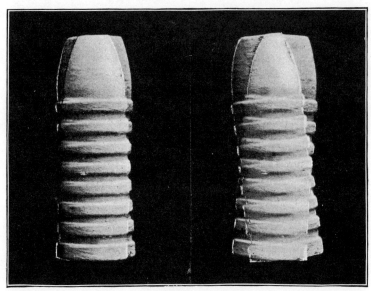

Fig. 100.

better to make the errors of ordinary observation apparent than any amount of explanation.

The No. 2 shot, shown in cut (Fig. 101) at right upper corner, made as perfect a print at 100 yards as possible to be made by any bullet, but it also shows that this shot was tipping 11 to 13 degrees when passing through screens placed at 7, 18, and 29 feet. It printed 4.5 inches to the right at 100 yards without a tip, and it was not tipping while passing through screens placed at 13 and 23 feet.

It was found necessary, where exact knowledge of tip was to be determined, to use a tough, thick, pure linen paper, backed by thin wood pulp board. The white paper will show the prints perfectly at point and the direction of tip, while the board will not deceive much in regard to its amount. It was also found necessary to place screens no more than 12 inches apart for a space of 10 feet, to discover where the bullet had its maximum tip.

If a rifleman overlooks the above caution in his experiments, he should be very cautious about reporting his tipping or non-tipping shots to his fellows because

THE BALLISTICS OF SMALL ARMS 231

errors are only multiplied thereby. Such methods, if persisted in, keep riflemen in an eternal dilemma regarding the influence a tipping bullet has upon an interesting score. There is no quibbling here. If the one shot in 10 which lowers the score and loses the prize was a tipper, and the marksman honestly believes it was not because it made a perfect print at the target, and persists in this error year after year, is he not seriously handicapped?

We repeat, that with the exception of "Plank Shooting," every tipping bullet over the homestead range during the period of six years that has left its record on screens, has been an oscillating one, and has made perfect prints at uniform distances varying from two to fifteen feet.

This constant and usual erroneous reading of the target prints in the past

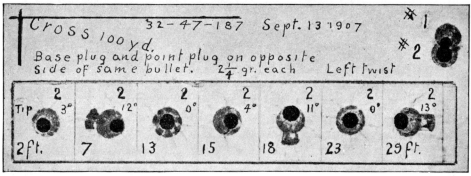

Fig. 101.

has caused us all to admit that it cannot be proved that an occasional tipper has any demonstrable effect upon the score. These tests disclose, however, that a bullet with a large tip or a moderate tip may make a fair, a fine, or a perfect print at the particular place where the target is set, depending entirely upon the phase in its oscillation at which it happens to find the target. For instance: in cut illustrating this article (Fig. 101) one would decide, when examining prints made by the two bullets in upper right-hand corner, that it did not matter whether a bullet tips or not since No. 1 was tipping 10 degrees when it struck, yet made as good a shot as No. 2 which shows no tip, but in fact both were excessive tippers, though at different places of their flight.

Bullet Tip Scale. In prosecuting experiments for determining the action of tipping bullets a necessity was soon felt for some convenient and accurate method of measuring the degree to which respective bullets tipped, and the scale illustrated by cut (Fig. 102) was made to answer such a purpose. Instead of being drawn by the artist, as directed, to fill requirements of a .32-caliber bullet, it fits a .34-caliber one, but principle of the scale remains the same.

Such a scale can be used whenever the center of bullet's point can be detected with certainty in its print, and where the print shows full marking of its base.

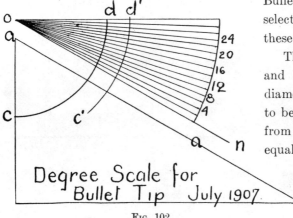

Fig. 102.

Bullets and paper were carefully selected while experimenting with these points in view.

The distance between lines *on* and *aa* should be one-half the diameter of bullet whose tip is to be determined. The distance from *o* to the arcs *cd* or *c'd'* should equal length of the respective bullets. The distance from center of the bullet's point to the extreme mark that its base makes should be laid off from the point of intersection of its arc with line *aa*. If the bullet did not tip, the measurement would come on line *on*, as it represents one-half the diameter of bullet. If the base of bullet stood out of line as it passed the target, the distance from its point to base would extend up the scale a certain number of degrees, which would be its approximate tip.

Measuring the number of degrees a bullet tips should be indulged in to a limited extent by every studious or interested marksman, not so much because the exact tip in geometric degrees is of any special value, but because he is liable to make valuable discoveries about the flight and amount of tip of bullets he is using and which may be giving more or less trouble. This he must do if ever expecting to obtain correct ideas of existing conditions in regard to tipping bullets. Theorizing will not answer in these scientific times.

A rifleman may choose one of several ways of measuring tip of his bullets, but if a flat-point one is to be measured, probably no easier method than that shown in cut will be found. This scale is not introduced altogether for show, but to put before the reader a comparatively accurate method which may be used to verify the author's deductions, or make his own observations upon the numerous photographic reproduction of prints which are furnished. A word of caution, however, in this connection may be of some assistance to the tyro; the print of a bullet is usually considerably smaller than its own diameter, which must be taken into consideration in measurements.

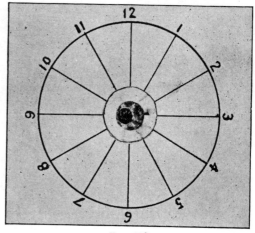

Fig. 103.

Speaking about the measurement, degree, or print of a tipping bullet, the cut (Fig. 103), which illustrates a scale executed on paper, will be found very convenient for marking the o'clock tip, which most riflemen like to know.

The central aperture, represented by a ring about the print of a bullet, should be larger than any marking a tipping bullet is likely to make at target. It will be noticed that radial lines represent the different phases of a clock's face, and by placing this paper scale over a bullet's print the approximate o'clock tip is immediately determined. It is well to carefully mark that side of the print towards which the bullet tips, as illustrated by short black mark in cut, before the scale is placed over it, or one's judgment will too often be found faulty; and this should be done with the aid of a good magnifying glass. If it was properly done in the illustration the fact is pretty closely determined that the bullet's trip there represented was at 2 : 45 o'clock.

234   THE BULLET'S FLIGHT FROM POWDER TO TARGET

Correcting Measurements.   In taking a measure of bullet prints throughout our screen work, from their center to the India ink lines, as was customary, it was discovered that an error was introduced which in some cases amounted to over .03 inch. The center of the prints as shown on screens does not represent the center of gravity of the tipping bullet which was passed through, and to show where this center belongs the diagram (Fig. 104) was drawn to scale of one to four.

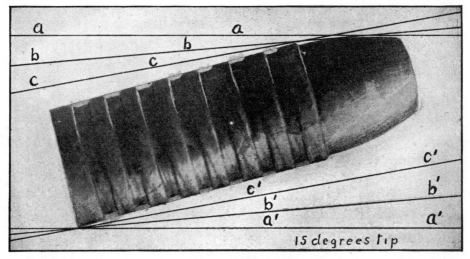

FIG. 104.

On this diagram of an enlarged bullet the center of gravity is shown by a white spot near its center, and all measurements should have been taken from this point, to show actual position of any bullet in question. The bullet in the diagram is represented as tipping 15 degrees, as marked by lines aa and $a'a'$; the lines bb and $b'b'$ showing 10 degrees, and cc and $c'c'$ five degrees. By measuring the white spot, or center of gravity, it is found to be nearer the line aa than $a'a'$ by .20 inch, and .16 inch nearer line bb than $b'b'$, and .12 inch nearer cc than $c'c'$, therefore one-half above measurement gives distance of the white spot from center of their respective parallels which represent the extreme edges of the bullet's print from its several tips.

Reducing the size of this diagram to that of normal bullets by dividing by four, gives the error which runs all through our screen work where measurements have been taken from the center of print of a tipping bullet. To illustrate: According to this diagram the measure of the print of a bullet tipping 15 degrees was taken .025 inch too far away from the point side of print; that of the 10-degree tip .02 inch, and the five-degree tip .015 inch.

We had been made aware of some error of this kind, but did not realize its magnitude until after screen work had been completed and put to test of this diagram. With this diagram interested students can make their own corrections throughout this whole screen work.

236   THE BULLET'S FLIGHT FROM POWDER TO TARGET

Flight of a Bullet. In an attempt to picture the phantom flight of a bullet, the artist produced Plate (23), hoping by appealing to the eye to aid the comprehension. The upper figure of the plate represents a .32-caliber bullet, full size, flying point on and therefore taking a straight flight, and its phantom path through the air is .32 inch in diameter.

The next figure shows the flight of same caliber that is tipping six degrees; though its flight is apparently straight, it is not in the direction of its long axis, which makes it a tipper. This bullet, by the air pressure on its under side, is being deflected upward towards the line of its axis though only slightly so. So slight is this deflection that its flight is actually straighter than the artist's rule. These two upper figures on the plate represent, to full size, the flight of a non-tipping and tipping bullet through seven inches of their respective flights. It will be seen that the bullet in the second figure makes a phantom path through the air much wider than the non-tipping one in the upper figure, which means more air resistance, a condition that should be fully realized.

The fourth, or lower, figure should be joined to the right end of the third, from which it was cut, then it will show the artist's attempt to represent a spiral with a diameter of $\frac{5}{16}$ inch, 14 inches long. Observe that this diameter of the spiral does not refer to the width of the white path, which represents the phantom flight of a six-degree tipping bullet, but is the diameter of the air spiral which the bending of this path was made to represent, although the curves of this spiral have been shortened by the artist from 45 feet or more to six or seven inches, to make them visible, thus presenting a gross exaggeration.

This spiral of $\frac{5}{16}$ inch is larger than a normal bullet, tipping six degrees, can make, and much larger than a fair shooting bullet would make, or one having tip admissible with modern ammunition.

The distance between the two bullets shown in third figure of the plate, in actual flight would be about 24 feet .instead of three inches, as represented by the artist. The first one is shown at top of its air spiral, and if from a left-twist barrel it is here being deflected to the left out of its course more rapidly than at any other place. It will be seen, however, that its point does not stand in this left direction, but at just 6 o'clock. At a position 12 feet farther on, or

PLATE 23.

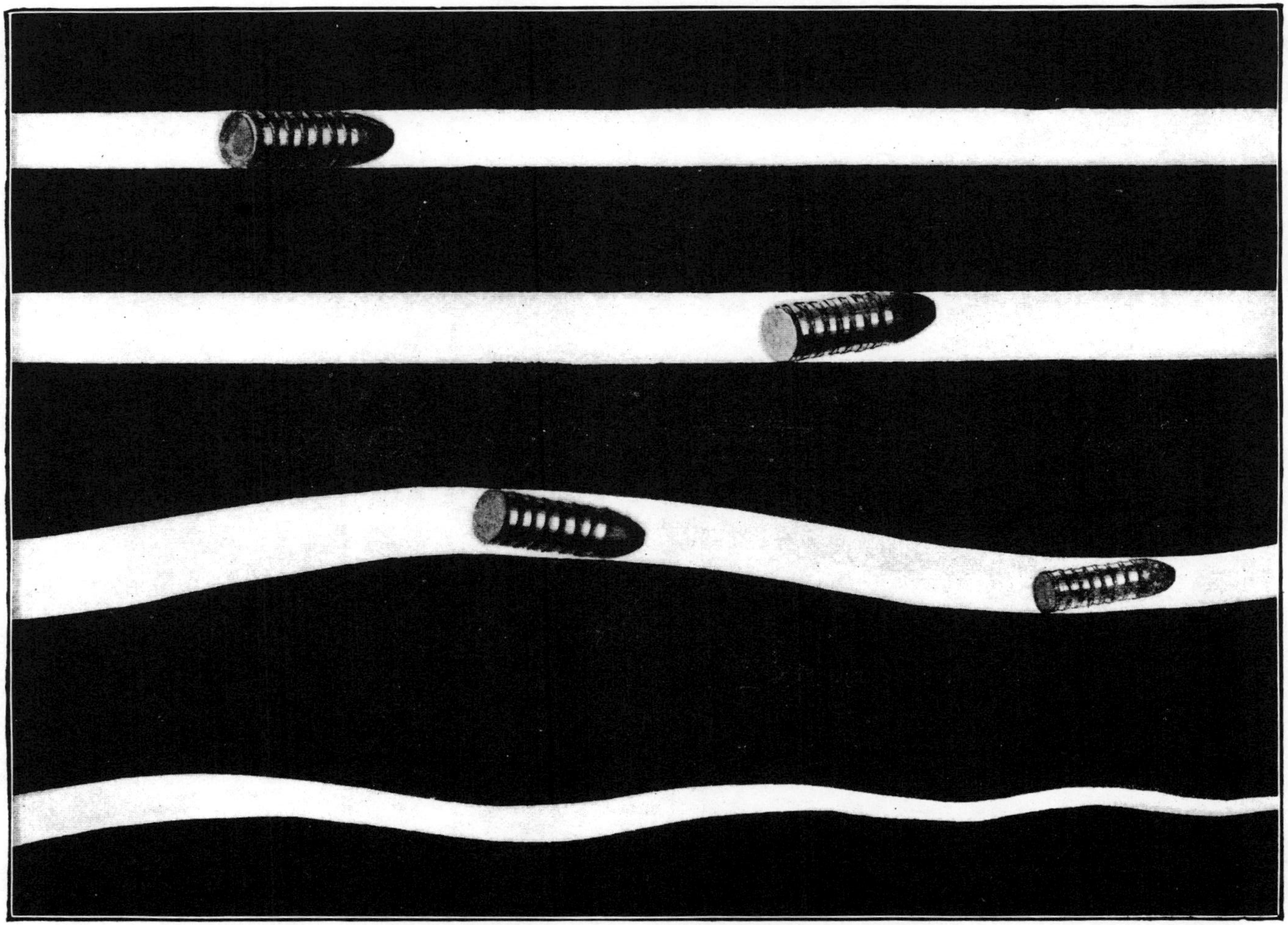

halfway between the two bullets, it will be going down in its spiral course faster than at any other place. At this position, however, its point does not stand at 6 o'clock, or the direction in which it is being deflected the fastest, but at 3 o'clock; thus the direction of tip is one quarter of a gyration in advance of its position in the spiral. The tipping bullet surges around in the air throughout its flight, forming a small spiral the turns of which are one quarter turn behind the direction of the bullet's tip.

The plate, however, does not fully represent a normal tipping bullet, because such a bullet always oscillates, while here it is only represented as gyrating, a condition exhibited only by our tests in plank shooting, where bullets gyrated without visible oscillation.

During many of our experimental years, the task to combine the several motions of a flying bullet was not found to be an easy one; and when finally comprehended, it may be still more difficult to make plain those movements to one who cannot possess equal opportunities. This may excuse us for constantly referring to the subject and still another appeal to the imagination.

If an ordinary rowboat 15 feet long was made of solid lead, as a bullet is solid, it would weigh eight and a half tons. Let us stretch our imagination a little farther and try to imagine such a boat weighing 6800 tons, equal to the weight of a solid block of marble 44 feet long, wide, and high. Since water is 800 times the weight of air, a boat 15 feet long of this estimated weight would bear about the same comparison to water that the bullet does to air, and would be deflected in its motion through water very much the same as air deflects a bullet. Such a boat would be expected to obey its helm to only a very limited extent. After its bow was deflected a few degrees in any direction, the boat as a whole would pass through a long distance before its direction was changed to an appreciable degree. The lead bullet in thin air, theoretically and according to actual measurements of bullet's flight, acts that way. A six-degree tip of a bullet deflects it about one quarter inch in 22 feet. A normal rowboat, if headed six degrees out of its course, would have its bow turned about 12 inches out, and if its weight was 6800 tons, it would probably only be deflected about the same as the bullet, one quarter inch in 22 feet.

Measuring Wind Drift. If a bullet be dropped from a stationary position through the air a given distance during a windy day, it will be drifted some distance out of plumb, and the machine illustrated (Fig. 105) was constructed to determine that distance.

The apparatus was composed of a vertical steel shaft, turn-buckles, cables, revolving platform or platens, each of the platens being graduated into 10 divisions; washers and collars, with set screws provided where necessary. The adjustable platens were machine made and a brass tube accurately reamed was supported vertically by the upper platen with its center $1\frac{5}{8}$ inches from center of the vertical shaft.

A number of 187-grain Zischang 1 to 20 bullets were carefully sized so their weight would just carry them through the tube, to be dropped upon a white paper circle, covered by a circle of carbon paper resting upon lower platen.

On November 20, 1907, when air was still, ten drops were made upon the circle of paper, represented in diagram (Fig. 106) on following page, turning the lower platen one division at each drop; the five

Fig. 105.

prints on the right semicircle were made from a drop of $9\frac{7}{8}$ inches, being the usual drop of a bullet before the .32–47 charge at 100 yards, and five prints on left semicircle were dropped 41 inches, the usual drop of bullets over 200-yard range.

Close examination of these prints shows that the apparatus worked well, as the whole 10 bullets dropped directly upon the line of circumference in calm air. The prints even show that some bullets tipped in the short space dropped.

240   THE BULLET'S FLIGHT FROM POWDER TO TARGET

After waiting 20 days for a windy one, December 10, the testing apparatus was again set up on brow of a hill, across which a gusty west December wind was blowing, though always towards center of the circle, and bullets were dropped at the moment when wind was in proper direction and at its maximum, as indicated by a flag posted for the purpose.

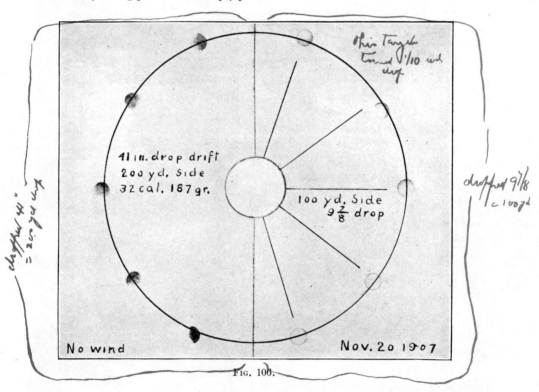

Fig. 106.

Prints shown in the diagram (Fig. 107) on opposite page which are in the direction of the radial lines were dropped $9\frac{7}{8}$ inches, while those nearest the lines which cut the circle were dropped 41 inches.

The wind was moving from twenty to twenty-four miles an hour, and the greatest drift of any one of the 10 which were dropped $9\frac{7}{8}$ inches was .08 inch, and at 41 inches it was .30 inch. If the wind was blowing 24 miles an hour, it would, by computation, have traveled 15.26 feet during the time the bullet was drifted .30 inch, or a proportion of 630 to 1; that is, the air traveled 630 times farther than the bullet was drifted.

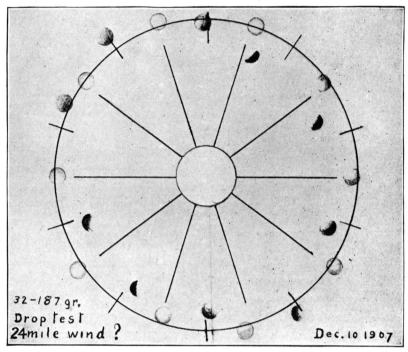

Fig. 107.

The results of this experiment give the rifleman an opportunity to compare the drift of a bullet dropped from a stationary position to the drift added to the wind deflection that occurs on a 200-yard range during a windy day.

242   THE BULLET'S FLIGHT FROM POWDER TO TARGET

Motions executed by Normal Flying Bullets.
Before attempting a statement of motions executed by a flying bullet, it is well to acquaint ourselves somewhat with a definition of terms, and in so doing correct some that have been constantly wrong as used in works on ballistics or writers on rifle practice.

The motion of a bullet which results from its being carried along by wind is a drift, not a deflection. The movement of a bullet out of its course by any other cause than that of being carried along by air currents is a deflection and not a drift. Webster defines drift as " to carry along as currents of air or water do a body, etc.," so the term is used here as defined, not only to be correct, but because better understood by the average reader.

Not every bullet will execute all the following motions, but some will, and all will make a majority of them: translation, spin, trajectory, gyration, oscillation, point deflection, center of gravity spiral (in rifle bore), center of form spiral (in the air), gyratory or air spiral, wind drift, wind deflection, trajectory deflection, tumble.

The swing of a pendulum may be designated as gravity oscillation, the vibrations of a rod which is rigidly held at one end may be designated as elastic oscillation; so we will designate the form of oscillation which a rapidly rotating or spinning body is observed to make as rotary oscillation, or oscillation of spin.

Gravitation actuates the swinging pendulum; elasticity, the vibrating rod; and spin actuates the oscillating bullet. In mechanics the terms "swing," "vibration," and "oscillation" are synonymous, therefore we designate one of the motions of a spinning body as oscillation of spin, which is always a curved motion, sometimes circular, at other times cycloidal, but probably circular with the bullet.

A pendulum at rest is forcibly held in this position by gravitation.

The non-vibrating rod, or a rod at rest, is forcibly held in position by its rigidity.

The axis of rotation of a rapidly revolving body is held in one direction by a force which is a result of its rotation or spin.

If the axis of rotation, the position of the pendulum at rest, or the position of a non-vibrating rod are altered by a momentary force, each will set up an oscillation characteristic of itself as soon as that force which moved them is removed.

Oscillation is not gyration. Oscillation of a pendulum, rod, or spinning body is produced by a temporary force which has been removed. Gyration of a spinning body is produced by the application of a constant force continually acting at right angles to the movement of gyration. These definitions will aid somewhat towards comprehending the following statements regarding the motions of flying bullets.

The first motion is that of translation, due to gas pressure from exploding powder; then it spins or rotates because of the spirally grooved rifle bore. Because of gravitation it has a trajectory, and this motion combined with air pressure causes deflection of the bullet, to the left if the twist of the grooves be left-handed, and to the right if right-handed. This motion is a trajectory deflection and will be further noticed in the next article.

Most bullets, being more or less unbalanced, begin to develop a tip and an oscillation immediately upon their exit from muzzle, and those that do not tumble in their flight will gyrate, due to air pressure on or near their points.

The 187-grain, .32-caliber bullet, in 14-inch twist, if unbalanced at base or at point, usually makes an oscillation after leaving the muzzle in every 9 or 10 feet, and a complete gyration in about 50 feet, as indicated by screens; that is, about five oscillations to one gyration.

Since the bullet with a tip is not flying in the direction of its long axis, it has a motion which may properly be termed point deflection, due to the glancing of the bullet as a whole on the air pressure, this pressure being more pronounced on one side than the other.

Because of a gyratory movement of the bullet the direction of its long axis continually changes around the line of bullet's flight, and because the deflected point must follow towards the direction it is pointing, or towards the direction of the bullet's axis, due to air pressure, it causes the tipping bullet to make a spiral flight.

When the bullet oscillates, it makes from three to seven oscillations during its period of one gyration, and during these oscillations point deflection occurs, as in gyration, but to a lesser degree owing to the shortness of time in which the point stands in any one direction.

The center of gravity of an unbalanced bullet, or one whose center of gravity is not in its center of form, is obliged to make a spiral flight while in the rifle bore. The center of form of the unbalanced bullet, which is obliged to make a straight flight in rifle bore, begins to make a circle around its center of gravity and, consequently, spiral flight immediately upon its exit at the muzzle, and the pitch of this center of form spiral in the air equals pitch of the rifle bore. This revolution of the center of form around the center of gravity is identical with the motion which produces the "hum" of a spinning top. The shifting of the center of form from a straight flight in the bore to a spiral flight at the muzzle occurs at the moment when the center of gravity changes from its spiral in the bore to its straight flight in the air.

Since the tipping bullet, due to point deflection, flies as a whole in a spiral after leaving the muzzle, we have this spiral joining itself on to the smaller spiral, made by center of gravity in the bore, within a few feet of its muzzle exit, which may still further increase the deflection from line of fire.

When a bullet changes its direction by being carried along by the wind, it is drifted, not deflected, and this drift cannot take place without some pressure upon the bullet, which will not only move the bullet as a whole but cause a slight deflection; deflection being due to its spin or rotation, or rolling upon the side of increased air pressure. A pressure of this kind may also change the direction of bullet's axis according to the laws of rotating bodies, and might be computed as to its extent if it could be known upon what part of the bullet a given wind was exerting its pressure. Thus far, however, it has not been determined.

There is no attempt in this article to make any suggestions or explanations of the cause of the major part of wind deflection which occurs during the bullet's flight. The wind drift test, page 239, is valuable on this line and stands as the first necessary test. No others have yet been devised to substantiate any theories that the author may have, respecting this deflection as observed on the range.

If a bullet does not tip sufficiently to tumble, it tends to follow its point, thus making a spiral flight. If it tips more than six degrees, E. A. Leopold determined by experiments that his bullets were liable to tumble and be lost, but our experiments do not coincide with his.

Trajectory Deflection. In the previous article the motion of a bullet termed "trajectory deflection" was defined, but should receive more extended notice because it has not, to the writer's knowledge, been as fully explained and comprehended as is possible. Works on ballistics designate this deflection as drift, improperly, because it is not a motion that corresponds to the definition of "drift" by any English dictionary. It is a motion one element of which is skin friction, due to a partial rolling and slipping of the bullet upon increased air pressure on its under side, or that side which is presented toward the center of the earth.

This increased air pressure, according to present works on ballistics, is partly due to falling of the bullet through the air in making its trajectory curve, but more largely to the axis of it not being in line of this curve or line of flight, but standing above it. Upon examining this statement a little closer, we will suppose a bullet commences its flight from the muzzle with its axis of spin and line of fire in a horizontal plane. Upon leaving the muzzle it enters its trajectory curve immediately, so its flight is not in a horizontal plane, while, by supposition, its axis of spin is. The bullet then becomes a tipper and according to tests on the homestead range it will, due to air pressure on its point, commence to gyrate as do all tipping bullets.

If the trajectory curve did not change, but was a straight line, this gyration would be equal on all sides of the line of flight, and the bullet, due to its tip, would make a spiral flight, but would have no trajectory deflection due to skin friction as far as its tip was concerned. The trajectory curve which, by previous supposition, was horizontal at the muzzle, leaves this plane for the curve immediately and leaves it more and more rapidly as the distance increases from muzzle, according to the law of falling bodies.

The supposed point-on bullet, therefore, becomes a tipper more and more rapidly, and its axis of gyration (for definition see page 221) is constantly striving to keep itself in line of the ever increasing trajectory curve, and would succeed for a space of 50 feet, or one gyration, if the curve would become a straight line for that distance; but at that place, as at all other places, the trajectory is making a sharper and sharper curve, so the line of the axis of gyration continually lags behind the ever increasing sharpness of curve; thus the bullet, due to its forward motion, receives more air pressure on the side presented downward, than on that which presents upward. This not only raises the bullet in its flight, but allows it to roll on the increasing air pressure below, toward the left if from a left-hand twist barrel, and to the right when the twist is right-handed.

The balanced bullet, due to its trajectory, becomes a tipper and makes a spiral, the axis of which is the trajectory curve or line of flight, but its gyratory axis never coincides with this line except at muzzle, where it does not gyrate because it does not tip. The cause precedes the effect, hence the bullet conically screws itself around a line slightly above the trajectory.

When the trajectory deflection is understood, and the cause of it mathematically accounted for, the author wishes to predict that the true cause of the major part of this deflection will be the "y" spiral of the tipping bullet, as discussed on page 253.

The skin friction due to the fall of the bullet through the air, or due to the axis of gyration lagging behind the trajectory curve, probably has but slight effect on the deflection in question, while the deflection which occurs at the time the non-tipping bullet is made into a tipper, and changes from a straight flight into a spiral one, is the deflection which we now call trajectory deflection.

To appreciate and illustrate this prediction, we go back at once to plank shooting. Here the bullet from left-twist barrel, leaving the plank from its right side, has a tip at 5 o'clock and prints far to the left at 100 yards. The bullet, going into its trajectory immediately upon leaving the muzzle and becoming a tipper at this point enters its y or air spiral at a 6-o'clock tip and would, as in plank shooting where the bullet has nearly the same direction of tip, print to the left. The 12-foot plank during its entire length produces a gradual tipping of the bullet towards 5 o'clock, while the trajectory curve produces a gradual tipping of the bullet, more and more towards 6 o'clock, during the whole time of its flight. It is the relation which the tangent of the spiral bears to its axis that causes the deflection at plank shooting, as experimentally demonstrated on the homestead range, and sufficiently discussed later on under the head of Mathematical Verification. It is this same relation between the tangent of the trajectory air spiral and the axis of this spiral, which induces the author to predict the true cause of trajectory deflection.

More Reflections. When Daniel Webster made his celebrated reply to Hayne of South Carolina, in the United States Senate, he very pertinently prefaced his speech as follows: "When the mariner has been tossed for many days in thick weather, and on an unknown sea, he naturally avails himself of the first pause in the storm, the earliest glance of the sun, to take his latitude, and ascertain how far the elements have driven him from his true course. Let us imitate this prudence, and before we float farther, refer to the point from which we departed, that we may at least be able to conjecture where we are now."

Some fifty years afterward a late Georgia senator clipped Daniel Webster's ponderous and dignified method by inquiring, "Where are we at?" as a prelude to another famous speech.

So, before we advance farther in our search for the elusive x-error, or burden the reader with other dry experiments and multiplied detail, it may be well to "refer to the point from which we departed," or pause to explain "where we are at."

Rifle practice and expert riflemen are multiplying throughout our own country as well as all civilized countries. With most the practice is simply a pastime, a fad, to take the man of business on an outing away from the grind or monotony of his everyday life. But there is a largely increasing class who are experimenting, hoping to discover why, when every condition seems favorable, one out of five or two out of ten shots go wild, while the other eight form a spreading group the size of which millions have been expended to reduce.

We made this inquiry so earnestly nearly forty years ago that it compelled us to commence these long series of tests and experiments; and when satisfied that the difficulty was located, other more crucial tests were instituted, not only to demonstrate that the cause of x-error was found, but to prove it beyond the cavil of that class of riflemen who are also earnestly seeking.

The question of being able to correct the error when found, though having its influence as an incentive, has not predominated in our tests and experiments, or in tabulating them for the public to examine. The main and leading incentive has been *why* a bullet cannot be made to hit a bull's-eye every time, or which one or more of nature's laws were not being complied with by thousands who were so assiduously trying.

Our state of mind in this matter must be much the same as that of the voyager or searcher for the North Pole. No doubt scores of intelligent and scientific people who have been, or still are, seeking the Pole, do not expect to change a hair's breadth of the earth's motion when the discovery is made; neither may we be able to improve rifle shooting when the cause of bad shooting is absolutely shown and demonstrated, but it may be of some aid towards reaching as near perfection as the laws of nature will allow. Certainly if the cause of an off shot is known, the chances for riflemen to improve upon it are greater than if unknown or only conjectured.

In order to eliminate many of the causes attributed by riflemen for bad shooting, a fixed, mechanical rest was adopted where the rifle barrel, denuded of its stock and action, could lie with the line of its bore always pointing to a tack driven into the 100-yard target, and hundreds of tests have been carefully conducted to eliminate or confirm assumed causes other than personal with the marksman. These make up the bulk of tests and experiments tabulated in these articles and assembled into book form. Though over 300 in number, they are only a small part of the many attempted or carried out, but it is hoped enough have been given in detail for the student to anticipate the writer by locating the cause of x before the closing articles are reached.

Although the mechanical V-rest and concentric actions were necessary for the elimination of many assumed errors which could not otherwise be proved unreal, yet our tests indicate that as good shooting may be and is done at the old shoulder and muzzle rest as from an immovable mechanical one.

That there is an average spreading of groups in finest target work, averaging from two and a half to three inches at 100 yards, not including fortunate groups or those from extra shooting or gilt-edged barrels, is well known. These large groups show up in the absence of wind, mirage, or irregularities of sighting; they show up when powder charges are of exact weight and ignite so uniformly that the perpendicular prints enlarge groups no more than horizontal or side ones; they show up when bore of the rifle is chemically cleaned, oiled or not oiled between shots, or when dirty shooting is practiced; they show up when bullets are lubricated or unlubricated, swaged or unswaged, if from a new barrel or an old one, front-seated or fixed ammunition, muzzle or breach loaded; they show up if a double rest, mechanical, or V-rest is used, open sights, telescope sights, or no sights.

If making alterations in any of the multiple forms of target work would uniformly eliminate these enlarged group sizes, or the x-error, riflemen would have known it years ago, for the time and experiments and money spent in the effort have been enormous.

Although many of our experiments and tests were seemingly haphazard ones, as in fact some of them were, something was learned from all, many times leading up to systematized tests that proved very valuable; for instance, snow shooting, short-barrel work, and plank shooting. Much was expected from vented barrels which only proved delusive, while the unexpected was constantly in evidence to illustrate how poorly our reasoning faculties were working. In short, the solution of our problem was so simple as to be for years overlooked or only half guessed, a common human failing.

Cause of x-Error Located. After years of patient research and a multitude of experiments extending over these years, eliminating one assumed cause after another and disproving the popular theories of the day which at one time and another appeared plausible, the real cause of the error at target, termed x-error, we have demonstrated mathematically and experimentally; that is, so far as any bullet is concerned which flies from the muzzle with its point directly on, in other words, the non-tipping bullet.

Dealing in this article solely with the non-tipping bullet, the difficulty is found to lie in the fact that the bullet which does not make a center print is unbalanced before discharging, or is unbalanced before it reaches the muzzle, thus preventing its center of gravity from coinciding with its center of form; neither can its center of gravity coincide with the axis of the rifle's bore. This must and does cause this center of gravity to fly in a spiral around the axis of the bore and to describe a circle around the center of bullet's form. When the bullet is released at the muzzle, being no longer forced to keep its spiral, it ceases to keep its spiral course, and takes a straight course which can only be a tangent to that spiral, as first mathematically stated by E. A. Leopold. That is, the substance of this statement was first given to me by him, after having had access to all my tests and theories.

It should be fully comprehended that this spiral form of flight which an unbalanced bullet must take in the rifle's bore is described by its center of gravity or weight, not its center of form, and that weight will throw form immediately upon the release of form from confinement, as is illustrated by swinging a weight attached to a string in a circle; as soon as the weight is released by letting go the string it flies at a tangent to the circle it has been describing. If the bullet did not take this tangent course at the muzzle, it would have to make a sharp change in its direction with no force present to cause it to make such a change, and contrary to all laws of moving bodies. When a bullet literally goes off on a tangent, it does not change its direction in doing so, but would have to change its direction if it kept in the line of fire.

A cut (Fig. 108) is introduced on opposite page to make this matter plainer. If a rapidly moving object, like a bullet before the powder blast, was following the spiral here represented through the confined space of a rifle bore, immediately upon its release from confinement it would not change its direction, but take the tangent represented at right of the cut. No power on earth could make it change its course and take the direction of the straight line to the left, which represents the line of fire, without reducing the bullet to a fine powder.

THE BALLISTICS OF SMALL ARMS

The spiral which an unbalanced bullet describes in a rifle bore might have a diameter of a part of a thousandth of an inch to two or three thousandths, depending upon the amount it was unbalanced, in most instances not larger than a hair, while our illustrated spiral has a diameter of nearly half an inch. If the rifle bore has a pitch of 12 inches, the hair spiral described within it by unbalanced bullet has a pitch of 12 inches, while pitch of illustrated spiral is less than quarter of an inch. The comparison between the illustrated spiral and the really described spiral of the bullet, therefore, requires a fearful stretch of one's imagination.

A bullet by following the spiral, as shown by cut, when released from its confinement in rifle bore would fly at a tangent from the muzzle at an angle

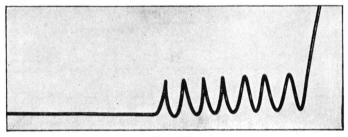

Fig. 108.

of over 80 degrees, but if a spiral no larger than a hair with a pitch of 12 inches can be imagined (for it cannot be photographed to be visible to the eye), instead of taking an angle of over 80 degrees, its angle would be so small that if carried in a straight line to 100 yards from muzzle it might not be more than an inch out of line of fire. This is exactly what happens to the unbalanced bullet that flies with its point on, and if we knew just where the bullet was unbalanced, it could be told which side of the bull's-eye it would print. It could also be computed accurately how far from the bull's-eye it would print if the amount and place of unbalancing of bullet was known, provided bullet flew with point directly on.

It is essential that this spiral flight which an unbalanced bullet must make through the rifle's bore be fully comprehended, and its resulting tangent deflection, before any explanation is attempted of the tipping one, therefore another spiral (Fig. 109) is introduced.

In this a little observation will convince that, although its pitch is kept the same throughout, its diameter and tangent angle is constantly changing. If a flying body was released from the left extremity of this spiral, it must take a tangent making an angle of 60 degrees to its axis; but if released from the right extremity, the tangent angle is so small that the eye cannot distinguish it from a straight line.

Again making good use of one's imagination by supposing the pitch of spiral described by an unbalanced bullet in bore, is 12 inches instead of half an inch,

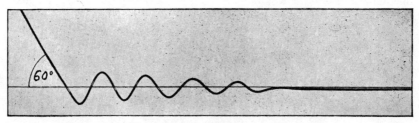

FIG. 109.

as in cut, and its diameter is equal to the line made by continuation of its axis to the right, and the real spiral we are attempting to describe may be comprehended. When this is understood, the tangent that an unbalanced bullet must take when leaving its confinement of bore at muzzle will also be comprehended, also that the angle described from the axis of the spiral, or from line of fire, by flying bullet is dependent upon size and pitch of spiral, which also determines the value of x at the target because a non-tipping bullet flies in a straight line, or with the tangent of its bore spiral.

This is a demonstration, oft repeated by experiments, of the cause of x-error at the target, and, though extremely simple, it had eluded our grasp for years.

**Cause of
y-Error
Located.**
As we began to congratulate ourselves that the cause of *x*-error was solved, the whole cause of off shots discovered, as explained in previous article, screen shooting in certain tests began to indicate that there was another cause besides that of the bore spiral, produced by an unbalanced bullet, which was in some way connected with a tipping bullet.

Plank shooting, first indulged from curiosity, eventually demonstrated an unmistakable error at target of this kind. The large deflection from a straight line which the bullet made immediately after leaving the plank, and its straight course afterwards, remained unexplained for five years in spite of all attempts at solution, so this error at target was very naturally termed the *y*-error.

Bearing in mind the fact that the instant a bullet flying in the air begins to tip it also begins to gyrate, and if it did not gyrate it would follow the direction in which the point began to tip, and by thus following direction of tip the resisting air would persist in deflecting it until such a bullet would describe the arc of a circle instead of moving towards the target, therefore it will be readily comprehended that a tipping bullet must gyrate in order to reach the target.

It has been admitted for years that a tipping bullet as a whole makes a spiral flight after leaving the bore, but its gyrating and oscillating motions have not been understood or absolutely ignored as completely as its tangent flight from bore spiral at muzzle. Plank shooting, however, demonstrated an unmistakable *y*-error at target resulting from a tipping bullet. The plank transformed a balanced point on bullet into a tipping one, and one that did not gyrate during time the plank held its influence. After the plank had been passed and its influence upon the tipping bullet lost, the latter immediately takes up its normal gyrations and mathematical air deflection spiral.

At the muzzle the causes of *x* and *y* errors add to each other and subtract from each other in a complicated and almost indeterminate manner, but by the fortunate curiosity of plank shooting the cause of *x*-error was left 21 feet to the rear, or the 12 feet of plank plus the nine feet at which plank was placed from muzzle, and the cause of *y*-error left standing alone for solution.

The too obvious fact which for years had eluded our grasp was this: The bullet when passing from a straight flight along the plank into a spiral flight after leaving it, must make a deflection from this straight line for the same reason that it is deflected from the line of fire by passing from a spiral flight in bore to its tangent or straight flight in the air, as explained in previous article. Thus it will be seen that the spirals which cause the x and y errors have the same influence upon the bullet, but the spirals are reversed.

The manner in which this y-error, y-spiral, or the deflection of the flying bullet, takes place in the air is easily comprehended by one who has followed out the experiments which so plainly indicate the cause of the error, but its

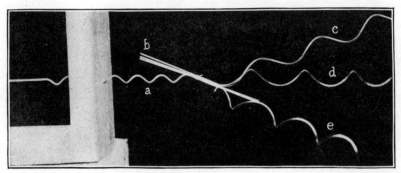

Fig. 110.

explanation, though made as definite as possible, may require a little study and thought before being understood; though by going slowly, step by step, and illustrating where possible, we may be able to make it plain.

Taking it for granted that the spiral described by the unbalanced bullet in the rifle bore, and the tangent from it which the bullet must take at muzzle, as explained in previous article, is fully understood, an illustration of y or air spirals is introduced in an attempt to show how a tipping bullet takes its y, or air spiral, from tangent of x, or bore spiral, but imagination must be freely exercised to overcome the great exaggeration presented.

Spiral a, passing through the beam, represents the spiral described by center of gravity of flying bullet through the rifle bore, the x-spiral. It must be imagined as no larger than a hair and with a pitch of 12 or 14 inches, while the air spirals, c, d, or e, may be about the right diameter, but should be imagined to be elongated to a pitch of about 45 feet, instead of less than an inch as represented.

The tangents of both bore and air spirals are represented by b bound parallel to each other by a string, and three air spirals by c, d, and e, any one of which, or a hundred more extending in as many different directions, the tipping bullet may take from the tangent b of bore spiral a. It will be observed that air spiral (d) has an axis which is a continuation of the axis of bore spiral (a) and directly in line with it because both spirals are joined in same phase somewhere within 24 feet of the rifle's muzzle, and such a joining of spirals by the tipping bullet, as has often been done on the homestead range, makes a center shot at target. Hereafter, in order to economize in repetition of words, we will designate bore spiral as x-spiral and air spiral as y-spiral.

The diameter of y is largely dependent upon amount of tip and shape of bullet, and its length of pitch upon its speed and number of gyrations it is making per second. By computation, equation (2), page 351, if diameter of x-spiral was .003 inch, that of y .125 inch, pitch of x 14 inches and of y 45 feet, the bullet would, after the first 24 feet from muzzle, take a spiral whose axis is true to line of fire or axis of x, provided the bullet should commence to tip in the right direction to cause the x-spiral to join y in the same phase. In other words, we should have the same conditions as presented by spirals a and d in the cut. If tangent of x-spiral did not leave x in the same phase in which it entered y, we should have the same conditions as presented by a and e, or a and c, or a hundred other deflections from axis of x, or line of fire, depending upon the direction of tip of bullet as it went into the y-spiral.

It is quite important that the manner in which spirals with different pitches, and diameters described by a flying bullet, join themselves to each other be well understood before the line of flight of unbalanced, tipping bullets can be comprehended. For further assistance, obtain two spiral springs with different diameters and pitches, similar to those seen in (Fig. 111); keeping the tangents of each bound together in the same line, as represented by cut where the tangents are thus joined, then observe how many directions may be given to

256   THE BULLET'S FLIGHT FROM POWDER TO TARGET

the larger spiral while still keeping the tangents of both bound in the same line.

Remembering that an unbalanced bullet must take the direction of the tangent from x-spiral, which is bound to the tangent of y, then it can be understood that if the bullet develops a tip it will commence a spiral of its own in

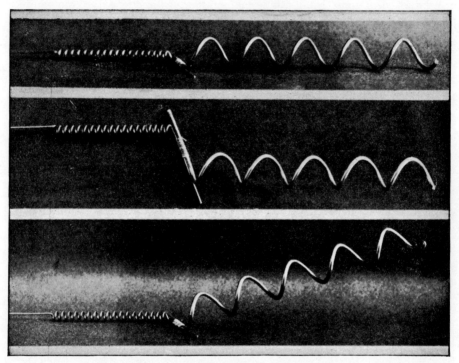

Fig. 111.

the air, or y. The axis of this y, or right-hand spiral, makes an angle with its own tangent, which is also the tangent of x-spiral, therefore the axes of the right-hand spirals of (Fig. 111) form an angle with either tangent, both of which are made to coincide by their binding.

By manipulating the two spirals seen in the cut (Fig. 111), it will be observed that about any direction desired may be given to the axis of y, or larger spirals, which are supposed to represent air spirals of a flying bullet. A direction may be given so as to preserve the axes of both x and y parallel to each other, as shown in upper two figures, which were reproduced from photographs of the same pair of spirals where no change was made except one is a vertical and the other a side view.

As has already been mentioned, a flying bullet may develop its tip in a direction to annul the tangent of x and so make a center shot, as may be demonstrated by the two upper figures in the cut, in which the tangent leaves the x in the same phase in which it joins the y. The third figure of the same pair, where the tangent enters the right-hand spiral in a different phase from which it leaves the left spiral, indicates clearly a deflection of the axis of the y-spiral from the axis of the x-spiral, and shows in an exaggerated manner the cause and position of the deflection of the unbalanced tipping bullet that makes an off shot at the target.

The experimenter will find it difficult to make his two wire spirals with such a diameter and pitch that when tangents are bound together the axis of the larger can be made parallel to the smaller. It is not a difficult problem mathematically, but its chance solution experimentally would hardly be probable.

While manipulating these wire spirals the experimenter must try to imagine that the smaller, or x, has a diameter of from .001 to .003 of an inch and a pitch of 12 to 14 inches, and the larger, or y, a diameter of $\frac{1}{16}$ to $\frac{1}{8}$ inch, and pitch of 45 feet, because these figures approximate very closely the spirals described by flying bullets, such bullets and with such rifles as expert marksmen select for their best work.

(Fig. 151) and explanation on page 352 indicate the practical working of the cause of the x and y error as shown by tests on the range. It would be well to examine this diagram in connection with the above.

Experimentally, it takes about 6 feet for a bullet to develop its full tip after exit from muzzle, and much more than this to develop the full size and direction of its air spiral, but our recorded tests indicate that it occurs within the first 24 feet. It takes time to deflect a solid body in thin air. Our recorded tests indicate that a bullet flying directly point on keeps the line of its tangent from

258   THE BULLET'S FLIGHT FROM POWDER TO TARGET

x, and the tipping one keeps the axis of its y in a straight line after direction of y has been determined or completed, and, as mentioned above, its completion is within 24 feet of muzzle, and the two tangents of their respective spirals, as shown in second figure of Fig. 111, may fairly represent this 24 feet. The above explanation makes it appear that conditions which produce the y-error are similar to those which produce the x-error, only reversed. In one case the bullet, due to its inertia, passes from a spiral into a straight flight; and in the other case, due to air pressure, the bullet passes from a straight into a spiral flight.

$x + y = 80\%$.
The Rifleman's Rainbow.
All tests but one recorded throughout this book are from machine rest in calm air, accurate powder charges, and same line of fire from shot to shot. They disclose the average spreading of the group at the target. They disclose the fallacy so prevalent among rifle shooters, that the bullet flies true to the aim of the arm at time of discharge; and if it does not strike at the desired spot, some one or more of the sixty and odd erroneous reasons are brought forward by them to account for the error. They illustrate the bane of the careful rest target shooter who year after year strives after the impossible, and illustrates the cause why the rifle crank has never caught his rainbow. They show up the x-error, the cause of which has not hitherto been stated or published as far as extended search can determine. The tests, though numerous and tedious and composed of much dry detail, tests performed under novel conditions and with novel results, should be read and re-read, until the careful rifleman shall appreciate the trouble he labors under, and understand the cause of his off shots. Prolonged experiments disclose that careful rest target work yields practically the same size group at 100 yards as machine rest, when all other conditions are the same. These experiments disclose that in calm air and careful target work, with modern ammunition, whether fixed or front-seated, at 100 or 200 yards, 80 or 90 per cent of the error at the target is the $x + y$ error, and is due to the unbalanced bullet. This same error was illustrated on all ranges of the country, when it was discovered that the refinement of the telescope sight, with its accuracy of

vision on the target rifle, gave only one point of advantage over the comparatively indistinct non-telescopic sights. The strongest argument at that time for the rifle scope was less eye strain and particularly as an aid to older riflemen where the eye had lost its accommodation. The argument was not that it made any marked reduction of the error at the target.

Before the cause of the x and y error was disclosed, constant observation and experiment for a long period decided us to state that 80 per cent of the error at the target in careful rest target work, under favorable shooting conditions, was due to some one unknown cause. To determine this cause was our self imposed task. Having accomplished this to a good degree of satisfaction, it becomes an easy matter to define and thus separate this error at the target from a multitude of other errors which may add to or subtract from it for any one shot or for any one of a series of shots. Given normal or modern ammunition, whether factory or home made, fixed ammunition or front seating, and a rifle, whether new or old, with clean, dirty, or pitted bore, to which the ammunition belongs, given a calm day and uniform line of fire from shot to shot, then, with these conditions 80 per cent of the distance that any one shot prints from the center of the normal group at 100 yards, comprises the x and y error. If the proper grade of Hazzard's old black or Laflin & Rand's high-pressure smokeless with accurately weighed charges is used, then about 90 per cent of the total error is the $x + y$ error. With modern fixed ammunition and cast bullets of any standard make in calm air and fixed rest, the average group is about twice the size of front-seating group, and the $x + y$ error with this fixed ammunition is from 85 to 95 per cent of the whole error.

As the average group becomes smaller from any cause, a smaller per cent of it is the $x + y$ error. In other words, in rifle shooting, in calm air, at short range, with weighed charges and uniform line of sight, as given by many different forms of machine rests, all of which conditions are easily obtainable, 80 or 90 per cent of the ever present error at the target is the $x + y$. The largest part of the remaining 10 or 20 per cent is due to the trajectory, which varies from shot to shot, and this slight error must always be present, since it is against all probability that any ammunition can throw every shot with absolutely the same velocity. If shots are made at different times of the day, the trajectory error may not be so slight, but this supposition takes us away from the ordinary ten-shot target group, and away from any subject treated of in this book. Our desire should be to understand the errors which do or do not compose the x.

From the foregoing, it is clearly evident that in calm air and careful rest work, the lateral error of a shot from the center of the group is entirely composed of the $x + y$; and it immediately follows, that where the lateral variation of a number of ten-shot groups is equal to the vertical variation as is often the case, then the whole error, vertical as well as lateral, at the target for these several groups is the $x + y$ error.

The above statements are made as a help to understand the relation which the unbalanced bullet bears to the average spreading of the group in rifle work, rather than a scientific statement of facts.

x and y Epitomized. Repeated experiments, as tabulated in these pages, indicate clearly that when $\frac{3}{4}$ grains of lead are removed from the base edge of a normal .32-40, 187-grain bullet, the x plus y error at 100 yards will be about 2 inches. When $1\frac{1}{2}$ grains are removed from the same position on the bullet, it will tip in its oscillation 9 degrees and make an air spiral $\frac{5}{32}$ inch in diameter, and the x plus y error will be about $4\frac{1}{2}$ inches, in either case printing at 4 o'clock when the mutilation emerges up at the muzzle. The y-error in these cases is practically $2\frac{1}{2}$ times the x-error, and always in a direction to add the full amount of one to the other. In plank shooting, where the x-error is always absent, the above bullet gyrates with no apparent oscillation. If the plank makes this bullet tip 9 degrees, its air spiral will be approximately $\frac{7}{16}$ inch in diameter, and its y-error at 100 yards 10 inches. When the mutilation removes the lead from one side of the body of the bullet as far from its base as the center of gravity, the bullet does not tip, the y-error at the target is therefore absent, the full x-error remains, and the bullet, if we omit its slight trajectory deflection, flies in a perpendicular plane from muzzle to its print on the target. When the mutilation is made towards or at the point, the x-error throws the bullet away from the target in the same direction as before, but since the bullet in this case is always a tipper, the y-error is present and deflects the bullet in the opposite direction from the x-error, because, owing to the direction of its tip, it enters its air spiral on the proper side to make it do so. The amount of tip is governed as before by the position and amount of mutilation, and when the y-error equals the x-error, as is often the case, the bullet makes a center print at the bud. If the y-error exceeds the x, as it may easily be made to do, the bullet will print at 10 o'clock away from the bud, while if the bullet is mutilated to make y less than x, it will print at 4 o'clock but near the bud. If mutilations are made, one at the base and one on the opposite side at the point, and the amount of lead removed be balanced respecting the center of gravity, such a bullet will fly with no x-error, but with an excessive tip and y-error, and will print at 4 o'clock if the base mutilation emerges up at muzzle, and from construction the point mutilation emerges down at muzzle. Either the x or the y error varies from zero to its full amount both depending upon the position and amount of the mutilation.

The principles upon which the x and y errors depend clearly have no connection with each other, and so the value of each error varies independently, although the spirals upon which both errors depend take their origin in the unbalanced bullet.

All the above statements are made up without guesswork or conjecturing from targets, made with cast and metal cased bullets under known conditions, and photographically reproduced.

Unbalanced Bullets; how Produced. The mold may be imperfect in that it will not cast a cylindrical bullet of uniform diameter, or it may cast one with an oblique base, that is, oblique to its cylindrical sides whether grooved or ungrooved. In the manufacture of molds it is difficult to produce one that will cast a cylindrical bullet with a square base. If the bullet is not tested in a properly made chuck by an expert, it cannot be known that the mold is not producing an oblique-base bullet.

We may make a cast bullet round and with a certain diameter by the usual method called sizing, by pushing it through a die, but this does not square the base and may make it even more oblique. This sizing process will usually make the base of a cast bullet oblique, even if it was cast square by the mold, and the obliquity will be about in proportion to amount of reduction caused by passing through the die. Again, the bullet will not go through central with the hole, and, if it is grooved, more metal is forced into grooves on one side than the other.

A rifleman may shoot oblique-base bullets if he cannot obtain those with a square base, but the harm comes in thinking he is shooting square bases when he is continually using oblique. Because a mold is finely finished or is working well does not insure that it will cast a square base.

Fig. 112.

Blowholes, or imprisoned air in a bullet will unbalance it, and there is plenty of evidence that such is frequently the case in bullets which are supposed to be well cast. The cut here introduced (Fig. 112) is sufficient proof without mentioning more which can be produced.

It shows four .28-caliber bullets, full size, from a Pope mold, and six .22-caliber ones from an Ideal mold, all of which were presented to the author by Dr. H. A. Baker. Four pieces of wire, varying in diameter, fit the air cavities and indicate their direction in the several .28-caliber bullets, the wire having been inserted after the point of each bullet had been cut away to expose the hole.

By splitting the .22-caliber bullets, from points down through the air cavity found in each, as represented in (Fig. 112), Dr. Baker exposed the imperfections so well shown by the camera. Bear in mind that these ten bullets were not selected from some old collection or museum, but from freshly cast ones, and all were found in the first batch examined.

The lead may not fill the mold, particularly at the edges of the base or at each groove, but this can and must be overcome. The alloy of tin and lead may not be homogeneous, and it cannot be known or ascertained by appearances when the tin is evenly alloyed throughout the contents of melting pot.

When casting a bullet, either at point or base, the lead may solidify heavier on one side than the other. To illustrate: for several years the castings for brass spinning tops, though made by a foundry of highest reputation, were very much unbalanced for a spinning disk until the metal was poured in a certain way and place. To balance these first disks three $\frac{1}{8}$-inch holes were drilled on one side nearly through the $\frac{5}{8}$-inch rim, because worthless until so balanced.

Imperfect lubrication, or where some of the grooves are not entirely filled with grease, may cause the projectile to unbalance at time of upset, since grooves containing no lubricant may fill with lead, while others, well filled with grease, the weight of which is only $\frac{1}{11}$ of lead, cannot be so filled. This throws extra weight of lead to one side of the bullet.

The swage should produce a round bullet with a square base, but a finely finished one may not do so; the supposition that the maker did his work properly does not alter the case. If a bullet be driven by a hammer out of a perfect swage, as is the universal custom, it is almost certain to be deformed, because a very light stroke of a hammer will shorten and usually crook the body of a lead bullet, especially if as long and heavy as the .38-caliber, 300-grain one. A .32-caliber, 200-grain bullet, which has a still smaller diameter, is easily bent in body by driving out of swage, no matter how carefully done. The bending, of course, takes place after the bullet has moved towards its exit from swage, when a part of its body is not supported by the die.

Swaging a grooved and lubricated bullet will deform it if all the grooves are not filled with hard grease, and that the grease absolutely fills each groove on all sides, on all bullets in a batch, is hardly supposable. If the grease filling the grooves was the soft variety, or temperature at time of swaging was sufficient to make it soft, results of swaging would be uncertain even with the press described and illustrated on page 51. The operation of swaging with a hammer is entirely out of question if balanced and unmutilated bullets are wanted.

By swaging a lubricated bullet is meant a slight pressure under a press to straighten up point and square up base and settle lubricant into place. Even this operation, though necessary, has its objections and must be performed with judgment guided by tests to know how much pressure is required to form the soft bullet as indicated above. For the lead, .32-caliber, 225 pounds direct pressure on plunger is nearly the correct amount.

If by chance none of the unbalanced bullets mentioned above are produced, and a balanced one is entered in the chamber, the explosion acts as a hammer, upsetting it before starting, the rear plunger in this swage being a wad of powder or paper or grease driven violently by a stroke many times heavier than is required for the proper swaging of any bullet. The barrel of this swage, instead of being polished steel, takes on many forms, brass, steel, and powder dirt, with a sharp shoulder midway. The front plunger of this swage is inertia of the bullet's point, which retards its forward motion but has no power to hold it central with the bore, as a steel plunger can. The consequence is at least oblique bases, as was shown in our snow-shooting tests, where 120 of the 122 bullets fired were found by micrometer measurements to have oblique bases.

A bullet patched with paper or an alloy of copper may become less unbalanced by upset than the regular form of lubricated ones. If so, it will make a better group at careful rest target work than the grooved bullet. Since patched bullets have no grooves, they may be swaged in a solid manner in a steel die, and if proper means are taken to force them out, a perfect bullet as to form may be produced. It is possible to produce perfect non-grooved bullets on a commercial scale, and if such bullets do not do fine work under fine shooting conditions, it would be due to the fact that the rifle unbalances them somewhere between breech and muzzle.

If one wishes to obtain even a little practical information in regard to the effect of burning powder behind a jacketed bullet as the explosion forces it through a steel bore smaller than itself, let him try to size the bullet down a few thousandths of an inch; for instance, reduce one of regulation size to bore diameter, requiring a change of only .008 inch.

Some of our first attempts to make such a reduction, with the swages at hand, resulted in such deformities as shown in the cut (Fig. 113), figures 1 and 2 being the normal and unreduced bullets, and the other five show manner in which they were mutilated by sizing. The swage was smooth and well lubricated, as were also the bullets, which refused to enter the swage when placed under the bullet press and subjected to a pressure of several hundred pounds.

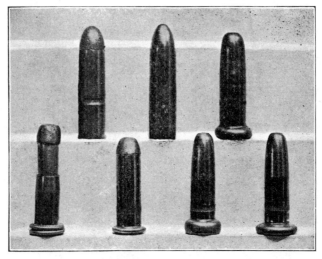

Fig. 113.

If the bullet is thus mutilated by a pressure of a few hundred pounds, what should we expect would happen to one subjected to the sudden impact of several tons? The cut represents, exaggerated to be sure, the conditions under which the regulation bullet is placed when put in front of 35 grains of high-pressure powder and shell inserted ready for discharge. The bullet is .008 inch larger than rifle's bore, as was the case with bullet and sizer which produced the above mutilations, and the only exaggeration here displayed is because the shell and chamber within the rifle will not allow the base of bullet to assume so excessive an enlargement, so the excess of metal must proceed to enlarge and change the form of the bullet in some other direction, probably by elongating it, and 99 times out of 100 producing an oblique base from the square base which was inserted.

The shell which is inserted into chamber is usually .005 inch smaller than cavity into which it is entered, made so by the factory to insure ease of insertion, therefore when the explosion comes and bullet receives its impact of several tons, and before the 220-grain bullet enters the bore, the shell has enlarged .005 inch and the bullet with it. Now the bullet being .313-inch diameter instead of .308, must be, and is, forced through a .30-caliber bore, which is even more of a mechanical proposition than the swaging as exhibited above by (Fig. 113).

The condition to which the oversize bullet is subjected in the rifle bore is even worse than in the swage which made the above mutilations, because the swage and bullets were well lubricated, while the bullet is forced into a bore without lubrication and into more or less dirt. Under such conditions, why should the rifle be expected to deliver a well-balanced bullet at its muzzle or one with a square base?

A gilt-edged rifle, while it remains so, will not deform a bullet, or it deforms it in the same place every time. Such a rifle will do fine work with an oblique-base bullet or a plugged-base bullet if such mutilations are uniform and emerge in the same place at the muzzle; but an ordinary rifle will not do fine work with any kind of a bullet.

While experimenting with swages for these metal jackets, the wonderful ductility of the metal from which the jackets are made was noticeable. With the excessive expansion of bases of the bullets, as shown in (Fig. 113), no rupture or appearance of rupture took place. The metal, which is an alloy of nickel and copper, expanded well. Though the particular form of sizer used would not successfully reduce bullets to required size, another form was found which would accomplish the desired result, but such a sizer, though producing the correct diameter, would cause the bullet to take a curved form as it was being pushed through, in some cases amounting to .013 inch.

Difficulties with Rifle Twists. A bullet must have sufficient speed of spin to keep it point on, even though a finely balanced one, and this speed is nearly up to the tables given by different authors upon ballistics. If the twist is increased above this normal to keep an unbalanced bullet from tipping, it fails to do so, as our experiments indicate, since the increased twist magnifies the angle of tangent to bore spiral, thus throwing axis of the unbalanced bullet out of line with increased force when set free at the muzzle. This additional force is somewhat proportional to the added rigidity of its axis of spin, due to the increased twist.

Increasing the twist also shortens up time of oscillation of an unbalanced bullet and lengthens its gyration, and a slow-gyrating bullet makes an air spiral with a long pitch, thus giving the air a longer time to act in any given direction before its axis is altered to allow air to deflect it in another direction. This enlarges the diameter of air spiral and the error at the target is increased.

To obtain any practical results the elongated bullet must spin, and its speed of spin should be the mean between enlarging the tangent angle of bore spiral and air spiral on one hand or, on the other, causing it to make an unsteady flight over the range. The numerous experiments on homestead range plainly indicated the difficulties under which riflemakers labor in determining the best pitch for any particular model.

PART II

Verification. During the winter of 1906–7, while tabulating and arranging the work of previous years with rifles and ammunition, theories were suggested which demanded verification, and during the spring of 1907 the homestead range was rearranged by building a 100-foot platform from the V-rest, immediately under the rifle's muzzle, running it two and a half inches below line of fire. Sixty small frames for holding paper screens, and 40 larger ones, were constructed, any or all of which could be supported by the platform to cover line of fire, and when placed one foot apart they reached the whole length of table or platform, thus forming 100 sliding and adjustable frames for the paper screens. A similar platform was also constructed five inches below line of sight, commencing at the 306-foot butt and building 30 feet towards the shooting stand or V-rest.

Fig. 114.

The cut (Fig. 114) is from a photograph of the range, as thus provided, looking from the rear and above it, and gives some idea of its appearance when finished and ready for business. It shows the rifle as laid by a spirit level, and suggests some of the trials the .32-caliber bullet must undergo before reaching the 100-yard butt.

Several brands of white-linen paper were tested to discover one suitable for screen work, until the whitest and strongest that could be found was finally selected, having a thickness of .006 inch, which proved satisfactory. From this the screens were cut, $2 \times 3\frac{1}{2}$ inches for the smaller frames, and 3×7 for the larger. A three-power telescope was fitted with rings identical with those on the 16-power glass, to be used for distances less than four feet.

Two persons were usually required to place screens expeditiously, and towards the close of the summer's work, placing of screens, manipulation of the two scopes, and the freak tricks the different rifle barrels enabled us to perform, became quite an art. The reliable Pope 1902 barrel and its straight bore, combined with previous experience, enabled us to compel the flying bullet to answer all questions we chose to ask, and do it expeditiously.

The test could be prepared in the shop, taken to the range, and the answer, printed with India ink and bullets for type, brought back in the evening. It was not until summer was half gone, however, that we fully realized over again the necessity of eliminating the small errors from our work if accurate answers were to be had.

The flight of the rifle bullet had been so much surmised for years, it seemed most desirable to determine some facts in the case and let theories dovetail where they fitted the fact. We went over the same ground as during the years past with new tests, but found no cause for change in the records of those years. The work had been conscientiously performed.

The weather for this season of 1907 was most remarkable, as 38 selected days, during which recorded work was performed, were practically free from air movements, excepting only three. Writing paper, screens, and even tissue paper would stay in place throughout nearly every shooting day. Twenty-eight rifle barrels were at our disposal, with tools to vary ammunition at will. Brother William was always on hand to assist, and the sawmill and workshop on the homestead could be drawn upon at will, but there were no visitors, competitions, or prizes.

The cut (Fig. 115) is from a photograph of the range at this time. No better shooting shed was required, or one would have been erected. This cover was not built for sociability during stormy days, but for work during fine weather.

We hope the experiments and tests which follow will not prove tedious to the reader, because verifying our past work, as they so uniformly did, made them extremely interesting to us.

Fig. 115.

***y*-error disclosed by Plank Shooting.**

Test 138. On June 12, 1907, the first experiment of the year upon the homestead range was made to trace the track of a bullet after leaving the 12-foot plank, and its results are shown by plots on Plate (24). The upper one, showing track of the shot to 83 feet, was taken from the original screens; the horizontal scale, representing the line of fire, is $\frac{1}{16}$ inch to the foot, or 1 to 192, but the vertical scale, which represents the deflection of the shot from line of fire, is full size.

Screens were placed at the respective distances indicated by ordinates, or vertical lines, and shown in feet upon the plot. The actual distance of bullet's print from line of fire is laid off full size as an ordinate, and the flight which the bullet made to the right of line, its lateral deflection, is shown by curved broken lines, while its flight below the horizontal plane of fire is marked by curved dotted line.

PLATE 24.

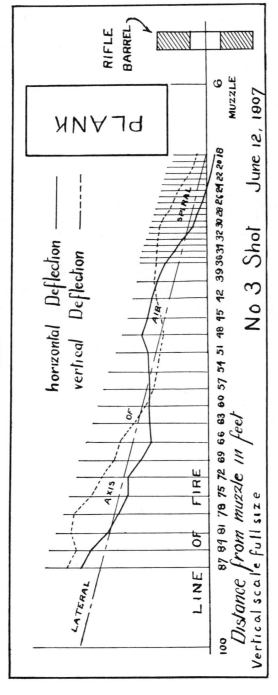

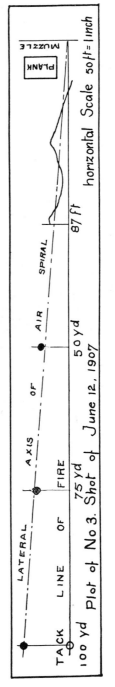

At 18 feet, where the bullet leaves the plank, its flight is started at the left and below the line of fire; following the dotted and black lines on plot it will be seen that its deflection to the right, up to 87 feet, was about the same as its deflection downward, added to its trajectory. The radius of this spiral can be measured from its lateral axis, as drawn on plot, showing its diameter to be between $\frac{3}{8}$ and $\frac{7}{16}$ inch, indicating the full-sized air spiral of a bullet tipping nine degrees without exercising one's imagination.

The apparent irregularity of the curves of flight is due to the fact that the cross on different screens was not correctly placed in line of fire. This error looks magnified upon the printed page, but it should be observed that the objective of telescope must be moved to focus at each distance at which the respective screens were placed, amounting to a movement in its tube of four inches while setting, and because the tube of scope was not straight, no placing of the cross at any distance was attempted without rotating the scope through 180 degrees. This rotation of scope in placing crosses, however, was conscientiously carried out during the entire summer.

The lower diagram of the plot shows where the lateral axis of this air spiral, made by the nine-degree tipping bullet, reaches the 50, 75, and 100 yard targets. The diagram was drawn to horizontal scale of 50 feet to the inch and made up from the original plot, which was 40 inches long. Measuring from the original plot, this lateral axis passes $\frac{5}{32}$ inch to left of 50-yard print, $\frac{5}{32}$ inch to right of 75-yard, and $\frac{1}{16}$ inch from 100-yard print. This seems either a coincidence or that the axis was purposely drawn to meet the 100-yard print.

It will be observed that the spiral from 18 to 87 feet, represented on the lower diagram, is grossly exaggerated; for if this spiral, with its prints shown by black dots at 50, 75, and 100 yards, were shown to scale, the whole flight of the bullet, spiral and all, would be contained within a line on the plot considerably smaller than one thousandth of an inch in width, and this proved to be a straight line.

Referring again to the upper diagram on the plot, it will be seen that the bullet was flying to the left away from the line of fire during its passage along the 12-foot plank, between six and 18 feet from muzzle. At 18 feet, or the farther end of plank, it is deflected quite sharply to the right instead of left, and prints four inches to right at 100 yards. It is at this point in its line of flight at end of the plank that it passes from its straight course, the tangent of its spiral, into its y-spiral, which causes the y-error at the target. See y-error discussion, page 253.

Test 139. — June 17, 1907, five days later than the preceding one, a more complete flight of a plank shot was mapped, and Plate (25) was plotted in same manner and to same scale. In this test the bullet tipped 10 degrees as it left the plank and shows the same amount of tip through every screen. There was less difficulty in placing crosses for this test, so screens were extended to full length of platform. Not much attention was given to vertical or dotted curve in this or preceding test because it contains the trajectory as well as spiral motion.

Particular attention is invited to the fact that at 33 feet the bullet is in low side of its spiral and, as plot shows, is being deflected away from its axis to the right at its greatest speed, and its tip stood at 6 o'clock. At 78 feet, where next position of greatest lateral deflection occurs, the tip is again at 6 o'clock, thus showing one gyration in 45 feet. The lesson to be taken from this condition is plainly indicated, that the position which bullet occupies in its spiral is 11 feet behind its tip. In other words, when bullet is in the top of its spiral, it is tipping directly downward; when going downward the fastest on left side of its spiral, being left twist, its point stands towards 3 o'clock, where it would naturally be expected its deflection to be most rapid to right instead of downward as found.

The action of bullet in its spiral is entirely mechanical and every part of a spiral is a curve. A tipping bullet is deflected by air towards the direction of the line of its long axis, but only slightly so. At any part of its spiral it is being deflected in the direction of its point, but only slightly so. When in the up side of its spiral it is beginning to go downward, but its point seems to be 90 degrees in advance. The bullet surges around in its spiral of flight something as a river sweeps around a bend of its course.

This lesson is an important one because upon it depends the question of which side of its air spiral the tangent enters when a flying bullet goes from a straight course into this deflecting spiral.

On the original large plot, from which the small diagram or lower one on Plate (25) was taken, the axis was drawn through the center of plotted spiral and found to pass $\frac{5}{32}$ inch to right of the 50-yard print, $\frac{5}{16}$ inch to left of 75-yard print, and $\frac{3}{16}$ to left of 100-yard print; and here the whole flight of this bullet, spiral and all, if drawn to scale, would be represented by a line having the width of one-third thickness of ordinary newspaper, and in this case, like the preceding one, this line proves to be a straight one.

This bullet reached the target 7.25 inches to right of line of fire, but this deflection was not because its air spiral had huge proportions. It was because the last two feet of its flight along the plank was a tangent to and therefore at an angle with the axis of the air spiral which the bullet went into immediately upon leaving the plank. This and the preceding test are not selected ones as they were the only ones made for verification of the proposition in hand.

PLATE 25.

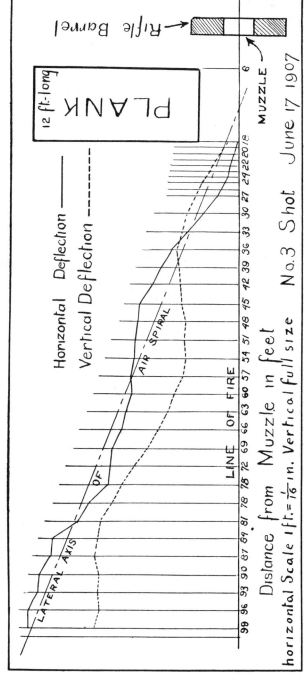
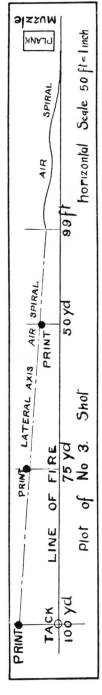

276 THE BULLET'S FLIGHT FROM POWDER TO TARGET

Cause of y Illustrated. Test 140.—August 27, a test with .32–47-caliber, left-twist barrel, was made to illustrate clearly as possible by half-tone reproductions, diagram, and description, the cause of y-error at target, and Plate (26) should be studied. As has been previously suggested, plank shooting was first attempted from mere curiosity, but while experimenting with that unusual sport it became apparent that by this means the y or air spiral might be demonstrated.

Sport soon developed into work with this kind of shooting, and as we proceeded day by day, errors grew less and exactness of the bullet's line of flight, after leaving surface of the 12-foot plank, was pretty closely determined, sufficient to enable us to make necessary computations. The main difficulty encountered lay in the fact that most normal shots would not follow surface line of plank, even though that surface exerted no effect, because very few shots start off from muzzle in same direction.

Four plank shots were made through 50-yard screens to 100-yard target, as shown on lower part of the plate, and shots 3 and 4 were made through seven plank screens. The weather was fine and rifle performed well.

The plank screens show that shot 3 was not deflected either away or towards the plank's surface, while 4 was deflected .02 inch towards it, or did not start from muzzle in line with it, and its print at 100 yards is to the right. Our experiments had already satisfied us that the pressure exerted between the bullet's point and surface of plank was equal to the " suction " at bullet's base, and this phenomenon seems to coincide with Professor Boys' photographs of flying bullets.

Since there is no deflection to right or left with shot 3, its deflection downward as it passes the plank is what our observations and computations must be based upon while tracing this y-error of flight, and the diagram (Fig. 116) on page 278 will aid us. Its horizontal scale is drawn $\frac{1}{4}$ inch to the foot and vertical to full size. Zero feet on Plate (26) corresponds to 6 feet on (Fig. 116).

The center of bullet at 6 feet from muzzle was in the line of fire and .40 inch from surface of plank. Between 2 and 12 feet of plank's surface, corresponding to 8 and 18 feet from muzzle, its direction downward is noted as the line of flight. Along the last two feet of plank the bullet was moving downward at the rate of

PLATE 26.

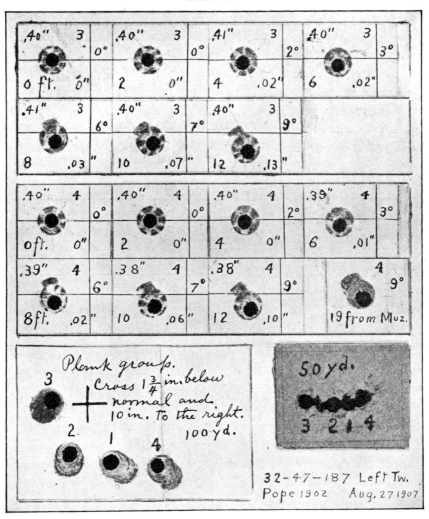

.06 inch in every two feet; and if it continued this direction, it would print at target 150 × .06, or nine inches directly below normal group. But it printed 9 inches to right and nearly in a horizontal line with it, as shown by Plate (26).

The tipping of bullet, causing more air resistance, combined with obstructions afforded by 10 screens and two cardboard backings, also combining with trajectory, should be separated from the deflection obtained at end of plank when bullet passed from its straight to its air-spiral flight while attempting to determine the question in hand.

We have thus gathered the following data from this No. 3 shot: the amount of tip at each two feet of its flight along the plank, as indicated to the right of

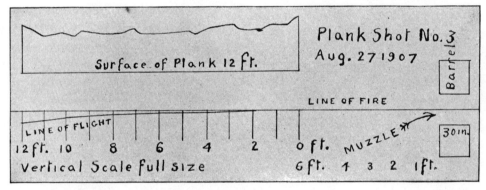

Fig. 116.

each print shown on Plate (26); second, the direction of tip; third, course of the bullet is positively known to within the limit of error of observation, especially its direction as it left the plank; fourth, previously determined by tests 138 and 139, the diameter and shape of the air or y-spiral; fifth, also previously determined, the pitch of this spiral or its distance in feet from one of its turns to another; sixth, the direction of axis of this spiral as indicated by its 100-yard target.

The tangent of this air spiral is the straight line taken by the bullet as it passes the last two feet of the plank and, extended, would reach the target, as before stated, nine inches directly below a normal group. In leaving the plank its tip is 9 degrees, 11 o'clock, where it goes into a gyration without apparent oscillation, forming a spiral $\frac{7}{16}$ inch diameter whose axis, barring trajectory, is a straight line, meeting the target $9\frac{1}{2}$ inches directly to right of a normal group.

It will be found difficult to make this air spiral, which causes y-error at target, more explicit or better comprehended except the reader is sufficiently interested to work the problem out personally, the numerous tests and illustrations furnished being ample for that purpose.

**Cause of
x and y
Disclosed at
Muzzle.**

Test 141. — July 18, a test was made to show flight of a bullet which both gyrates and oscillates, to distinguish between it and those which gyrate without apparent oscillation, and Plate (27) was accurately drawn from the flight of a normal bullet which was mutilated at base by a hole made with number 39 drill, removing 2.8 grains metal, and plugged. The plot was drawn to same scale and in same manner as two preceding plates. It was the only test made for the indicated purpose, being so successful in every particular that no more were needed, but it required the conscientious work of two men at range for eight hours.

The maximum tip of this shot was nine degrees at six feet, towards 10 : 30 o'clock, while at 12, 33, 54, 63, and 75 feet it made a perfect print. It was found to be at right .03 inch at six feet, which would be equivalent to 1.50 inches at 100 yards, but it printed there 4.60 inches to right. Since it was point on at about every 12 feet, and much of the way only tipped from one to three degrees, its air spiral, by actual measure on plot, is $\frac{5}{32}$ inch in diameter, being less than half the diameter of either plank shot recorded in two previous tests where bullets kept their respective nine and ten degree tips throughout their flight. One observation from the original plot shows a gyration in 51 feet; taken from another place it shows 48, and from still another only 30 feet, which latter is clearly due to error in reading the plot.

While the two previous shots, shown on two previous plates (the plank shots), show the y-spiral, they are out of the run of normal rifle shooting because neither shot oscillated. The plot accompanying this article, however, applies strictly to normal rifle work and shows to the eye the spiral and deflection of a gyrating and oscillating normal shot. It also shows the causes of x and y errors which result from an unbalanced bullet. That it was unbalanced intentionally does not remove it in practice or theory from a normal shot, since 120 of 122 shots were unbalanced even from a fine rifle bore, as was determined by mechanical tools and micrometer measurements.

Again, in this plate, as in the two previous ones, the diameter of air spiral is seen full size, so riflemen need not exercise their talent of imagination. The ordinates in this plot are full size; that is, the actual distance of bullet from the line of fire

PLATE 27.

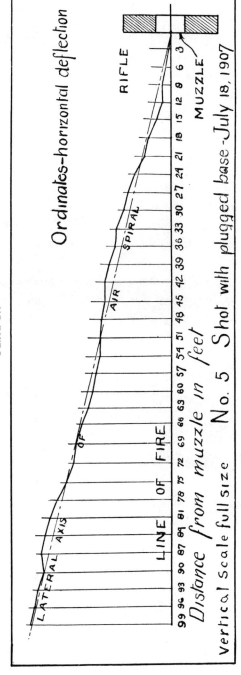
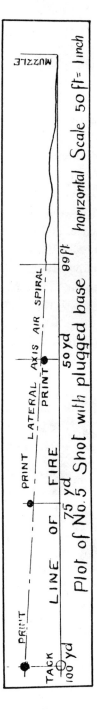

282   THE BULLET'S FLIGHT FROM POWDER TO TARGET

is laid off upon the vertical lines. On the original plot, which was 40 inches long, the lateral axis of air spiral was drawn through this spiral, regardless of its direction, being plotted in the writer's study by a student of the Massachusetts Institute of Technology, and the lower section of plate, not drawn to scale, shows the results. The axis when drawn was found to pass $\frac{3}{16}$ inch to the right of the 50-yard print, $\frac{3}{8}$ inch to left of the 75-yard, and $\frac{3}{32}$ inch to left of 100-yard print, as indicated by black spots.

The same student also plotted the two previous ones, and the small distance of this axis from the 50, 75, and 100 prints in plot of June 12 he thought was a coincidence, and the second plot surprised him. When this plot was finished he exclaimed, " Mr. Mann, just come over and see how this thing works out. It seems impossible!" Here again it may be remarked that if the flight of this bullet and its spiral was drawn to scale in the plot, it would be represented by a line one-sixth the thickness of ordinary newspaper and it proves again to be a straight line. It can be plainly seen from this plot, that this line, or axis of the air spiral, is not in line with the rifle bore, showing the y-error of 3.10 inches added to the x-error of 1.50 inches, making 4.60 inches at the target.

Test 142. — June 19, an experiment was made to verify test 34 a, page 43, in regard to the deflection of a shot by plank and muzzle blast combined, with which it should be compared.

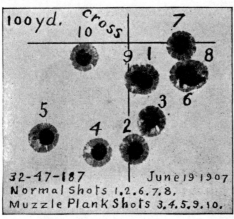

Fig. 117.

The cut (Fig. 117) shows 10 consecutive shots at 100 yards, five normal and five made over an oak plank 6 inches in length that was placed $\frac{1}{16}$ inch below the bore and against the muzzle of rifle. The results of the group are well shown in the cut and there described, and they tally with test 34 a, made seven years previous.

Test 143.— June 19, with the .32–47 Pope, left-twist, 187-grain bullet, five shots were made through screens to show direction of tip when a bullet was mutilated at base and emerges from muzzle with its mutilation up and down.

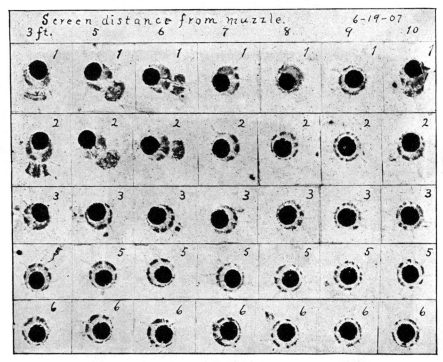

FIG. 118.

The cut (Fig. 118) shows original screens, numbers over the columns giving their respective distances from muzzle and figures on screens giving number of the shot.

Shots 1, 2, and 3 were mutilated by a different size cut on base of each, and entered to emerge from muzzle with cut down. Shots 5 and 6 had 1.5 grains cut from edge of base and entered to emerge with cut up. The tip of 1, 2, and 3 at six feet is about 3 : 30 o'clock, while with 5 and 6 it is at 10 o'clock. They tip in opposite directions to within the error of their emerging exactly up or **down** at muzzle, and this was well established by all tests and from all rifle barrels during this summer.

Observe that shot 2, which tips 14 degrees at five feet, makes a perfect print at nine feet; that 3, 5, and 6 make perfect prints at 10 feet; 1 tips at eight feet in opposite direction to its tip at nine, showing that its point-on position was between these two screens. No. 1, at nine feet, emerges from position of no tip in opposite direction from its last tip at eight feet, all six shots showing that amount of tip has little if any effect upon time required to make one full oscillation and, also, that the point of greatest tip is about midway between the muzzle and its first point-on position.

Nothing occurred during the entire summer's experiments to indicate that any elongated bullet shot from a normal twist bore, however badly mutilated, failed to assume a point-on position somewhere within the first 15 feet of muzzle.

Test 144.— June 24, with .32–47-caliber, 187-grain bullet, making three normal and six mutilated shots, to determine by correctly placed screens the relative position of a bullet at three and six feet from muzzle with its position at 100 yards, targets and screens shown in Plate (28). The figure on the right of each screen indicates the shot, and the decimal in the left gives inches that the center of gravity of the bullet passed to the right of cross.

The bullets in shots 1, 5, 6, and 7 were drilled in the side near as possible to center of gravity, removing 4.5 grains of lead, hole extending to center of bullet and plugged with wood. Taking our measurements at the six-foot screen, shot No. 1 should print on the 100-yard butt three inches to right, but it printed less than two; No. 5 should print 5.5 inches to right, but printed 2.5; No. 6 should print 2.5 inches, which it does; No. 7 at 3.5 inches, but prints at 2.75. Since the crosses were correctly placed an explanation cannot now be avoided, as was done with our work of 1903.

The above shots all started from muzzle towards the right because they were unbalanced and entered so that heavy side of bullet emerged down and plugged side up, the twist being left, or opposite to movement of the hands of a watch.

PLATE 28.

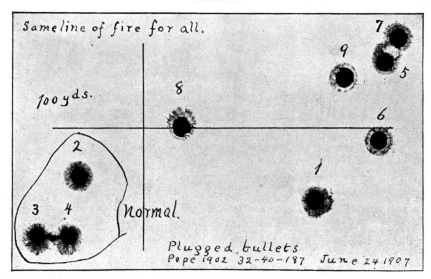

It is a fixed law that a moving body will not change its direction from a straight line without the application of some external force.

Shots 1, 5, and 7 did not fly in a line but changed their course after passing the six-foot screen. No. 6, judging from its six-foot screen, did fly straight and printed at its computed distance 100 yards away, but it will be noticed that No. 6 traveled point on, while 1, 5, and 7 were all tippers.

Since it has been believed for years that a tipping, elongated bullet forms some sort of spiral in the air, and all know that a spiral is not a straight line, it was to be expected that shots 1, 5, and 7, being tippers, would change their direction after leaving the six-foot screen, while No. 6, not being a tipper, should keep its direction in calm air.

Passing to No. 8, which was mutilated by a .10-inch drill hole near its front band, removing 2.5 grains lead, only half the mutilation in previous bullets, we find print of its center .09 inch to right on six-foot screen, which should cause it to print 4.5 inches away at 100 yards, but it is off only .37 inch, showing a discrepancy of over four inches. No. 9 was mutilated with same weight plug at its base instead of front as with No. 8, and by computation should print three inches instead of 2.20 out. Observe that 8 and 9 have opposite tips and, in common with 1, 5, 6, and 7, all emerge from muzzle with plugs up and drove to the right regardless of where plug was placed in the bullet. Unless the lesson is well learned from these tests and the article commencing on page 249, the flight of a rifle bullet will still remain a matter of conjecture.

Test 145. — June 29, 1907, with .32–47-caliber, 187-grain bullet, left twist, a test was made to further determine the relative positions of a flying bullet in calm air at six feet and 100 yards from muzzle, making four normal and eight mutilated shots, illustrated by Plate (29). The plate is self-explanatory after stating that the six-foot screen with its cross is pasted upon the 100-yard target close to its corresponding print, and distances are given below it, R for right and D for down, as case may be.

No. 5 bullet had one grain cut from its base, No. 6, 1.5 grains; and one-third the point was cut from bullets 7, 8, and 9, as will be noticed in their print;

PLATE 29.

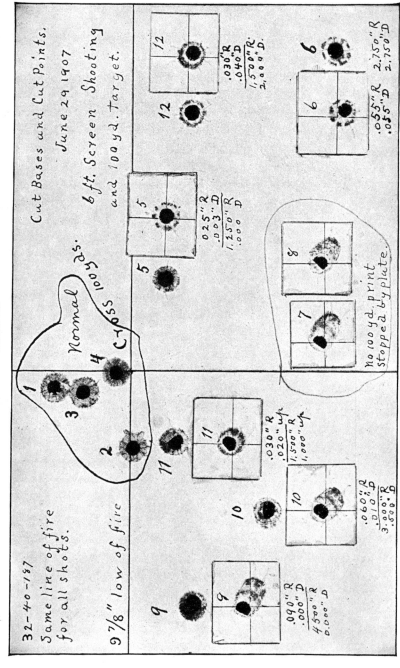

288 THE BULLET'S FLIGHT FROM POWDER TO TARGET

10 and 11 were mutilated by cutting less than a third from their points, and 12 had 1.5 grains cut from its base.

No. 5 printed one inch out at 100 yards while its computed distance was 1.25; No. 6 printed 3.5 inches to right instead of 2.75 as computed; 9 should have printed 4.5 inches to right, but went 7 inches to the left of this place; 10 printed 4.5 inches to left of its computed position; 11 printed 2.25 inches to left and 12 printed 2 inches to right of computed positions.

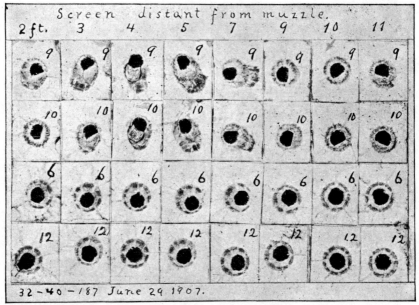

FIG. 119.

The advantage in this test and its accompanying plate over our work of 1903 lies in the accurately placed six-foot screens which show, as in previous test, that all bullets emerging with light or mutilated side up, from a left-twist bore, start off sharply from muzzle towards the right regardless of position of mutilation or plug in their side.

Observe that the tip of cut-point bullets is 4:30 o'clock at six feet, while mutilated bases show it opposite, or 10:30 o'clock; also, that the 100-yard butt caught No. 9 in nearly point-on position while its tip at six feet was 12 degrees, thus teaching a very important fact, viz. it cannot be told how much of a tipper any given bullet is by examining a fine print at 100 or 200 yards.

The cut (Fig. 119) shows eight screens for each of shots 9, 10, 6, and 12, as preceding plate (29) only gives the six-foot screen pasted on 100-yard target. Observe that all the shots pass the 10-foot screen with their points perfectly on but commence their tip again at the 11-foot screen; also that shots 9 and 10, with their mutilated points emerging up at muzzle, tip at each screen opposite to shots 6 and 12, the latter having their mutilations at base.

Test 146. — July 9, with .30–30-caliber, 175-grain lead bullet, barrel with 8-inch twist, same as tested during 1904, to test action of mutilated bullets shot from a twist shorter than length of bullet demands, and the notes upon cut (Fig. 120) explain themselves, the cut being reduced to half size of original photograph.

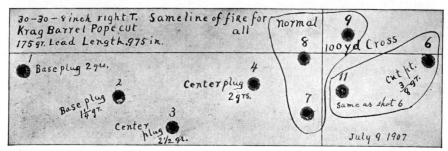

Fig. 120.

The bore being right twist the base-plugged shots go to left and point-plugged ones to right, opposite those emerging from a left twist. Shots 7, 8, and 9 were normal or unmutilated, and as the reduced cut of target fails to show amount of deflection of the mutilated ones, it will suffice to say that it was greater than with the .32-caliber 14-inch twist, using same amount of mutilation. This increase of x-error was expected, though Plate (30), where degree of tip is represented to be also magnified by the quicker twist, shows up different from what was hoped of it. This plate of screens, natural size, show how much all bullets went to left of cross on the six-foot screens when their mutilations emerged from the muzzle up.

Shots 3 and 4, mutilated in such a manner that they did not tip, flew straight to 100 yards in line with their six-foot prints, as computation indicated, and any of the others may be computed by multiplying the decimal on their respective six-foot screens by 50. Shot 5, which tipped 16 degrees at $2\frac{3}{4}$ feet, exceeded all other recorded tips; observe, however, that this, like all other recorded shots, goes point on at regular intervals though it did not reach the butt.

These 170-grain bullets oscillated perfectly every 5.5 feet, even the one that tipped 16 degrees, all showing their first point-on position at the $5\frac{1}{2}$-foot screen, indicating that the quicker twist shortens the period of oscillation.

Plate 30.

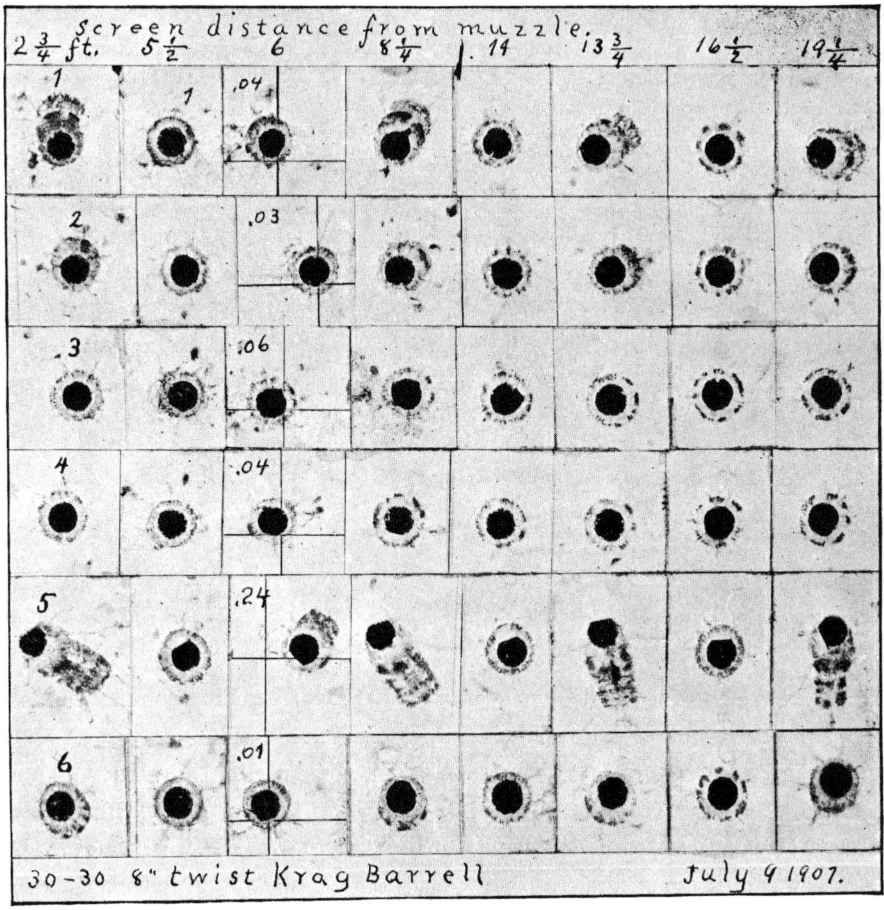

Screen distance from muzzle.
2 3/4 ft. 5 1/2 6 8 1/4 14 13 3/4 16 1/2 19 1/4
1
1
.04
2
.03
3
.06
4
.04
5
.24
6
.01
30-30 8" twist Krag Barrell July 4 1907.

Test 147.—July 24, .32–47-caliber, 187-grain bullet, 14-inch twist barrel, and one shot was made to demonstrate that the same bullet, barrel, and ammunition used in tests 144 and 145 will throw a bullet with a different period of oscillation, the cut (Fig. 121) being a very fine one for the purpose. Figures upon their respective screens give distance from muzzle in feet.

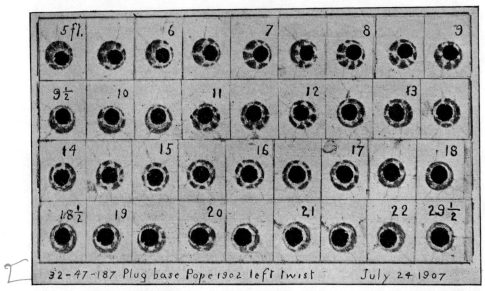

Fig. 121.

A marked feature in this cut is the vividness with which the enlarging lead point of bullet is shown, beginning at five feet and batting its way through tough paper every six inches up to $29\frac{1}{2}$ feet. All the bullets used in this test were point on at about 15 and $29\frac{1}{2}$ feet, while in tests of June 19 and 29 they found their point-on positions at 9 and 18 feet. During this test particular attention was given towards securing fine prints at short distances, and to perfect a test of same character made September 7, 1902, to study the tipping bullet throughout one oscillation. Others of like character had already been made, and still others are recorded farther on.

One of several questions to be answered by this test was a determination of what part of a bullet forms a turning point or center while making its tips, oscillations, and gyrations. Was it the point about which it tips, or does it oscillate and gyrate about its center of gravity, or some other point along its longitudinal axis? The question regarding direction of tip when its period of oscillation is 15 feet is well answered in the accompanying cut.

The question of curiosity which instigated so many tests of this nature, was about this: What are the exact motions of the bullet just as it is passing through its no-tip position? The reproduction, by half-tone process, of original targets are so perfect that from them one with a little ingenuity can determine for himself the precise movements of the oscillating bullet, bearing in mind that the majority of all bullets from all rifles in all places are oscillating ones. The cut before us shows the last distinguishable tip at 14 feet, 3 o'clock, and the first succeeding one at 17 feet, 6 o'clock.

In test 143, page 283, where period of oscillation was 10 instead of 15 feet, the bullet emerged from its no-tip position opposite to its entrance instead of three hours ahead, as in this test.

The gyroscope, and the laws which govern spinning bodies, plainly illustrate the motion which an oscillating bullet makes, and the student should apply them to our statements and illustrations, for these last are set forth as impartially as possible.

Test 148. — On July 25, with .32–47-caliber, 187-grain bullet, 14-inch left twist, was a continuation of previous test, to determine the motions of an oscillating, gyrating bullet; screens illustrated by Plate (31).

Screens were placed every six inches from four to $32\frac{1}{2}$ feet and every foot from 90 to 100 feet, as indicated upon respective screens. One shot was fired in which the bullet was mutilated by removing $2\frac{1}{2}$ grains from its base and plugging, and it will be noticed that this shot straightened up at $11\frac{1}{2}$ feet instead of 10 and 15 feet as in previous tests, and no change was made in ammunition to produce this result. At 22, $32\frac{1}{2}$, and 91 feet it was point on, and its maximum tip at 98 feet was nearly equal to its first maximum at six feet.

Our observations indicate that the maximum tip of an oscillating bullet decreases a little in each successive period of motion, but our screen work has not yet shown this to our entire satisfaction. It has not been determined positively in any test that oscillation diminishes sufficiently to allow a bullet to pass into a gyration without visible oscillation, though at 100 yards there is usually every appearance of oscillation holding its own motion.

Another movement which attracted attention in flight of this bullet, was the fact that last tip occurs at 11 feet, 5 o'clock, and appears again on opposite side at 13 feet, 10:30 o'clock, but if this screen had been absent, its first tip would have been recorded at $13\frac{1}{2}$ feet, 8 o'clock.

Observe again that where the tip is clearly at 2 o'clock on 21-foot screen, in the next, $21\frac{1}{2}$-screen, it is 3 o'clock, clearly a backward movement of bullet from left-twist barrel. Another of these backward movements is shown between the 23 and 24 foot screens: at 23 feet tip is at 4 o'clock; at $23\frac{1}{2}$ less if any tip; and at 24 feet its tip is 6 o'clock, but after this print it goes on its regular left-hand motion to 93 feet. At 91 feet it is point on and at 92 a tip shows at 11:30 o'clock; point on again at 93 feet, then takes its regular backward motion.

Irregularities of this kind are frequently occurring, which seem to indicate that the bullet has yet another motion besides gyrating and oscillating; though slight, it makes its appearance in nearly every test where screens have been carefully placed to catch its movements through its no-tip position.

PLATE 31.

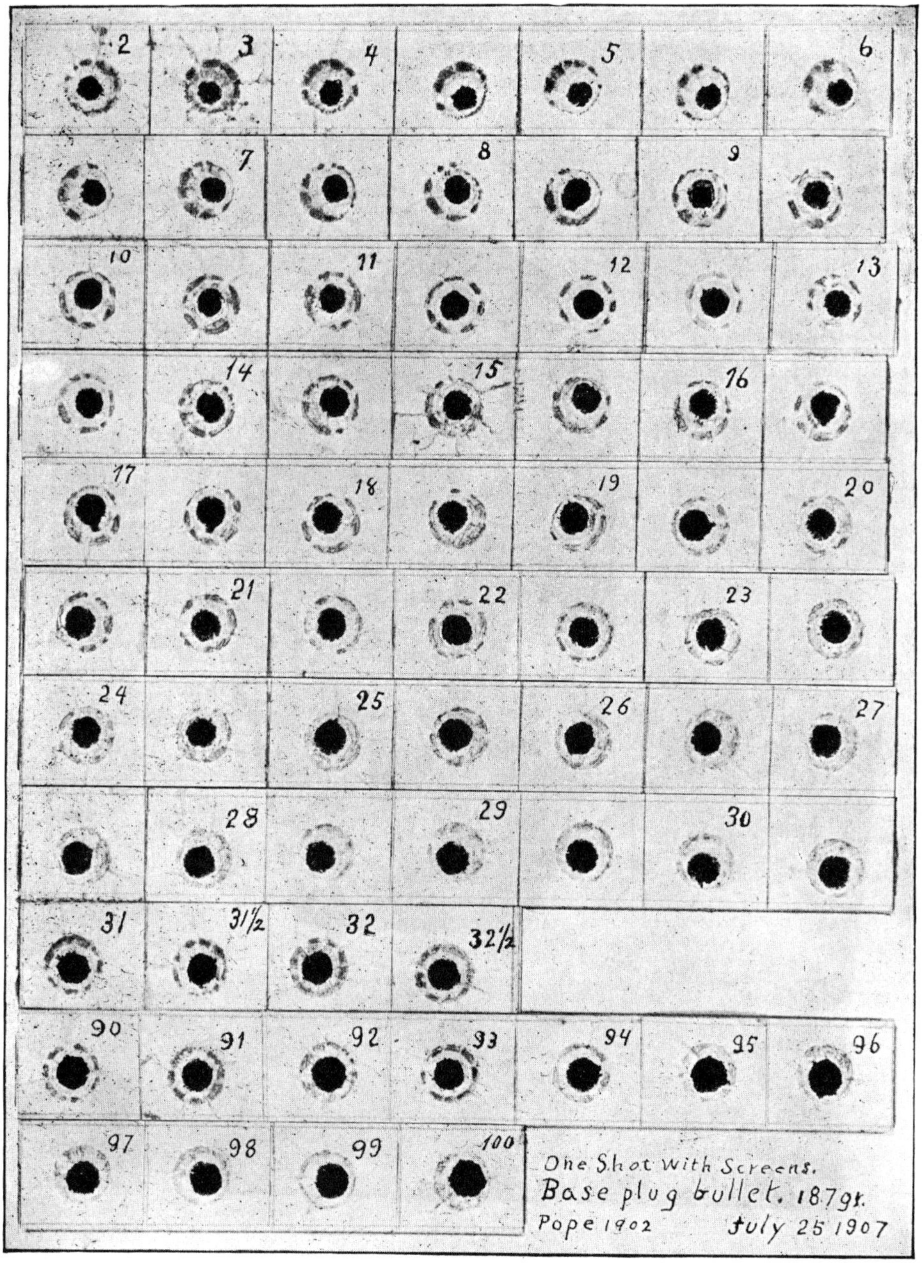
2
3
4
5
6
7
8
9
10
11
12
13
14
15
16
17
18
19
20
21
22
23
24
25
26
27
28
29
30
31
31½
32
32½
90
91
92
93
94
95
96
97
98
99
100
One Shot with Screens.
Base plug bullet. 187 gr.
Pope 1902
July 25 1907

296   THE BULLET'S FLIGHT FROM POWDER TO TARGET

There is no physical reason why a flying bullet should not take on any motion that might be given to it by an external force, and we know that the external form of all these plugged ones have a motion which in a spinning top produces the sound called " hum," caused by the center of form revolving around the center of gravity. This motion in the bullet is so small, however, that it hardly accounts for the irregularities noted above, though they might result from this " hum " motion in combination with some other.

Test 149. — July 25, 1907, a shot was made to indicate the movement of a bullet mutilated at its point, through two oscillations, to supplement several

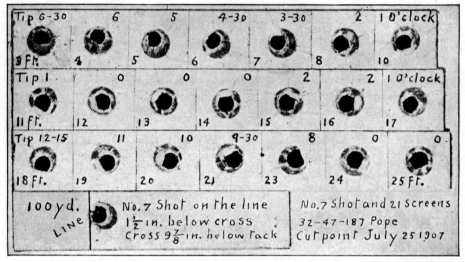

Fig. 122.

other tests where mutilated bases had been used; and the cut (Fig. 122) presents fine prints through each foot screen, from 3 to 25 feet from muzzle.

At six feet, this bullet mutilated at its point tips directly opposite to one with mutilated base, and its period of oscillation is 13 instead of 12 feet. Its last tip before 22 feet is at 1 o'clock, and first tip after 14 feet is 2 o'clock, indicating that with this arrangement of screens it emerges from its no-tip position one hour later than it entered. The hour of tip is marked on the respective screens, the first maximum being at 7-foot screen, 3 : 30 o'clock; next at 19 feet, 11 o'clock; and if this reading be correct, we have $2\frac{3}{4}$ oscillations of 13 feet each, making 36 feet to one gyration. This computation indicates the chief trouble present in all tests where time of gyration was sought after.

Test 150. — July 25, same day as preceding, but extended to 100 yards to indicate that regular oscillations continue up to this distance, a test was made and illustrated by Plate (32).

The shot straightened up at 13, 25, and 100 feet, and nearly so at 282 feet. It goes point on somewhere between 286 and 291 feet, though no screen was there to show it, and is practically point on at 100 yards. The last tip, 11 feet, is 1 o'clock, at 24 feet, 8 o'clock, a difference of five hours, showing $2\frac{2}{5}$ oscillations of 13 feet which makes the gyration 31 feet, but this computation is quite clearly in error. The maximum tip at 7 feet is 4 : 50 o'clock; at 19 feet it is 12 o'clock, making $2\frac{3}{4}$ oscillations for one gyration of 35 feet, giving the same result as in preceding test.

The student who may be scanning these tests and illustrations may discover a more correct solution for this distance, for he has here the original screen prints and all the data to measure from that has been obtained.

PLATE 32.

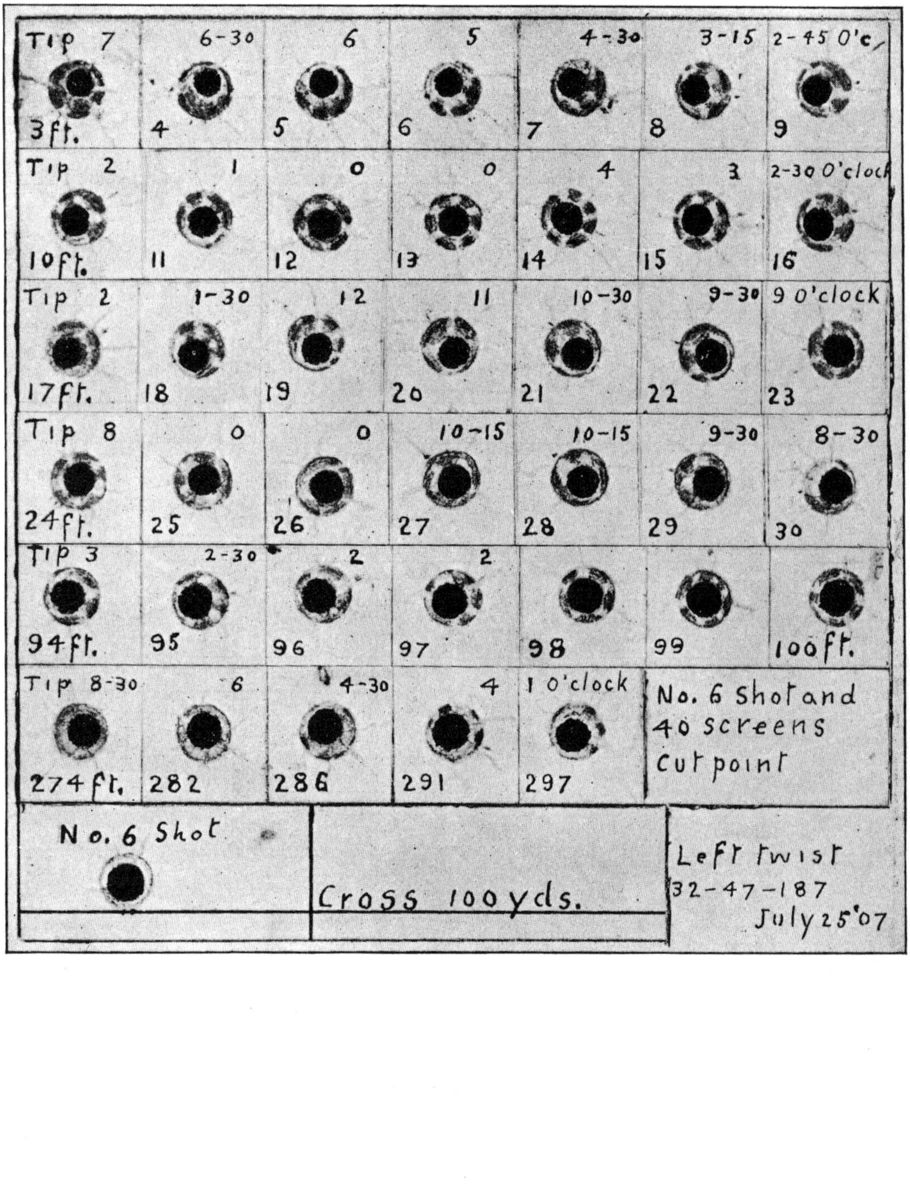

Tip 7
6-30
6
5
4-30
3-15
2-45 O'c
3ft.
4
5
6
7
8
9
Tip 2
1
0
0
4
3
2-30 O'clock
10ft.
11
12
13
14
15
16
Tip 2
1-30
12
11
10-30
9-30
9 O'clock
17ft.
18
19
20
21
22
23
Tip 8
0
0
10-15
10-15
9-30
8-30
24ft.
25
26
27
28
29
30
Tip 3
2-30
2
2
94ft.
95
96
97
98
99
100ft.
Tip 8-30
6
4-30
4
1 O'clock
274ft.
282
286
291
297
No. 6 Shot and
40 screens
Cut point
No. 6 Shot
Cross 100 yds.
Left twist
32-47-187
July 25 '07

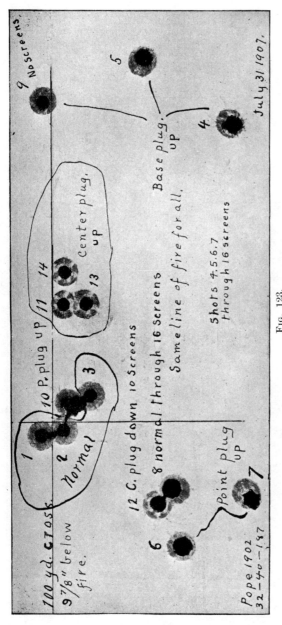

Fig. 123.

Test 151. — On July 31, .32–47-caliber, 187-grain bullet, 14-inch twist, this test separated, for the first time on homestead range, the x-error or error of unbalanced bullet, from y-error, or error of flight. Four normal shots, 1, 2, 3, and 8, were made, and eight mutilated ones, all shown in cut (Fig. 123), while Plate (33) gives the screens for six of the shots.

A little more than a month before this test was made we were able, by plank shooting, to clearly separate the y from x, as already shown. In this test success was obtained by fortunately locating the center plugs for shots 11, 12, 13, and 14 in a position which allowed the greatly unbalanced bullet to fly strictly point on, so there could be no y-error. This was the attempt made in test 144, page 284, where the records are puzzling and only partially explained.

Eight of the bullets used in the test now under consideration were drilled from

PLATE 33.

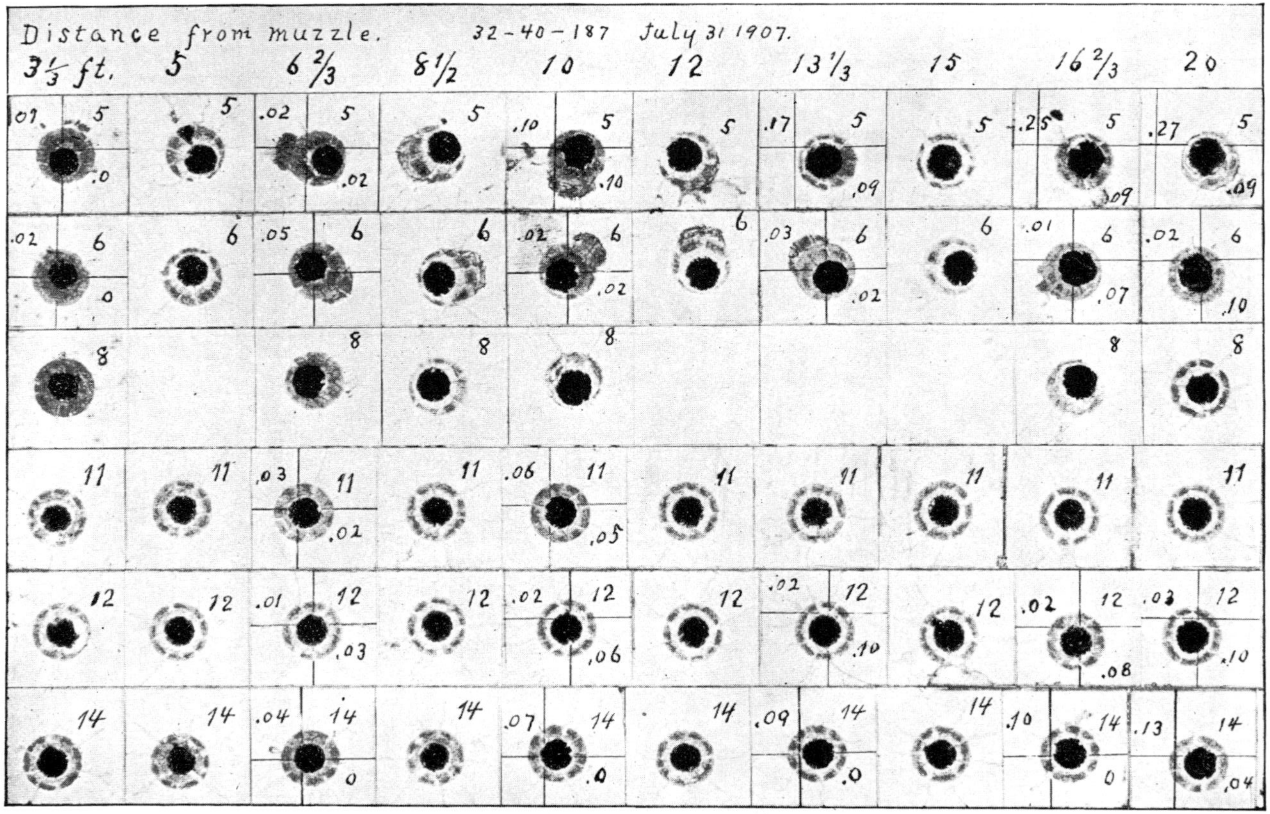

Distance from muzzle.   32-40-187   July 31, 1907.

| $3\frac{1}{3}$ ft. | 5 | $6\frac{2}{3}$ | $8\frac{1}{2}$ | 10 | 12 | $13\frac{1}{3}$ | 15 | $16\frac{2}{3}$ | 20 |

their side to center with a 39 drill, removing 2.8 grains lead, afterwards plugged and swaged.

The bullets entered to emerge with plugs up printed at 11, 13, and 14, with plug down at 12, the former starting off from muzzle to the right true to their prints at 100 yards, as exhibited by the $6\frac{2}{3}$-foot screen, and 12 started to the left. All the other mutilated bullets emerged with plugs up and started off to right at the $6\frac{2}{3}$-foot screen at about same angle for all in common with the center-plugged ones; but the base-plugged shots printed 2.5 inches to the right of center-plugged ones at 100 yards, and point-plugged ones printed 2.5 inches to left of those plugged in center.

While the center-plugged bullets did not tip, the base-plugged ones tipped at 6 feet, 10 o'clock, and point-plugged on opposite side, or 4 o'clock. The base-plugged ones must have entered their air spiral soon after leaving the muzzle on the opposite side to which the point-plugged ones did.

The diameter of air spiral for these particular bullets was not far from .20 inch, and the angle which the tangent of spiral of this size makes with its axis subtends an arc at 100 yards of about 2.5 inches, which is shown by the distances upon the 100-yard target and also by computation based upon the algebraic formula, equation (2), page 351.

Referring to Plate (33) of screens which were made up from this test, we note that shots 1, 2, 3, 10, and 9 were made without screens so their flight was unchecked. Shots 11, 13, and 14 passed through 10 tough paper screens; 4, 5, 6, 7, and 8 were passed through 12 paper and 4 thin strawboards. Brother William suggested that No. 8 shot be made through 16 screens with a normal, unmutilated bullet, which printed low also, indicating fairly well what he had surmised, that the increased air pressure on a tipping bullet, combined with the obstructions received from screens and strawboards, might have caused shots 4, 5, 6, 7, and 8 to fall 1.3 inches low while all others kept up.

An examination of the center-plugged shots, 11, 12, 13, and 14, indicates that the x-error alone throws them in a circle whose center is the normal group, and that the y-error added throws them in a circle whose center is not the normal group, but the x-error group.

In this connection a close study of illustration on page 300 is here suggested.

Test 152. — August 6, the Lee navy, .236-caliber, or 6 mm., 10-inch right twist, was dismounted from its bolt action and placed with rings upon the V-rest, to test its capability at 100 yards with regular Winchester ammunition, 32 grains W. A. powder. This rifle, with two others like it, was obtained after being discarded by the government and, so far as could be seen, was a new one. After being stripped of 37 separate pieces, Mr. Pope lapped it out. A new complete

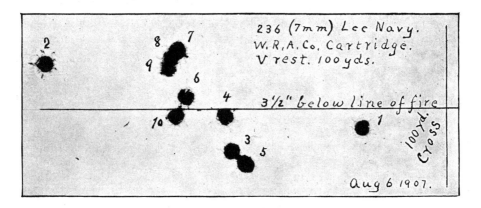

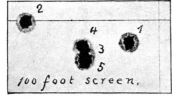
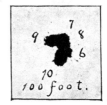

Fig. 124.

concentric action of great strength was made to support chamber pressure and so avoid danger. It carried a 112-grain, metal-cased bullet with a velocity of 2500 feet per second, in diameter about the size of the .22-caliber, and, at short range, could be passed through half an inch of steel plate. It was with considerable interest that this much-talked-about small bore was taken to the range.

The cut of first 10 shots made with this barrel, commencing with a clean bore (Fig. 124), speaks for itself.

304   THE BULLET'S FLIGHT FROM POWDER TO TARGET

The 100-foot screens for first and second five shots are shown in lower part of the figure. After grease was well removed from bore by first five shots, the next five made a group of .7 inch.

Observe that with this barrel, as with all others, the prints bear the same relative position to each other at 100 yards as at 100 feet; that the group at 100 feet is just one-third the size of that at 300 feet.

Since all tipping bullets, as indicated by tests, make a small spiral in the air, we know that they cannot be in their perpendicular plane of flight only at those

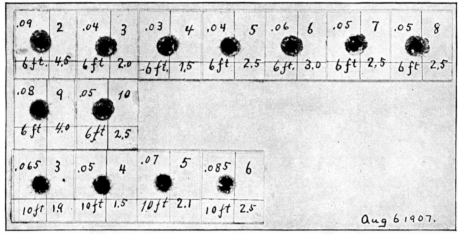

Fig. 125.

points where this plane cuts the spiral in which bullet travels. This is illustrated in the cut (Fig. 125) of 6 and 10 foot screens. Upon it the fractions of an inch are given which each shot passed to the left of vertical line of fire, and below each print is given the computed distance at which these shots should print to left at 100 yards.

The distance of each screen in feet is marked at bottom of each; the number of shot at upper corner and the computed distance at lower corner.

Observe that these perfect bullets, for they were before being shot, make tipping prints in most cases at six feet. No. 2 did not tip at six feet and printed true to its computed distance at 100 yards.

THE BALLISTICS OF SMALL ARMS 305

Judging from the action of bullets intentionally mutilated during our past work, these bullets would not have tipped if their center of gravity had coincided with their center of form when they reached the muzzle. Neither was there any evidence that these bullets tipped more at 100 yards than at six feet, and this is precisely how bullets act which are intentionally mutilated.

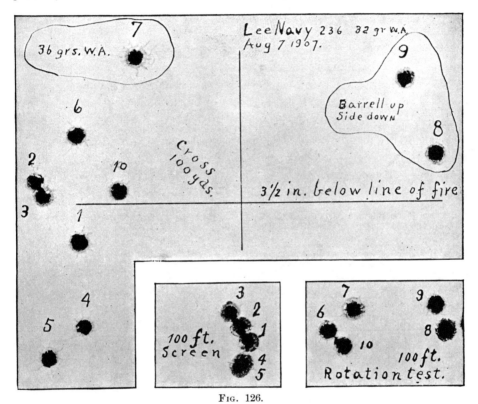

Fig. 126.

Test 153. — August 7, 1907, with Lee navy, 6 mm., 112-grain bullet, 100-yard target, and 100-foot screens, to test bore-diameter bullets in this barrel and show rotation group. (Fig. 126) will explain itself by adding a word or two.

The warming shot was sent into the pond; shots 1, 2, 6, and 10 were made with regular Winchester ammunition; shots 8 and 9 with same ammunition and barrel upside down; shots 3, 4, and 5 with bore-diameter or two-cylinder bullets, $\frac{1}{16}$-inch base band, front-seated, leaving $\frac{5}{16}$ inch more air space than the Winchester ammunition; shot 7 with bore-diameter bullet and 36 grains powder, an increased charge to reduce air space.

The accuracy of this small-bore rifle was surprising, as exhibited in this and previous test. The bore-diameter bullets did well considering that the charge used was not adapted for this bullet, but for an oversize one and high-chamber pressure; not counting No. 7 with its excessive charge they would all hit a ramrod $\frac{5}{16}$ inch in diameter at 100 yards.

The rotation test shows a hollow group of about four inches diameter, and a line drawn through center of this group cuts the perpendicular .5 inch above the cross, indicating that this bullet with its Winchester ammunition drops only three inches in 100 yards, while its trajectory taken at 50 yards is .75 inch.

Test 154. — August 15, ten shots were made from the Lee navy rifle, all in the same line of fire, and screens were placed a foot apart for bullets 6, 8, 9, and 10, to illustrate the effect of mutilated bullets with this small bore and high pressure.

Plate (34) gives the 100-yard prints and direction of tip of four bullets up to 15 feet from muzzle. Maximum tip of shots 8, 9, and 10 was at 4 or 5 feet; and No. 8 was point on at 9 feet, No. 9 at 10 feet, and No. 10 was point on at 11 feet, while 8 was lost.

This tally is almost identical with the old reliable .32–47–187 ammunition with its 14-inch twist, practically showing how closely the twist of the two rifles had been proportioned to their normal projectile by the manufacturers.

Shots 5 and 6 were notched on edge of their bases with a file, removing .2 grain of metal, and entered to emerge from muzzle with notch up; and it will be observed that from this right-twist barrel they printed to the left, opposite those from a left twist.

Shot 8, with a .06-inch drill hole .06 inch deep near the base, removing .7 grain metal, had excessive tip and was lost. Shot 9 with same drill hole less deep, removing only .5 grain metal, printed as shown. Shot 7 was filed slightly

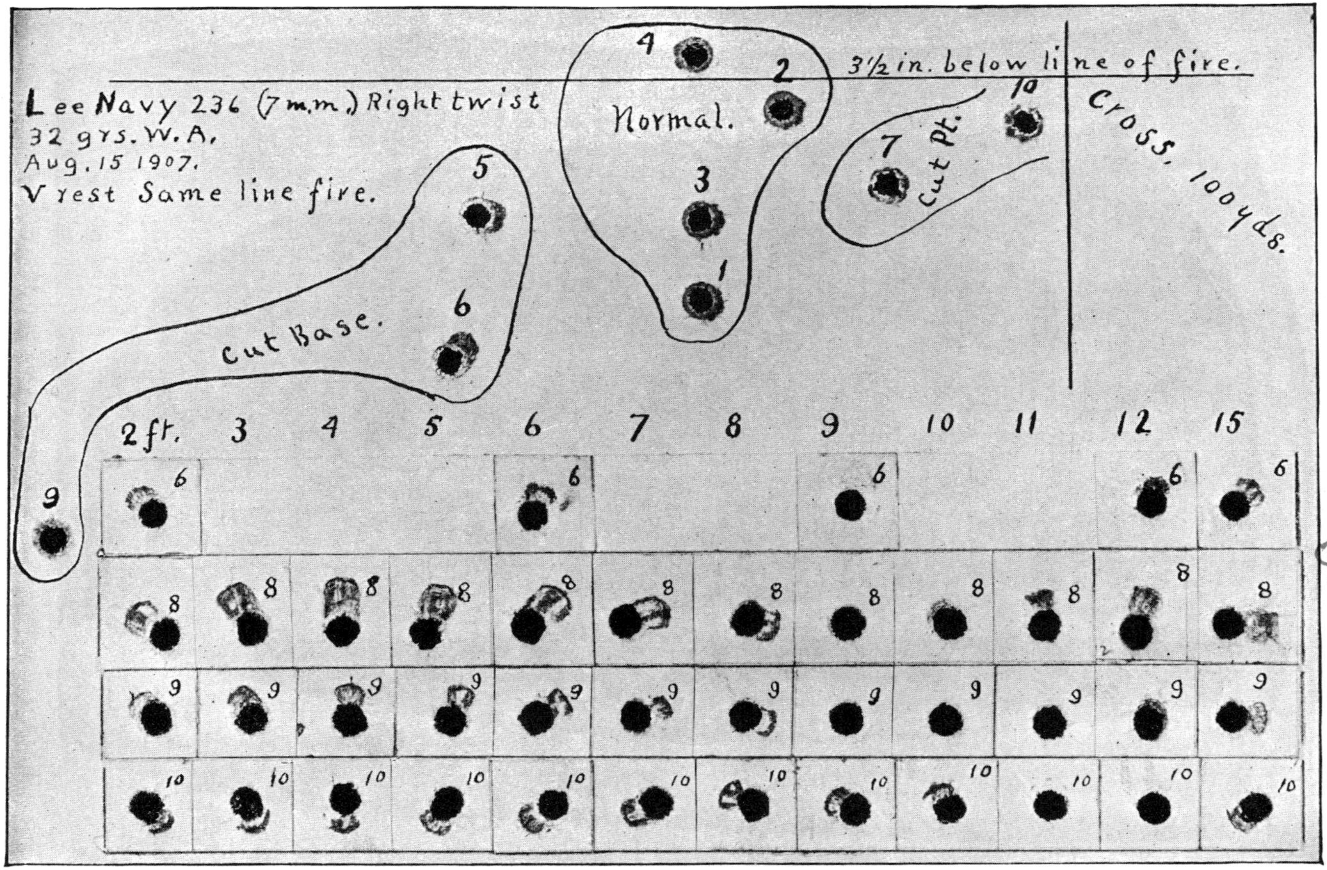

PLATE 34.

Lee Navy 236 (7 m.m.) Right twist
32 grs. W.A.
Aug. 15 1907.
V rest Same line fire.

Cross. 100 yds.
3½ in. below line of fire.
Normal.
Cut Pt.
Cut Base.

(6) mottled case corr. 0.2 gr
(8) .06 dr. new iron .15 gr lost @ 100
(9) do ½ gr
(10) Puni filed .02 gr

on one side of its point and shot 10 more so, removing about .2 grain metal. Observe how slight a mutilation is required to cause these small bullets to fly wide.

These tests with the government Lee navy may be of some interest to those working with high-pressure ammunition, also indicating that the extensive tests made with .32 black powder cartridge apply in principle to all modern arms.

Test 155. — August 17, with a .32-caliber, muzzle-loading, round bullet, 10 grains Hazzard F. G., oleo wad, from V-rest, and same line of fire for all shots, a test was made to determine the effect of plank shooting with spherical ball.

After theorizing in regard to effect imparted to a cylindrical bullet by plank shooting and comparing with Professor Boys' wonderful photographs of flying bullets, we were curious to test our theory that a spherical bullet would glance away from the plank, and continue its flight away from it after passing. A fine mold was procured and powder charge so gauged that good prints might be procured upon our screen paper; target placed at 100 feet, and four normal and two plank shots were made. (See Fig. 127 on opposite page.)

A black line to the right of each print upon the screens represents the plank surface along which the bullets flew. Shots 7 and 8, measuring from their centers at muzzle end of the 12-foot plank, were .4 inch from it; at farther end of plank shot 7 was .6 and shot 8 .5 inch from plank. That they were forced away from the plank by air pressure .2 and .1 inch respectively was to be expected, for our surprise with cylindrical bullets was induced by their not being thus forced away.

The interesting difference between the spherical and modern bullet lies in the fact that the former continues a straight course after leaving plank, while the latter makes a large deflection to the right. Between 10 and 12 feet shot 7 was deflected .05 inch, and computation shows that at 100 feet it should print out 2.5 inches, which is .18 inch from where it did; shot 8 works out in a similar manner.

Observe that the sprue end acts as a point and gives to the print the appearance of having been made by the ordinary cylindrical bullet. The plank converted these point-on bullets into apparent tippers, but they are tippers only in appearance, for they spin around an axis which is parallel to the line of flight, as is shown by the sprue point passing once around every 14 inches of its flight, the same as the twist in rifle bore.

THE BALLISTICS OF SMALL ARMS 309

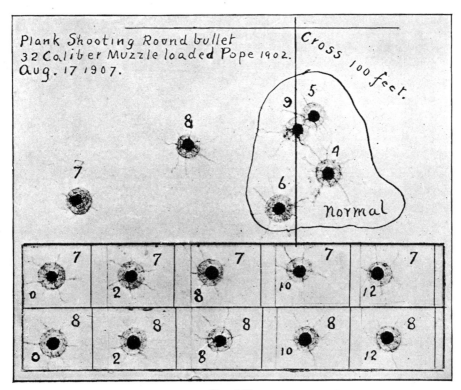

Fig. 127.

Test 156. — August 17, .32-caliber Pope, muzzle-loading, spherical bullet, and 10 grains F. G. powder, to test effect of mutilation with such a bullet.

The cut (Fig. 128) on following page represents screens through which three bullets, mutilated with side plugs, were shot with the cross placed on the 12-foot screens. The screens were placed 14 inches apart, representing pitch of rifle bore, from 36 to 106, while the 127-inch one was placed one and a half times the pitch of bore beyond the 106-inch one.

At 36 inches the bullet prints point on, but at 64 has nearly its full tip, if it does tip; and all shots print with point in same direction until reaching the 127-inch screen, where it is reversed, thus indicating that if screens had been

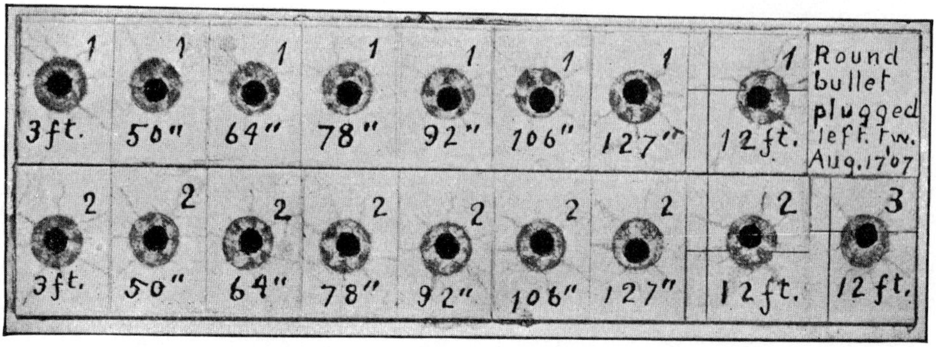

310   THE BULLET'S FLIGHT FROM POWDER TO TARGET

Fig. 128.

placed seven inches apart the point would have been reversed at each, so we have no tip here in the sense that the term is used regarding a modern bullet.

It is plain that these spherical balls spin around an axis which is in the line of flight, as suggested in previous test, the unbalancing caused the equator of rotation obtained in rifle bore to change its plane soon after leaving the muzzle, but this secondary plane of rotation which obtains during its flight is in same plane as was primary equator of rotation. The unbalancing caused the particles of the sphere to change their plane of revolution, while plane of revolution of the sphere as a whole remains unchanged. Probably this explanation will be as Greeky to the reader as it is to its author. The language of these spherical shot as they print upon the screens is perfectly intelligible even if the printed language is obscure.

No oscillation or gyration was recorded on the screens which corresponds to those motions made by cylindrical bullets, but crosses in 12-foot screens for each shot shows that as their plugged sides came from muzzle up they all drove to the right, thus indicating that their x-error existed in common with other projectiles and in the same direction.

Test 157. — August 22, with .30-caliber Krag, Pope cut, 8-inch twist barrel, the one covered with a steel tube in 1904 to show with screens the effect of mutilating the 220-grain government bullet in a bore cut for it by Mr. Pope.

Plate (35) exhibits a normal group of five shots and their 100-foot screen at

PLATE 35.

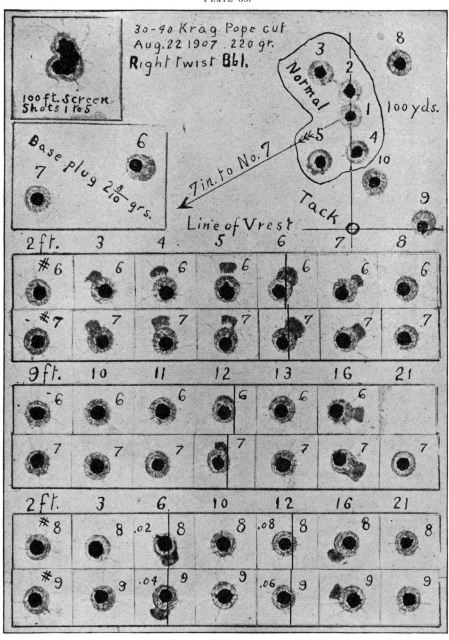

top, also screens for shots 6, 7, 8, and 9 with distances marked at top of each column. The arrow indicates position of shots 6 and 7 at 100 yards, being brought into plate to come within its limits. The points of these metal bullets were turned flat in a lathe, thus causing prints as shown and allowing the tip to be detected and measured with facility.

Observe that shot 7, which tipped 7 degrees at 6 and 16 feet, made a perfect print at 10 and 21 feet, and was caught point on at 100 yards, again indicating that by examining an ordinary target the tipping or non-tipping bullet cannot be recognized. Shot 6 passed its no-tip position also at 10 feet.

This was a crooked barrel but was revolved in its V-rest until caused to make a normal group on the vertical line of the V and there retained, which caused shots to carry above the tack. Being a right-twist barrel, the x-error carries all mutilated shots, 6, 7, 8, and 9, .03 and .04 inch to left at 6 feet, which would amount to two inches to left at target. The y-error carries 8 and 9, the point-mutilated bullets, 2.7 inches to right of this place, and plugged bases four inches to the left, same as observed with other tests and barrels. With this barrel the oscillations occur in the same period as with the Lee navy and the old .32-caliber rifles.

Test 158. — August 26, with same barrel as preceding, to show motion of mutilated 100-grain government miniature bullets, and their points were lathe-squared to make the print readable. The cut (Fig. 129) on opposite page gives screens for two of the three shots that were made.

No target was made, the shots being stopped by an iron plate, often used to save the paraphernalia on the range and make the surrounding country safe for its inhabitants. A 2.8-grain plug was inserted in each and made to emerge up at muzzle, and the screens presented a puzzle, though at the end of several hours' study the solution came.

Number 6 is point on at 4 and 8 feet. Number 7 passed through no-tip position between $3\frac{1}{2}$ and 4 feet, and the oscillations occurred every 2 feet instead of 10 or 15 feet, as with a bullet adapted to the twist, or twist adapted to the bullet. The gyratory motion proceeds right-handed about an hour for each

THE BALLISTICS OF SMALL ARMS 313

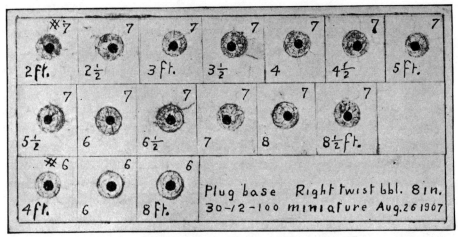

Fig. 129.

oscillation, thus making one gyration and 12 oscillations in 24 feet; clearly, this 8-inch twist was out of proportion for this short bullet. This test indicates that shortening the bullet or increasing the twist shortens the period of oscillation.

Test 159. — August 26, with the .30-caliber, 8-inch twist, and 170-grain U. M. C., soft-nose, .30–30 bullet, shots were made through screens which were placed as noted in cut (Fig. 130).

This bullet, mutilated by a base plug, shows maximum tip at about two feet,

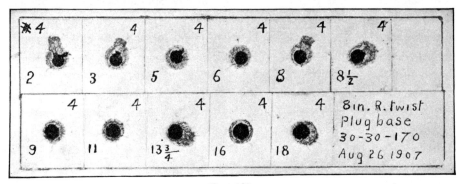

Fig. 130.

also 8 and 13 feet, and first no-tip position at 5.5 feet and last at 16 feet on screens furnished. This last was the only screen in exact position to show that this bullet, like all others tested, passed through its no-tip position for each oscillation.

Notice that the period of oscillation is lengthened when the 170-grain bullet is used in the same twist as the 100-grain one.

Test 160. — August 28, with .30-caliber steel-covered barrel, 8-inch right twist; more plank shooting to determine the action of the 100-grain miniature, 170-grain .30–30-caliber, and 220-grain government bullets. Screens were placed, commencing at 20 feet from muzzle, or 2 feet beyond far end of plank, extending out to 50 feet from muzzle; target was placed at 100 feet. Plate (36) gives the 100-foot target at top and screens with their several distances below. Number 6 shot, with 220-grain bullet, was the only one followed through the screens, and would have printed 10.5 inches to right if carried to 100 yards, in same direction as a left twist. In leaving the plank, however, it tipped at 6 : 30 o'clock as all right-twist bullets do, instead of 11 o'clock towards which the left-twist ones tip, thus either left or right twist bullets are deflected in same direction by the plank.

This 220-grain Krag bullet oscillated after leaving plank, a motion which is never visible at plank shooting in a bullet whose velocity is in the vicinity of 1400 feet per second, or one fired by the old charge of black powder, a phenomenon that accords with the laws governing spinning bodies. Number 6 shot flew $\frac{1}{8}$ inch from surface of the plank, and 1 and 2 were $\frac{1}{4}$ inch. It was noticed that this long, heavy, high-speed bullet was much less deflected after leaving the plank than the .32–47–187 ammunition.

Shots 1, 2, 3, and 4, prints of which are shown upon upper half of Plate (36), have their plank screens shown in the cut (Fig. 131, page 316). Shots 1 and 2 with 220-grain bullet left the plank about .06 inch and would have printed an inch to left at 100 feet, but they tipped towards 7 o'clock, the usual position for right twist, and air spiral more than carried them back into line at 100 feet.

Number 3, the 170-grain bullet, acted very similar to the .32–47–187 ammunition in not leaving the plank. Number 4, the short 100-grain bullet, acted

PLATE 36.

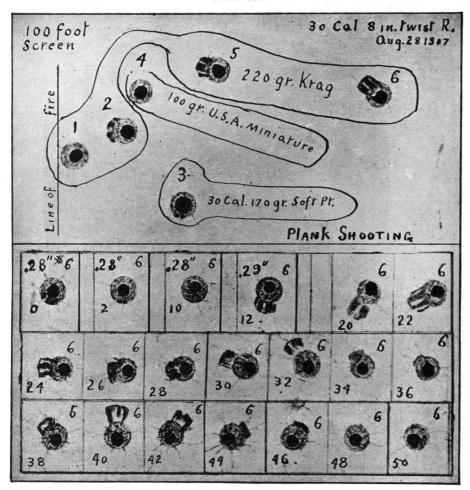

316   THE BULLET'S FLIGHT FROM POWDER TO TARGET

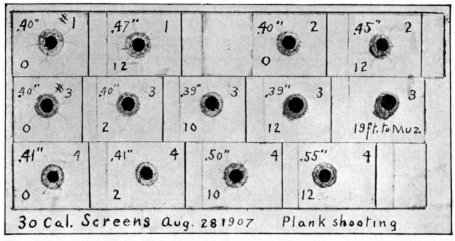

Fig. 131.

like the spherical base one in test 155, page 308, as it passed the plank, but its tip and air spiral made it print an inch to right; while the spherical shot, which could have no air spiral, kept its course and printed two inches to left at 100 feet.

Test 161, September 7, was made with .32–47–187 ammunition, 14-inch left-twist barrel, to perfect doubtful tests in our work of 1901 in regard to the influence of a plank when bullets flew a certain distance from its surface. Plate (37) gives one normal and seven plank shots, with screens for 2, 3, and 5 shots, and distance they passed from plank can be seen by the eye, bearing in mind that the black line at right of each print was drawn before the screens were removed from plank, using surface of plank for a rule.

This plank was 16 feet long, and all the bullets were pushed away from its surface. As they did not tip, the y-error was absent, so they continued a straight course, printing at 100 yards as computed from their direction along the plank, thus completely verifying observations upon this subject from tests made in 1901. This test not only shows that a shot flying along a plank, one-half inch or more from its surface, is deflected away from it at the target, but indicates the reason why, viz., because they do not tip, while shots flying nearer develop a tip.

PLATE 37.

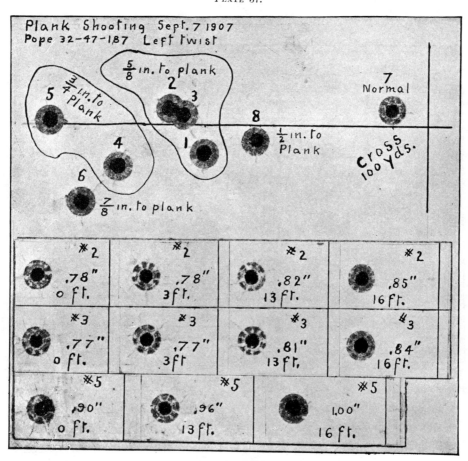

318   THE BULLET'S FLIGHT FROM POWDER TO TARGET

Test 162, September 10, was made with .32–47–187 ammunition, cutting the bullet's points, to determine rate of spin from the 14-inch twist barrel, at muzzle and 100 yards. The cut (Fig. 132) shows the 100-yard prints of three shots with points cut at an angle which removed only a fraction of a grain of lead from near its center of gravity.

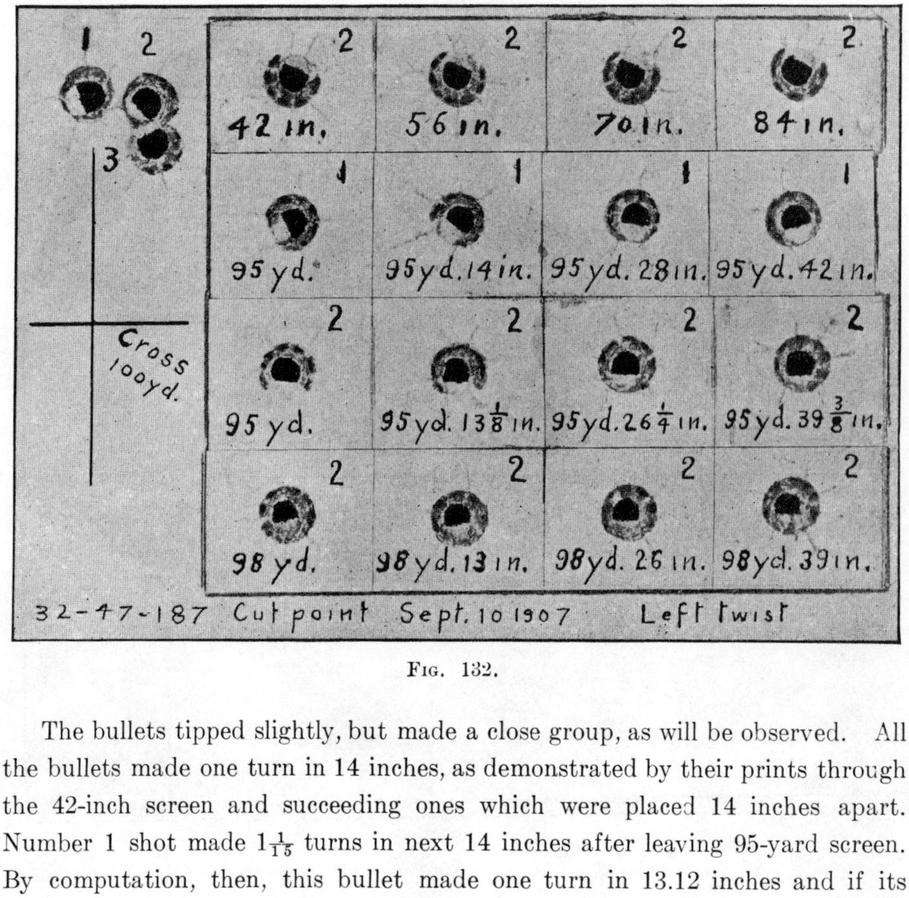

FIG. 132.

The bullets tipped slightly, but made a close group, as will be observed. All the bullets made one turn in 14 inches, as demonstrated by their prints through the 42-inch screen and succeeding ones which were placed 14 inches apart. Number 1 shot made $1\frac{1}{15}$ turns in next 14 inches after leaving 95-yard screen. By computation, then, this bullet made one turn in 13.12 inches and if its speed of revolution had not diminished its forward velocity at 95 yards had diminished 6 per cent, or about 84 feet per second. It is quite certain, however, that its rate of spin had decreased somewhat, so the above computation can only be true by supposition.

After making the above computation the screens for shot No. 2 were placed at 95 yards plus the 13.12 computed inches and so on; but the shot did not tally with the computation, as in that space it made one revolution and 2.2 degrees towards another. At 98 yards this same shot did not make one revolution in 13-inch spaces to within 4.6 degrees.

Tests of this character, made on other days, indicated that successive bullets had a different rate of spin at 100 yards; and Brother William expressed his conviction that the more a bullet tips the longer was its period distance of spin, notwithstanding its greater loss of forward motion due to its increased tip which, in itself, would shorten the distance period of spin.

This test again illustrates a statement made by E. A. Leopold, years ago, that a flying bullet loses a larger per cent of its forward velocity than of its rate of spin.

Success comes; x-error stands alone.

Test 163. — September 10, 1907. This experiment was made with .32–47–187 ammunition, 14-inch, left-twist barrel, to illustrate the action of x at the muzzle when no y exists.

After our long years of experimenting, the idea seemed to be suddenly born, like Topsy, without any fathering or mothering, that if a bullet plugged at its base tipped one way and plugged at point tipped the other, we would make both these mutilations on the same side of same bullet, and watch for results which, so well illustrated by the cut (Fig. 133) on following page, were eminently satisfactory.

This cut shows the cross at 100 yards in the left upper corner, which represents the same line of fire for all shots, and about it the prints of five normal shots. At the right upper corner is seen the four base and point plugged shots, which form a group about $3\frac{1}{4}$ inches from the cross. On the 6-foot screens below, their respective crosses appear, and a glance will show that each of the four plugged shots flew .06 inch to right. They printed very close to their computed distance, .06 × 50, to the right, making a better group than the normal ones.

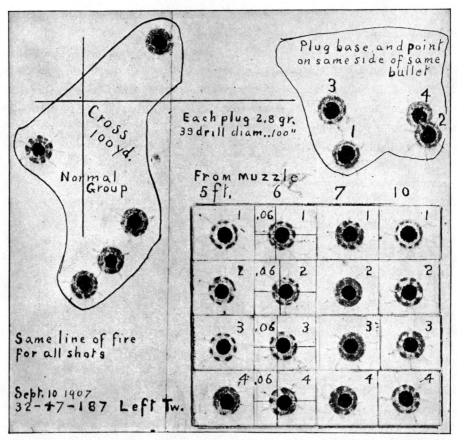

Fig. 133.

The y was absent or nearly so, since their tip was very slight at any place in their flight, the error of observation or measurement probably being greater than error of flight.

Only one who has spent years of diligent search, whose unwearied labor has been finally crowned with success, can appreciate the satisfaction this test afforded. The x was experimentally made to stand alone for the first time and print its own error at the target without interference from y.

THE BALLISTICS OF SMALL ARMS 321

Bullets oscillate about Center of Gravity.

Test 164.—This was with same ammunition and barrel and day as preceding, to show that a bullet in its oscillations turns about a point which is located nearly central between its point and base. Having failed to produce a photographic illustration of motion of oscillation which would produce a print, at the time the gyrating bullet was made to spin by our spinning machine and photographed, this test was devised to compel the powder charge to produce a visible print where spinning machines and photography failed.

Since a bullet mutilated on one side of its point tips the same way as another mutilated on its opposite side at base, we concluded to test a bullet with a

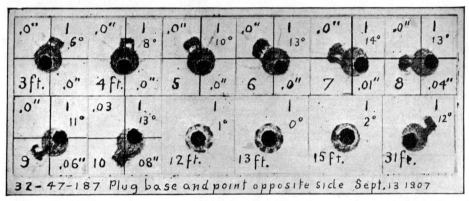

Fig. 134.

plug on one side at its point and another plug in its other side at base. The result was an exaggeration of the tip, the effect of one mutilation adding to the effect of the other, but it eliminated the cause of x-error and left a double cause for the y.

The cut (Fig. 134) shows the bullet with a maximum tip of 14 degrees at 7 feet, but it did not leave the line of fire, as shown by upper left figures on each screen, until reaching the 10-foot screen. If the measurements to determine position of this shot, as it passed the lines on screens, had been taken from the position which its point or base occupied, the result would have shown a curved course between the muzzle and its last cross, at 10 feet, which no swiftly flying bullet could make.

This bullet must have oscillated about some point located between its point and base. The drop of this shot, .08 inch in 10 feet, never could have been produced by gravity or by x, because all tests show that x never carries the shot downward when mutilation emerges either up or down at the muzzle. The fact is, the deflection downward was the beginning of the y-spiral, which is well shown in next test.

The result of this test was very satisfactory, as it had enabled our apparatus on the range to show more quickly and conclusively the picture we were after than could be done in a photograph studio, in combination with the perfected spinning machine.

y-error stands alone. Test 165. — September 13, 1907. This was a continuation of the previous test which separated at muzzle, for the first time, the y from x, although the causes of these errors had been separated at an earlier date by plank shooting. The cut (Fig. 135) gives the 100-yard print of

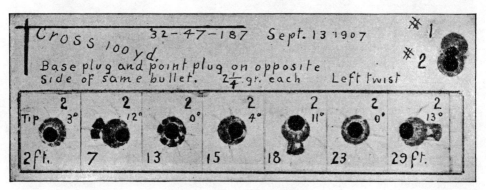

Fig. 135.

shot 1, with its cross in the left upper corner. The screens for this shot being shown in the cut which accompanied the previous test. (Fig. 135) shows also shot 2, made on same day and screens for this shot.

THE BALLISTICS OF SMALL ARMS 323

These twin shots, with same mutilations, showed 12 degrees tip at 7 feet and 13 at 29 feet, but were point on at 13 and 23 feet. Observe that shot 2, with its enormous tip of 13 degrees only six feet from muzzle, made a perfect print at the target. The deflection directly downward of .08 inch between muzzle and 10-foot screen of shot 1 is well shown on screen in previous test; the fact that there was no lateral deflection before reaching this screen eliminates x, but this shot 2, going into its air spiral rapidly downward, did not print downward at 100 yards, but 4.5 inches to the right; thus the axis of its air spiral was in one direction, while the direction of its flight as it went into this spiral was in another. This was also shown at plank shooting in test 138, page 270, and in test 139.

The principles which govern this y-error, here separated for the first time from x, are explained elsewhere.

Cylinder Bullets do Stunts. Test 166. — September 14, 1907, this experiment was made with .32–47–187 ammunition and left-twist barrel, using cylinder bullets mutilated at the base by removing 2.8 grains of lead and plugging. The two half-tone cuts (Figs. 136, 137) furnished with this article provide interesting and important information.

Prints made in screens in cut (Fig. 136) were from a flat-end, cylinder bullet with its base plug emerging up at muzzle of the rifle. Its first and second maximum tips at 2 and 8 feet were 12 o'clock, an important observation, as this does not occur with any other form of bullet tested. It oscillated every five feet, and following its oscillations through to 97 feet, its first tip, after its no-tip position at 96 feet, points to 12 o'clock, same as when the bullet was at 2, 7, and 13 feet.

After leaving the 13-foot screen it commences to gyrate backward, or right-handed, from a left-twist barrel and completes one backward gyration at 97 feet. It will also be observed that it has completed 19 forward or normal oscillations at its 96-foot screen. This bullet, with its mutilated side emerging up, starts to right at 6 and 12 feet, as bullets of any kind do, and at 37 feet it is .31 inch to right. Since it gyrates backward, or opposite to the gyration of a normal bullet, it enters its y or air spiral opposite to the ordinary base-plugged one; so at 75 feet it is found on left of the line of fire, instead of right, being .75 inch to left at 100 feet.

324   THE BULLET'S FLIGHT FROM POWDER TO TARGET

This test very plainly indicates, also, that the direction of the axis of y-spiral is governed by the side at which the bullet enters it by its tangent.   This bullet

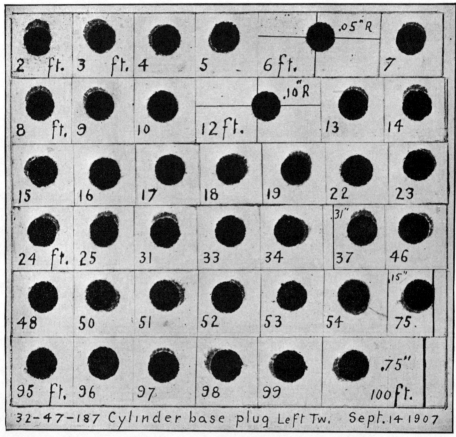

Fig. 136.

did not commence to gyrate before reaching the 25-foot screen, so could not enter its spiral before reaching that position.

Some other things were learned from this experimenting with a cylinder bullet by explaining the cause of some peculiar targets made in the past with them.

THE BALLISTICS OF SMALL ARMS 325

Test 167.—September 18, with same ammunition and barrel as the preceding; plank shooting again, to verify conclusions that a cylinder bullet gyrates back-

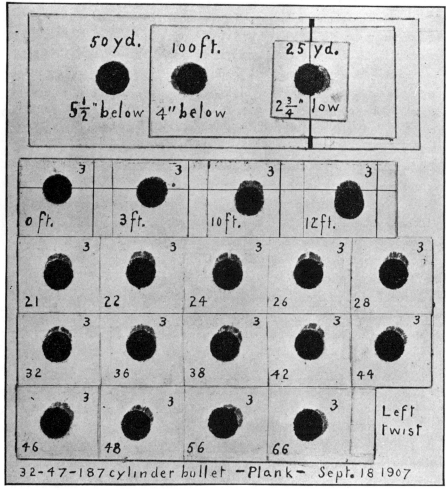

Fig. 137.

ward, and the cut (Fig. 137) gives the 50-yard, 100-foot, and 25-yard screens, so placed that the perpendicular line of fire for each is made to coincide. The bullet used in this test was not mutilated, and commences its flight along the plank's surface one quarter inch away. At 12 feet, or end of plank, it tips towards 12 o'clock, showing at once why normal-pointed bullets at this place tip towards 11 o'clock. This may be explained as follows: The pressure on the point and "suction" at the rear end of any elongated shot, due to the influence of plank's surface, moves the axis of rotation of bullet at right angles to these two forces, or toward six o'clock for the point and 12 o'clock for the base. The cylinder shot as tested, due to the form of its forward end, does not gyrate for the first 25 feet, and this allows the plank's influence to show its effect upon this form of bullet, uninfluenced by a gyratory motion. With the normal pointed bullet, its gyratory motion is combined with the plank's influence, and results in carrying the base to 11 o'clock from 12 o'clock, where the influence of the plank alone would carry it.

Since this shot does not begin its gyratory movement, only to a limited degree, until after 50 feet of its flight, it flies downward, owing to air deflection, but not sidewise, and reaches the 25-yard screen on the line.

Since it gyrates backward, it prints at 100 feet and 50 yards away from the line of the plank, quite different from the curious phenomenon of plank shooting with normal bullets, which have an opposite gyration, thus demonstrating that in either case it is the tipping bullet, after leaving the plank and its y-spiral, which causes its plank deflection and determines its direction.

Driving Tacks with Bullets. *Test 168.* — September 21, with .32–47–187 ammunition. Old-timers are good story-tellers, and their masterful relations of wonderful shots are extremely interesting to the boys. They tell of hitting a nail on its head and of driving tacks with bullets at some marvelous distance, though rarely, if ever, exhibiting such targets to the uninitiated. Perhaps, like the proverbial fisherman, the biggest fish that took the hook never accompanied the relation of the story to clinch its reliability.

It has often been stated by hunters that the contact of a flying bullet with a small twig deflects it from the game, thus accounting for a miss. Others have questioned that small twigs exert any appreciable influence towards deflecting a bullet from line of sight. It was with these conflicting statements in mind that the following test was instituted, viz., hitting the smallest attainable tack upon its point, thus discounting, though unintentionally, fiction with fact. In reality, however, this experiment was made to hit a tack upon its point, pick it up out of center, hoping by this means to unbalance the bullet by a known weight at a known position, after it had left the muzzle. A photographic illustration of results are given below.

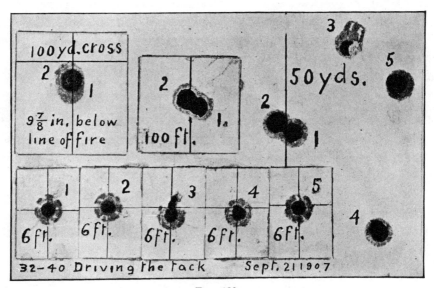

Fig. 138.

Selecting several tacks $\frac{1}{8}$ inch long, weighing .8 grain each, truing up and thickening their points with a file, one was made to pierce the 6-foot screen at the center of its India ink cross, bringing out its point towards the rifle's muzzle so as to stand directly in line of fire.

The cut (Fig. 138) shows the 100-foot, 50-yard screens, and 100-yard target for shots 1 and 2, and 50-yard screen for all five shots, and vertical line of fire is shown on all. It also indicates that shots 1 and 2 did hit their respective tacks on their points centrally, and drove them to the 100-yard target and through the same hole into the butt.

For shot 3 the 6-foot screen was intentionally raised to allow the bullet to pick up the tack as much as possible out of center, but the point of tack did not stick to bullet, and the inertia of this .8 grain of metal plowed a hole through the lead point to full size of tack's head, as indicated by its print on 50-yard screen. It also tore through the 6-foot screen as shown.

Shots 4 and 5 caught the tack, as intended, out of center, with results also exhibited at 50 yards, and all the prints present marked enlargement of the bullet's point, owing to impact of the .8-grain tack.

This relation of one of our numerous experiments is not expected to equal in marvelousness some of the old-timers, but it is accompanied by a modern target which may be examined at leisure.

Illustrated *x*-error. Test 169. — September 27, with .30–40 Springfield barrel, to show x-error at 6 and $12\frac{1}{2}$ feet. Two screens were placed 6 and $12\frac{1}{2}$ feet respectively from muzzle, and five shots were made with unbalanced cylinder bullets, mutilated at their center so they did not tip. The cut (Fig. 139) shows results.

The barrel was laid with a spirit level the same for all shots and a 79-grain, $\frac{3}{8}$-inch long bullet used, having two holes drilled into their sides to their centers which removed $\frac{1}{16}$ their weight or 12.2 grains of lead. These plugged drill holes were entered to emerge on the quarters at muzzle, and a charge of five grains black powder was the load.

The prints, as seen on 6-foot screen, are within .01 inch of a true circle. This circle if carried to 100 yards would have a diameter of six feet. This is not a selected test, as it was the only one made with this barrel; it was sufficient to make the x-error quite apparent.

Test 170. — October 2, with .30-caliber, 21-inch twist, Pope-cut barrel, same as was tested in 1904, to again illustrate the x-error.

The cut (Fig. 140) gives screens at 4, 6, and 12 feet for all four shots, the

THE BALLISTICS OF SMALL ARMS

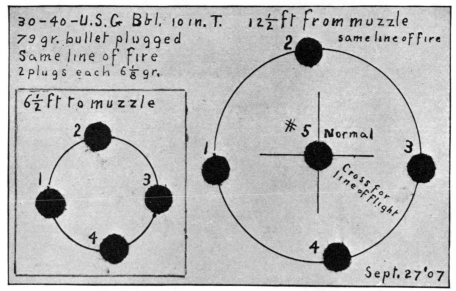

Fig. 139.

bullets and charge being the same as in last test, where barrel had a 10-inch twist instead of 21 as in this. The plugs were made to emerge as near as possible on the quarters at muzzle, without a graduated bullet seater, not having one at hand suitable for this particular barrel.

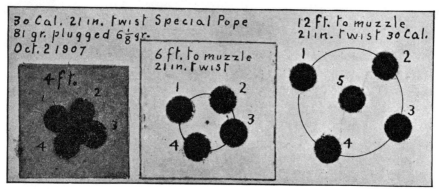

Fig. 140.

330 THE BULLET'S FLIGHT FROM POWDER TO TARGET

Diameter of the several groups, as shown in this and preceding test, are in exact proportion of the different twists of the two rifles, or as 10 to 21.

Test 171.—This was made on the same day as the last, with .30-caliber Pope, 8-inch twist, and the same kind of mutilated bullets, to show again the x-error and its proportional amount when compared with different rifle twists. The

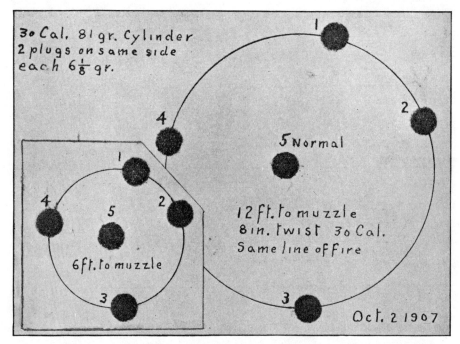

Fig. 141.

cut (Fig. 141) gives the screens at 6 and 12 feet for same shots, and shows that the prints are not more than .03 inch out of a true circle. This x-error would make a circle at 100 yards of over six feet.

It will be seen that size of groups are again in proportion to the several twists. For instance: as 8 is to 21 so is size of group in test 171 to size of group in test 170. Test 172, with the .32-caliber, 14-inch twist, having same propor-

THE BALLISTICS OF SMALL ARMS 331

tion of mutilation, also showed the same proportion of error as these, all working out true by equation (2), page 351.

Test 172. — September 21, with .32–47-caliber, 14-inch left twist, 77-grain cylinder bullet with a drill hole .16 inch diameter in center of one side, removing 6.5 grains lead, or $\frac{1}{12}$ weight of bullet; length of bullet was $\frac{3}{8}$ inch, powder charge 47 grains.

This test was undertaken to illustrate beyond any doubt the cause of x-error and its result at target, being planned some time after x had been separated from y by plank shooting, and the cut (Fig. 142) almost speaks for itself.

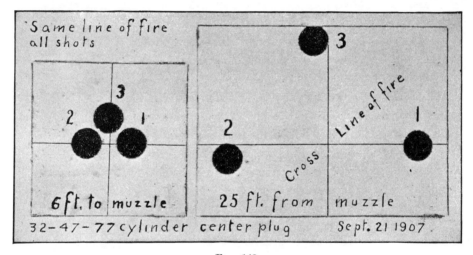

Fig. 142.

Shot 1 was entered to issue from muzzle with its plugged side up, shot 2 with plug down, and shot 3 with plug to the left. This form of bullet being easily unbalanced central to its longitudinal axis, did not tip. Screens show that shot 1 was moving directly to right as it left the muzzle, its center of gravity did not change its direction when leaving the muzzle, as it was forced to do while in the bore. This center of gravity, holding its line of flight, or the tangent of the x-spiral, could only print as seen on the 6 and 25 foot screens, which makes the cause of x-error apparent. Shots 2 and 3 behaved in same manner, printing as they were severally loaded to print, again emphasizing the fact that an unbalanced bullet will not leave the muzzle in the line of fire.

Jacketed Bullets throw Melted Lead. Test 173. — September 24, with Springfield barrel, .30–40 government Krag, 220-grain U. M. C., soft-nose bullet, to test the oscillations of an unbalanced bullet from a government rifle, Springfield barrel; and Plate (38), produced from this test, is quite interesting.

Number 3 bullet was drilled close to its base .1 inch in diameter, removing 2.4 grains metal, and Nos. 1 and 2 with larger drill, removing 4 grains metal. These mutilated bullets were entered by the shell as nearly as possible in same position, but what that position would be when emerging at muzzle was not known except as could be told by the tip.

The bullets oscillated regularly in the same period as the old .32-caliber black powder ones, with no tip at 11 feet. The maximum tip of No. 1 at 6 feet was 11 degrees and was deflected .23 inch at 12-foot cross.

The interesting feature of this test was the pouring of melted lead out of the drill holes, probably due to the excessive heating of the metal jackets by friction; this heat, passing through the jacket, melted more or less of the lead core before the first bullet reached the 8-foot screen, the 2d one the 2-foot, and 3d the 5-foot screen. The melted lead poured out of the drill holes of all three bullets to and past the 16-foot screen.

The three shots thus illustrated by markings of melted lead were made, commencing with a clean rifle, the bore being in very good shape and cleaned easily, though it had been slightly corroded before coming into our possession. After such a marked heating of bullets several other tests were made on succeeding days with different forms and makes of bullets, and with vents in jackets at various distances from their bases, but in no day or with any condition did they throw so much melted lead as shown on Plate (38); that the markings on the screens were made by lead and tin, we determined by chemical analysis.

PLATE 38.

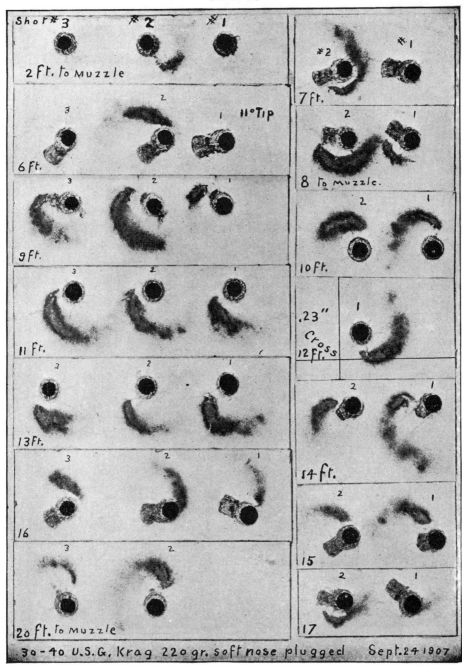

The lead melted sometimes and poured out of the drill hole when made $\frac{5}{16}$ inch from bullet's base; and a notch cut through the jacket at edge of its base did not throw more than when vented $\frac{1}{4}$ inch from base, indicating that friction and not combustion produced the melting heat.

This barrel with a clean, dry rifle bore usually produces heat enough to throw lead, but when entirely clean and thoroughly wet with soft water or oiled with pure lard oil did not throw lead with any variety of bullet tested, and after a few shots without cleaning no bullet that was tested melted.

Plank Shooting, Service Rifle. Test 174. — September 25, with a .30–40 government Springfield barrel, 10-inch twist, service charge of 35 grains W. A., similar to test 160, page 314, but with different barrel, to show action of plank upon government ammunition in a service rifle and determine, if possible, whether a bullet ever oscillates and does not perfectly straighten up somewhere between its positions of maximum tip. The cut (Fig. 143) gives results.

The target for this test was placed at 100 feet instead of yards, one normal and four plank shots being made. The number of feet distance of screens for shots 1 and 3 are measured from muzzle and marked upon respective screens. After leaving the plank, screens were placed every foot, but many are omitted from the cut.

Shot 1 straightened up perfectly at 30 feet, far as could be detected, but 3 and 4 were not caught at any screen in an accurate point-on position, thus leaving us in doubt because screens were too far apart to prove that these shots did not straighten up. Although the action of these two shots may seem of trivial importance, it was not so with us at this time, for, other than plank shooting, no bullet has been found which oscillates and does not accurately straighten up at every period. Neither have we found, outside of plank shooting, any normal bullet, mutilated or otherwise, fired from an ordinary rifle that gyrates without oscillation.

Plank shooting with low-pressure powder produces a gyrating bullet without visible oscillations. If a plank will not cause a high-speed bullet to give an imperfect oscillation, that is one where the bullet does not perfectly straighten up between its positions of maximum tip, it must be decided that an oscillation is one of the movements of a flying bullet which is quite regular and persistent in its form.

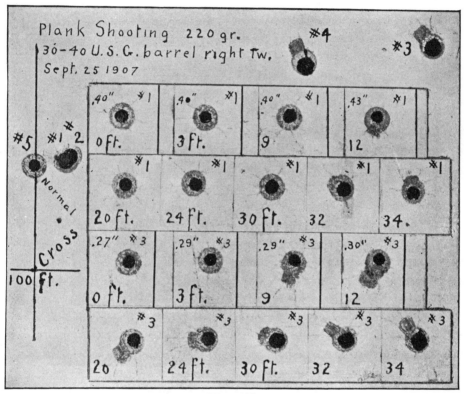

Fig. 143.

A gentle, continued, and equal force, such as a slow-traveling shot may receive as it passes the plank, does not produce a visible oscillation; while the quick impulse given to the high-speed shot by the plank, or the sudden movement in any bullet as it leaves the muzzle, caused by mutilation, produces a regular oscillation.

Paper Plank Experiments. Tube Shooting.

Test 175. — On September 27, 1907, a test was undertaken out of curiosity, to obtain markings of the melted lead which streamed through the drill holes of metal-jacketed bullets, and Brother William suggested placing white paper along the surface of our plank for that purpose.

The .30–40 government charge was used, and a strip of thin, white, poor quality paper, three inches wide, was stretched along the three planks from end to end of each, paper being removed from the plank's surface quarter of an inch by splines placed between at each end. These bands of paper formed a straight surface when set up edgewise and were sensitive to slightest motion of air. This paper plank was lined up to present 12 feet of paper surface .12 inch from line of fire from the 35-grain service charge and 220-grain drilled bullet, hoping to get good markings of the spiral trail of melted lead which might issue from the drill holes.

The paper was closely watched, not to see the stream of lead, but expecting the paper to rupture by air pressure from end to end. Nothing of this kind happened, however; the paper was neither marked with lead or made any visible movement; but to our surprise this slender, slimsy 12-foot strip of paper converted the 220-bullet into a tipper, in same direction and to same amount as when the wooden plank was used, and there was no sign of any melted lead.

This paper plank shooting was freely indulged in for two days, though not for purposes first intended. It was found that this fragile paper deflected the bullet in all cases the same as an immovable surface of a plank, and, so far as this high-pressure bullet is concerned, the paper was as immovable as a plank. The paper was brought nearer and nearer to line of fire until, at times, the ball touched the paper along its surface 12 inches, only blackening it a little without tearing. In one test the shot cut through the paper an interrupted channel two feet long, interruptions being from $\frac{1}{16}$ to $\frac{1}{8}$ inch in length. It could not be detected that the air played any part in cutting the paper, and the metallic mark showed plainly that it was done by touch of bullet.

This unusual and really accidental test threw much light upon the action of air motion around a flying bullet, eliminating some strenuously argued theories, and very neatly illustrated the property of inertia which all substances, even thin paper, possess.

Test 176. — October 25, with same ammunition as in previous test, to further experiment in regard to air pressure which was supposed to surround these high-speed bullets in their flight. The apparatus, as shown by cut (Fig. 144), was designed for this purpose, consisting of a wooden block adjusted vertically by screws upon which a V-rest for holding paper tubes was secured. Two of the

Fig. 144.

regular sliding frames for holding screens in our test work are seen at the right, all supported at six feet from muzzle of barrel (which lay in its own V-rest) upon the 100-foot platform described elsewhere.

A shell, two bullets, and three paper tubes are seen at the left. Tubes were made of one thickness of thinnest tissue paper, four inches long, $\frac{1}{2}$, $\frac{7}{16}$, and $\frac{3}{8}$ inch diameter, which were so frail that it was found necessary to lift and handle them by carrying upon a straw when laid in the V prepared for them.

Other tubes were made of writing paper, $\frac{1}{2}$ inch in diameter and six inches long, which were first tested, but it was found that unless the bullets in passing through touched them no visible effect was produced except to raise them very slightly vertically.

Placing the $\frac{3}{8}$-inch, 1 inch long tissue tube in its V-block so that line of fire passing directly through only left .03 inch between the bullet and the tube on all sides, and this high-speed bullet only moved the tube slightly upward, but not along the line of fire, when it passed through. It did not rupture the tube.

To protect this adjustable V-block from possible injury when the small tubes were used, where bullet must pass only .03 inch from the two surfaces of wood, supplementary V-blocks were laid in the larger one, as seen in the cut, and the little tubes laid gently into these. Though a supply of these blocks were at hand, not one was touched by all the shots made. Some four-inch tissue tubes were also made and tested; but were all ruptured beyond the first $1\frac{1}{2}$ inches of their length; but this air pressure in the long tubes was an abnormal condition, having no bearing upon the action of the bullet upon air in its normal flight.

These tests and all our screen work have failed to indicate any visible effect produced by air upon any object through which or near which a bullet of the high-pressure kind may pass. Some years ago Mr. Leopold detected air motion near a flying bullet by using a light dust, but not until the dust lay upon a surface within the fraction of an inch of the flying bullet.

Determining Rifle Twist. Test 177.— October 2, 5, and 10, with .30–40 government Springfield barrel, to determine its twist, and a novel method suggested itself. After making some cylinder bullets by cutting the points from the 220-grain U. M. C., soft-nose ones, two sides of front end were filed into a blunt wedge. Supposing twist of this barrel was nine inches, we tested it with a tight-fitting swab according to old-time custom, which verified our supposition; but a more novel test than this was undertaken.

Five screens were placed 9 inches apart, commencing six feet from muzzle, as seen in the upper row of screens in cut (Fig. 145), and the first wedge-pointed bullet was shot through, printing as shown.

The bullet did not make one full turn for each space to within 34 degrees,

THE BALLISTICS OF SMALL ARMS 339

and computation from this data showed that the twist of barrel was 10 instead of 9 inches. We were deceived by the swab test.

Two days later five more screens were set, placed 10 inches apart, and another shot tried, resulting as seen in second row of prints. These seemed to tally very closely, but did not look quite right, and close examination with glass and scale

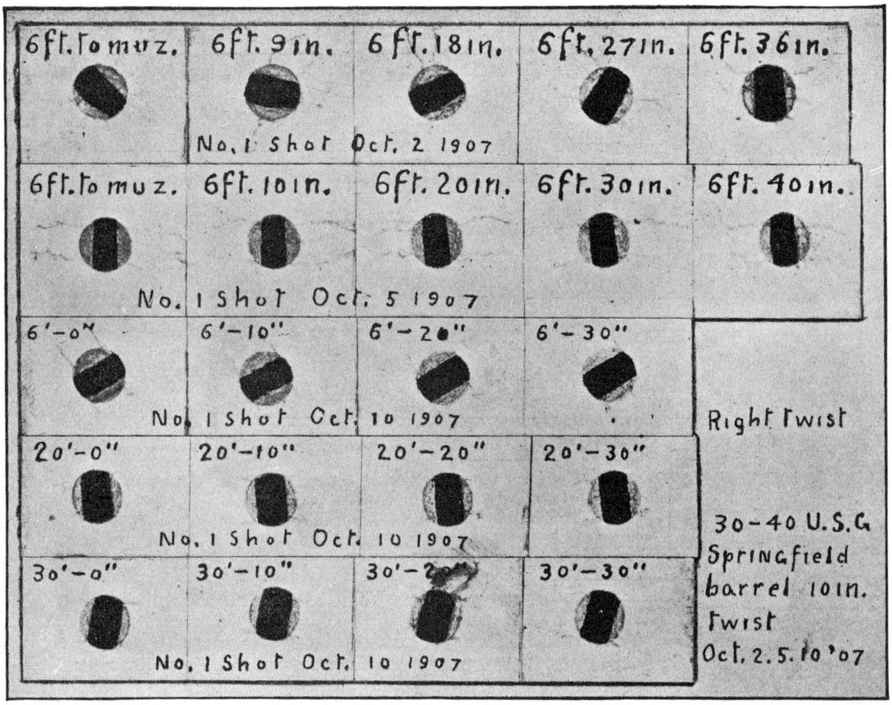

FIG. 145.

indicated that six degrees was lost in the first four spaces, or 1.5 degrees to each 10-inch space. Five days later these spaces were examined with a steel scale, and a total error in setting the four spaces was found of .032 inch, thus reducing loss in turn of bullet from 1.5 to 1.25 degrees.

With a template we were able to set screens for the third day's test fairly accurate, and prints in third row show results. Here, however, the shot lost one degree for each 10 inches, which tallied with previous test to within .25 degree. These tests therefore indicate that the Springfield armory cut a twist of one turn in 10.034 inches, or the bullet was traveling faster at six feet than when leaving muzzle.

Professor Bashforth of England, as mentioned before, demonstrated by his chronograph, to the satisfaction of his government, that a shot increased its velocity for 25-caliber lengths after exit from muzzle. Consequently, if the Krag barrel tested was 10-inch pitch, the ball, by computation, had .3 per cent greater speed, or an increase of 6.3 feet over its reputed 2100 feet velocity.

This bullet at about 20 feet, represented in fourth row of screens, made one turn in each 10 inches, and at about 30 feet, seen in fifth row, having lost speed, made one turn plus 1.5 degrees in each 10 inches.

Test 178. — October 4, another experiment was made, hoping to show movements of an oscillating bullet through its period of no-tip position, and Plate (39) gives results.

All conditions were made the best possible; powder charge was reduced to 30 grains, which makes best prints through screen paper; screens were placed $3\frac{1}{4}$ inches apart, close as wooden frames would allow; bullets were made more pointed in a lathe so that the paper punched out by them would not mutilate succeeding screens so badly. These paper punchings were strewn about the range all summer so that the path along the range was covered with them.

The four upper screens on the plate show three shots in the three upper ones, and four in the fourth without being cut into squares, the frames which held them being moved to right as usual after each shot. They show amount of tip at distances indicated. The lower half of plate gives prints of shots 3 and 4, 14 screens for each shot placed $3\frac{1}{4}$ inches apart, first being set 8 feet 9 inches from muzzle, and first no-tip position is at the fourth screen, or 9 feet $6\frac{3}{4}$ inches from muzzle. The direction of tip is given below each print on every screen.

Both shots enter no-tip position at 4 o'clock, one coming out at 3 o'clock and the other at 12:45, an irregularity that would not have been noticed if similar differences had not occurred throughout all tests. See remarks regarding this

PLATE 39.

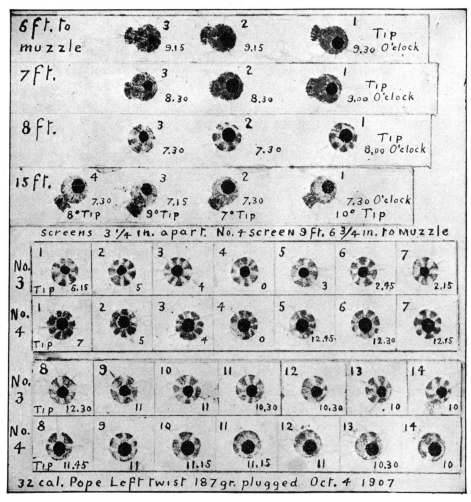

under test 148, page 294. The maximum tip of shot 3 at 6 feet is at 9 : 15, shown in upper uncut screens; its next maximum, at 15 feet, is at 7 : 15 o'clock, showing six oscillations for each gyration of 47.7 feet. This proved one of the most satisfactory determinations of length of a gyration for an oscillating bullet that has been made in any test.

We do not attempt to explain how the base of a bullet can come out of its no-tip position on the same side at which it entered, when six oscillations occur for each gyration, as many casual observations made upon the range indicate when screens are some distance apart. This and several other similar tests, with their perfect illustrations, can be studied during leisure hours of any ingenious or interested student.

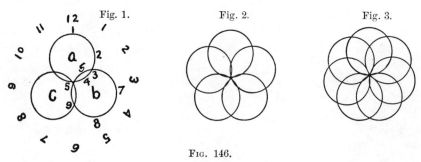

Fig. 146.

To make this subject more interesting, these geometric figures are brought forward, having been in part reproduced with a description of same on page 224. The outer circle of figures shows the hours of the clock. If the last observable tip in figure 1 occurs as the center of base of bullet passes 3 on the circle a, then the first observable tip will be at 3 on circle b. Here then the bullet will enter its no-tip position at 2 o'clock and reappear on the same side also at 2 o'clock. If its last observable tip is at 2 : 30 o'clock where the center of base passes 4 on circle a, its first observable tip will be at 6 on circle b, which is at 1 : 30 o'clock, being one hour late. Working out the motion in figure 2, if the last tip be at 2 : 30 o'clock it will reappear at 12 o'clock, being two and one half hours late. In figure 3, with seven oscillations for each gyration, if its last tip is at 3 o'clock its first tip after the no-tip position will be at 11 o'clock, being four

THE BALLISTICS OF SMALL ARMS 343

hours late. The reading of these geometric figures tallies so well with recorded tests at screen shooting, and with the action of the gyroscope illustrated on page 222, that we can assume they have a similarity with the actual motion of the base or point of the oscillating and gyrating bullet.

Plank Shooting, Spherical Bullets. Test 179. — October 10, with .32-caliber spherical bullet and 10 grains black powder, to test the action of plank upon this form of projectile, but particularly to show its form of gyration, if any, after passing plank. It will be noticed in the cut (Fig. 147), as in test 155, page 308, that the round bullet is deflected away from plank to a marked degree, and

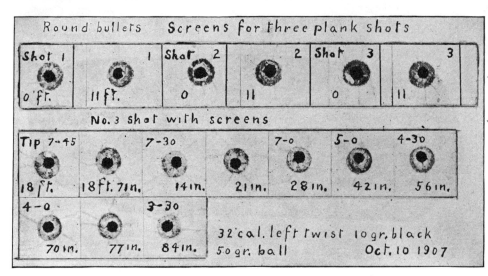

Fig. 147

the pressure which carried them away made No. 1 tip at 11 feet at 4 o'clock, Nos. 2 and 3 at 8 o'clock. This variation never occurs with a cylindrical bullet, even though it be a short one.

The object of this test is indicated at end of the plank 18 feet from the muzzle, 18 feet 7 inches, 18 feet 14 inches, and so on, where tip is shown to change to opposite direction in every 7 inches. This very plainly indicates that this round bullet is spinning about an axis that is in the line of flight, and the plank converted it into a tipper only in appearance, since a tipping shot is one whose axis of rotation is not in line of flight.

Strictly speaking, however, since the bullet had lost some velocity, its tip is not exactly the same for each 14-inch screen, as indicated by notes given upon each, but gains about one hour for each distance, which tallies with speed of all bullets, indicating that the per cent loss of spin is less than the loss of velocity.

These plank-shooting and other recorded tests with spherical bullets plainly indicate that, in any form of its apparent tip, it is spinning about an axis which is in the line of flight; and further experiments might be conducted to show why a round bullet, shot from a rifled bore with a twist, makes such curved flights in calm air as our tests have shown them to make, although these particular tests have not been recorded in these pages. This curved and non-uniform flight might be explained if it could be shown that the spherical shot ever takes on a true tip, as all cylindrical projectiles can be made to do; but the deflection of a spherical shot that tipped would be from a different action of air than that which deflects the cylindrical, the one acting solely on front half of the sphere and the other mainly on side of bullet.

Cause of Excessive Tips Disclosed. Test 180. — October 12, 1907. This was instituted to illustrate how a mutilated bullet is caused to tip to so large an extent in six feet of its flight from muzzle.

Computation indicates that a 187-grain bullet with four grains metal removed from near its base and plugged throws its center of gravity about .0016 inch out from its center of form, and it puzzled us for months to understand how this small fraction of an inch could be multiplied by a 14-inch twist to produce the large tip so universal at only 6 feet away. When this problem was solved, however, it was quite easy to institute the following experiment and produce the cut (Fig. 148) for critical inspection of any who may be interested. The attention of riflemen is particularly invited because this action of an unbalanced bullet is always present with all rifles.

THE BALLISTICS OF SMALL ARMS 345

Several bullets were carefully prepared with a four-grain plug in the side of each and close to the base. The bases of two of these bullets were sawed off to leave their plugs central with the side of section, thus producing two .32-caliber cylinder bullets out of these bases, each $\frac{3}{16}$ inch long. A special .326-inch swage was made for this experiment, since a bore-diameter bullet of this length would not take the rifling. Only three shots were made, being sufficient to complete the experiment.

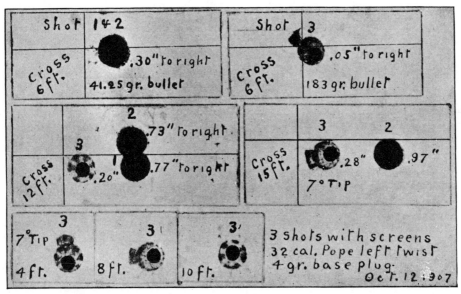

Fig. 148.

The cut gives results very well and will be readily understood if explanations are carefully followed.

The three lower screens give shot 3 at 4, 8, and 10 feet. At 10 feet it was point on as usual, and its plug came out up at muzzle, as did the other two shots.

The upper two screens should receive particular attention, where shots 1 and 2, at 6 feet, went .30 inch to right and through same hole. With shot 3, which was the normal cylindrical bullet plugged like 1 and 2, its center of gravity, at 6 feet, is .05 inch to right; but the base, which was shot directly to right, in common with shots 1 and 2, would have been, by the law of rotating bodies, carried up to 12 o'clock. By air pressure, which caused a gyration left-handed in this case, it was carried past 12 to 10 : 30 o'clock where this plugged base is well seen in the cut. The unbalanced base causes the oscillation, so well shown with gyroscope and explained on page 221. In the illustration of a gyroscope the gyration is absent because air pressure is absent; it is the air pressure upon a flying bullet from the front that causes it to gyrate.

The proportion by weight of shots 1 and 3 are .225 to 1. Multiplying .30 inch, distance of the first shot, by .225 gives .067 inch, which is the distance of third shot to right within .015 inch; but in the cut shot 3 at 6 feet is incorrectly marked. Making the proper correction (see page 234), it would read .065 instead of .5 inch, making the computed distance correspond to the actual distance within .002 inch.

Referring now to the third screen at 12 feet, in which all three shots were made without changing the screen, it will be found the same proportion and computation places shot 3 at .17 inch instead of .20, where it did print, and since it tipped so slightly at this screen a corrected reading is unnecessary.

The 15-foot screen works out nearly the same as the two computed ones. All this should plainly indicate to a careful student that a four-grain plug which unbalances the normal bullet only about .0016 inch unbalances the base end, as used in shots 1 and 2, so that its center of gravity makes a spiral in rifle's bore, the radius of which is .0016 inch × 4.5, the ratio of weight between the two bullets used, the diameter of spiral being twice its radius, or .014 inch. This size and pitch of spiral, by computation equation (2), page 351, produces an x-error at 6 feet with these two 41-grain bullets of .42 inch though their prints on screen show a .30-inch x-error.

Computations and tests plainly indicate that it should not be surprising for a bullet to tip 7, 10, or more degrees at 6 feet. It is a mathematical problem in common with all other motions of a rifle bullet.

Retrospect. As mentioned at the time and in its proper place, the record of much work with the 8-inch and 21-inch twists, .30-caliber barrels, was omitted and some of it is introduced here as more appropriate.

During September, 1905, continuous tests were made with these barrels to show that the x-error, as first stated to me by E. A. Leopold in 1904, was the only one for which search was being so assiduously made. In other words, we were trying to discover if the error caused by an unbalanced bullet at 100 yards was in proportion to the twist, other things being equal, and some of the tests made at that time are illustrated by Plate (40), made one quarter size of originals. Three days' experiments, September 11, 29, and October 4, 1905, are represented in the plate, which will not be difficult for a student to comprehend, the different forms given to the various circles representing bullet prints are indicative of which were shot on same day.

Eighteen grains lighting powder was used in the 8-inch twist and 23 grains in the 21-inch, and twist of the rifle used is given at lower left-hand corner on each target; same line of fire for all shots. Normal groups and others are clearly indicated, the two upper targets made with 170-grain bullets plugged, and the two lower with 100-grain with lighter plugs.

Observe that the plugs in these 100-grain bullets bear a close proportion to those in the 170-grain ones, or as 100 is to 170. The error caused by the 6-grain plugs in two upper targets, where 170-grain bullets were used, is in proportion of 2 for the 21-inch twist to 3 for the 8-inch; and for the $2\frac{3}{4}$-grain plug the proportion of error is again about 2 to 3, though not in proportion to the twists, which is 2 to 5.

In two lower targets the 100-grain bullet produced an error at 100 yards with 8-inch twist $2\frac{1}{2}$ times greater than with 21-inch twist, or in proportion of 8 to 21. Thus the two lower targets show that the x-error is the only one, while the two upper ones hint strongly of some other existing error.

As has been fully explained in tests of previous years, the 21-inch twist could handle the 170-grain unmutilated bullet with 23 grains powder, but there was no margin in the twist to take care of imperfections or mutilations, while the 8-inch twist, which was much sharper than demanded by either bullet, contained a margin of safety.

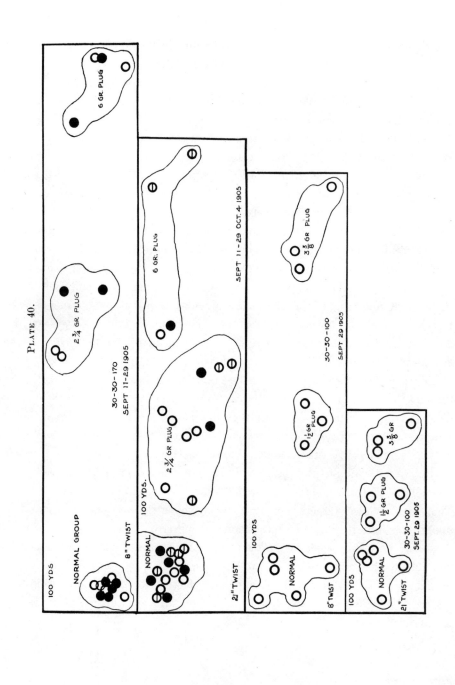

PLATE 40.

Although the third and fourth group show the proportion of error as 2 to 5, compare the second and fourth group, both with same barrel, where mutilations were in close proportion to weight of bullets. If the deduction from the third and fourth target be correct, then second and fourth should contain nearly same amount of error and be practically of the same size, but it is nearly three times as large. Here again it is clearly indicated that some other than x-error must exist. Our ignorance of it during those days made progress slow and very discouraging. Persistent experiments convinced us that the x-error was not all, but let in very little light for over two years.

It was not until the winter of 1907 that light dawned; then it was not through experimenting, but seemed to filter in through the medium of the brain in some unconscious manner. No doubt the brain had been prepared, through the years of close rifle study and experimenting, to solve the problem. Anyhow, it seemed to arrive in its own good time. Then tests were devised one after another in rapid succession and without difficulty for the summer of 1907, to prove that the problem had been solved.

No hindrance or drawbacks occurred, and each test, as has been related, stamped its proof of our solution upon the target and screens day by day with astonishing regularity; even the wind and weather gave of their best.

PART III

Mathematics of *x* and *y*. So far in this work algebraic formulæ have been omitted, and clearness has been attempted without bringing in extended mathematical deductions. It is expected that the tests speak for themselves, experimentally, accompanied as they are by explanatory notes. It is hardly possible, however, to clear up many interesting questions that have risen, without a few simple equations, which apply directly to some of the work gone over, especially those equations which express the properties of the helix, or spiral. In shop practice and among riflemen, the word "spiral" is used in place of the term "helix," and since this is allowable, we shall continue this usage.

To illustrate a spiral (Fig. 149) is introduced, showing a brass wire made into the form of a spiral spring. The end of the spiral *a* is straight as the wire was before being manipulated, and forms a tangent to the spiral. The straight wire *cc* bound to the spiral by a cotton thread illustrates a tangent drawn to it at the point of contact. The tangents *a* and *cc* are parallel to each other because they are tangent to the same side of the spiral at two different places. Tangent *cc* would not be parallel to tangent *a* if it was tangent to any other part of the circumference of the spiral. *dd* is a straight wire representing the axis of this spiral, and indicates the direction of the spiral as a whole. The radius of this spiral is the perpendicular distance from the above designated axis to the wire which represents the spiral, and the diameter of this or any other spiral is two times the radius. The distance from one coil *s* to the next coil *s* is the pitch of the spiral. The angle that the tangent to a spiral makes to the axis of the same spiral is dependent upon the diameter and pitch of the spiral.

In a mathematical spiral for determining the angle that the tangent to the spiral makes with the axis, we have the formula: —

THE BALLISTICS OF SMALL ARMS

$$\tan \alpha = \frac{\text{circumference of spiral}}{\text{pitch}}$$

or $\quad \tan \alpha = \frac{\pi d}{P} \quad$ (1)

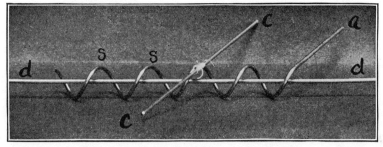

Fig. 149.

EXAMPLE. Given a spiral with a diameter .003″ and a pitch of 14″, to find the angle which the tangent makes with the axis.

Substituting in (1), we have

$$\tan \alpha = \frac{3.1416 \times .003}{14}$$

Angle $\alpha =$ about 2 minutes.

Since in gunnery the error E at the target is of more interest than the value of the angle α, and since we have determined the cause of the x and y error in rifle shooting, the following formula, which is easily obtained, gives the relation between the x-spiral in the rifle bore and its resulting error E at the target, also the relation between the y-spiral in the air and its resulting error E: —

$$\frac{\pi d}{P} = \frac{E}{R \times 36}$$

where P equals the pitch, d the diameter of the spiral that the center of gravity of the bullet makes while in the rifle bore (or d may equal the diameter of the spiral that the bullet makes in its flight over the range), and R the range in yards. Clearing and transposing to find E, we have, for a working formula,

$$E = \frac{\pi d \times R \times 36}{P} \qquad (2)$$

EXAMPLE. Taking as before a diameter of .003″ and a pitch of 14″, range 100 yards, to determine the x-error in inches at the target.

Substituting in formula (2), we have

$$E = \frac{3.1416 \times .003 \times 100 \times 36}{14}$$

Solving,

$$E = 2.4 \text{ inches.}$$

The cut (Fig. 150) indicates how the formulæ (1) and (2) were obtained by developing the spiral and drawing triangles.

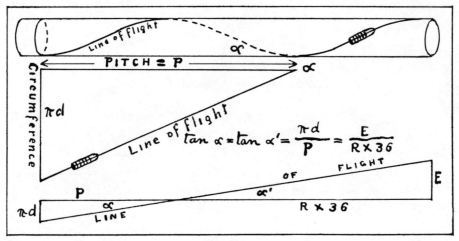

FIG. 150.

In connection with formulæ (1) and (2) just derived, Fig. 151 shows at a glance the effect of the x or bore spiral, and that of the y or air spiral upon the flight of the bullet and the consequent x and y error at the target. The rifle being at the left and the target ad at the right, b represents the center of the normal group, Ob the line of fire, which is the axis of the bore spiral. OM is the tangent leaving the x-spiral at the muzzle at O. If the unbalanced bullet does not develop a tip, it will continue on the tangent OM and reach the target at c. The angle bOc is angle α in equation (1) and is the angle that the tangent of the bore spiral makes with its axis, or line of fire. If the bullet is mutilated at the point or base, it will tip and go into its air spiral at M somewhere within 24 feet of the muzzle.

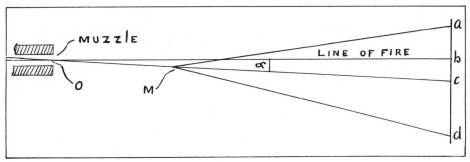

Fig. 151.

The tangent of this air spiral is OM, which is also the tangent of the x-spiral. If the mutilation is at the point of the bullet, its tip will be in a direction to deflect it into a y-spiral whose axis Ma will be towards the line of fire. If mutilated at the base, the tip will be in the opposite direction, and the axis Md of this y-spiral will be away from the line of fire. The deflection at M for either case is away from the tangent Oc of the x-spiral. The angle dMc or cMa is also angle α in equation (1), and thus solved when the pitch and diameter of the y or air spiral has been determined by properly placed screens.

cb is the x-error at the target. ca or cd is the y-error. With a plug-base bullet the x-error bc is added to the y-error cd and produces the total error bd. With a point plug we have the error at the target ab, which is the y-error ca less the x-error cb.

In some of the tests where the errors have been clearly separated, the y-error has been about $2\frac{1}{2}$ times the x-error, and the drawing has been made upon this basis. In theory and practice, however, there is no relation between the values of these two errors.

When working for the x-error with equation (2), P is always equal to the pitch of the rifling, and the range R in yards must be known; d and E are the only two unknown quantities. To find d, or two times the distance between the center of gravity of a mutilated bullet and its center of form, we take three special cases. When d is known, E can be determined.

1st. Where the mutilation is a drill hole, commencing on the side of the bullet and extending to the center of form.

2d. Where the mutilation is caused by removing a known weight of metal at the surface upon one side of the bullet.

3d. Where the plane of the whole base is oblique to the long axis.

For the first case, let M equal the mutilation, or the weight of lead in grains removed by the drill hole. Let W equal weight of bullet, $d =$ diameter, $r =$ radius, $y =$ distance of the center of gravity of one-half of the unmutilated bullet from the longitudinal axis of the whole unmutilated bullet, $z =$ distance from the center of gravity of the half of the bullet containing the drill hole, to the longitudinal axis of gravity of the whole of the mutilated bullet. (See Fig. 152.)

We then have the following equations, (3), (4), (5).

Equation (3), in which y equals the center of gravity of a semicircle from the center of the whole circle, is obtained by calculus. It is y in Fig. 1 (Fig. 152).

Equations (4) and (5) are obtained from the proportions which exist between the weights and lever arms in figure 1, where each half of the bullet is represented as suspended from its center of gravity. Noting that the lever arms in one case are each equal to y, and in the other case where the weight M has been removed, one arm has a length of z and the other $2y-z$.

To lengthen the lever arms and make room for lettering, the parts of the bullet have been drawn widely separated. Line BO represents the center where the two halves would join, if placed in their normal position as in figure 2, and represents the center of gravity of the unmutilated bullet. Shifting the fulcrum from O to C on the lever LE, establishes a balance after the mutilation M has been produced.

The distance x between BO and AC is to be determined. This distance is constant, irrespective of the position at which the hole is drilled along the length of the body of the bullet, since it only represents the distance of the center of gravity from the long axis.

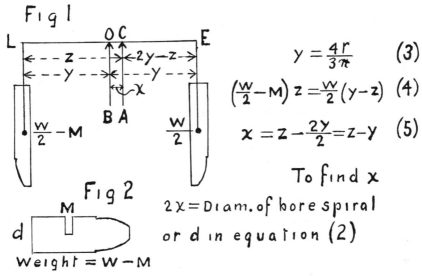

FIG. 152.

The above equations are only approximately correct, a slight error having been introduced when we assume that the center of gravity of half of bullet does not change its position when we mutilate it by a drill hole. If this error was corrected, the value of x would be slightly reduced.

Simplifying the above equations, we have for working purposes the following:

$$y = \frac{4r}{3\pi} \quad (3)$$

$$z = \frac{Wy}{W - M} \qquad (4)$$

$$x = z - y \qquad (5)$$

where x equals the distance between the center of form and center of gravity of the unbalanced bullet or the radius of the bore or x spiral.

EXAMPLE. Given 187-gr. .32-cal. bullet with drill hole extending from one side to the center, thus removing 7 grains of lead. We then have $W = 187$, $d = .320$, $r = .160$, $M = 7$.

Substituting in equation (3), $y = \dfrac{4 \times .160}{3 \times 3.1416}$

Solving, $\qquad\qquad\qquad y = .0679$

Substituting in equation (4), $z = \dfrac{187 \times .0679}{187 - 7}$

Solving, $\qquad\qquad\qquad z = .0705$

Substituting in equation (5), $x = .0705 - .0679$

Solving, $\qquad\qquad\qquad x = .0026$ approx.

Second special case, where the entire mutilation is supposed to be in one place on the surface of the body of the bullet. Since the weight is removed from the surface instead of the center of gravity of the mutilated half of the bullet, we can dispense with equation (3) in the first case, by making the lever arm y equal to one-half the diameter of the bullet, or $\dfrac{d}{2}$, this being the length of the lever arm of the center of gravity of the mutilation. Formula (4) will then give approximately the value of z in this second case, by writing instead of y, $\dfrac{d}{2}$, or half the diameter of the whole bullet.

For this case, then,

$$z = \frac{W\dfrac{d}{2}}{W - M} \qquad (6)$$

$$x = z - \frac{d}{2} \qquad (7)$$

where $x =$ radius of the bore spiral.

A complete solution of this problem shows that for any practical mutilation the simple formulæ above give results well within the degree of accuracy required.

EXAMPLE. Given 187-gr. .32-cal. bullet with 7 grs. removed from the surface upon one side of the body.

Substituting in (6), $z = \dfrac{187 \times \dfrac{.320}{2}}{187 - 7}$

Solving, $z = .166''$

Substituting in (7), $x = .166'' - .160'' = .006''$

Third special case, where the base of the bullet is oblique. Let T equal the number of thousandths of obliquity of the base, or the number of thousandths that one side of the bullet is shorter than the other.

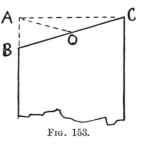

FIG. 153.

This mutilation then is in the form of a circular wedge ACB, see (Fig. 153), where $AB = T$. Its volume is one-half that of a transverse section, T thousandths long of the bullet. So we have the equation,

$$\text{Volume } M = \frac{\pi r^2 \times T}{2}$$

where πr^2 equals the area of a transverse section of the bullet.

Since one cubic inch of lead weighs approximately 2881 grains, we have

$$\text{Weight of mutilation } M = \frac{\pi r^2 \times T}{2} \times 2881 \qquad (8)$$

Although M represents the weight of lead removed from the base of the bullet, only a fraction of M acts to unbalance it. Drawing the line AO in (Fig. 135) separates from the total mutilation ACB, AOC, which is not effective, and leaves the effective mutilation AOB. This effective mutilation then is a section of a semicylinder, and takes the form of a wedge with a circular base.

Since T is always a small quantity, we can, without introducing any appreciable error, consider this wedge-shaped mutilation as composed of pyramids, the sum of whose bases form the base of the wedge, and whose altitude is r, and since the volume of a pyramid equals the base times $\tfrac{1}{3}$ of its altitude, we have

$$\text{Volume of effective } M = \frac{AB \times \frac{\pi d}{2}}{2} \times \frac{r}{3}$$

Simplifying, substituting T for AB, and reducing to weight by multiplying by 2881, we have

$$\text{Weight of effective } M = \frac{T\pi dr}{12} \times 2881. \qquad (9)$$

The center of gravity mechanically obtained of this wedge-shaped piece is $.58\,r$ from center of cylinder.

Going back to the first special case, we can again omit equation (3) if we make $y =$ to $.58\,r$. Taking equation (4), making the lever arm equal to the distance of the center of gravity of the effective mutilation from the axis of the bullet or $.58\,r$, we have

$$z = \frac{W \times .58\,r}{W - M} \qquad (10)$$

$$x = z - .58\,r \qquad (11)$$

where x equals the distance of the center of gravity of the unbalanced bullet from its center of form, and here is the actual error in careful rest target work, the primary cause of both the x and y error at the target.

Heavy Ordnance. The author is not unmindful of the fact that the same laws of ballistics which govern the flight of the rifle bullet from small arms also govern the flight of the shot from heavy ordnance. "If the center of gravity of a shell or solid shot does not coincide with its center of form, this shot will be deflected from its line of fire at the muzzle." Leopold. If this unbalancing is situated in front of or in the rear of the center of gravity of the projectile, it will become a tipper as certainly as the laws of inertia act. No amount of spin can annul this motion, since shortening the pitch of the rifling in the bore of the cannon to increase its spin will make the force which produces the tip greater. The tipping projectile will at once take up a spiral flight from the line of its first deflection, and the axis of this spiral can only be another deflection.

The value of these two deflections can only be determined by those who have access to the necessary data, and on the proving grounds of heavy ordnance. Here the x and y errors will appear at the target.

The Spitzer Metal-cased Bullet. After the foregoing pages were in type, while reading the proof, it was discovered that nearly all past experiments had been performed with what may be termed "bluff-pointed" bullets. It is noticeable also that screen shooting, upon which our deductions in the ballistics of small arms should be based, was performed in every case with the normal or bluff-pointed target bullet. This is not to be wondered at, since all target men and even manufacturers of ammunition have universally settled down to this form of point both for hunting and target work. Hardly any other form of point will be found used in normal modern ammunition, therefore the advent of the long, sharp point and its adoption in the 1903 model army rifle necessitated careful tests with this bullet, since it is unsafe to decide upon the flight of a projectile, unless the flight of a similar-shaped one has been determined.

Having obtained the use of a new Springfield rifle, Model 1903, a special machine rest for this arm was built on the homestead range in the bleak

Fig. 154.

month of December. We were fortunate in getting this rest constructed properly upon the first attempt. We believed, judging by our past successes with the bluff-pointed bullet, that we could compel the long, small-pointed projectile to yield up its secrets by carefully placed screens. If we had known the disappointments in store for us, even the following meager experiments might not have been forthcoming. The new machine rest behaved kindly and did all that could be asked of it during the six days of actual use.

Upon December 12, 1908, two shots were made; December 14th, two shots; 15th, two shots; 16th, three shots; 17th, three shots; 19th, four shots. The ammunition and material were prepared each morning in the shop and tests made in the afternoon. It was not feasible to make more than three shots per day, owing to the time consumed in placing the screens 4 inches apart for considerable distances. The 150-grain 1906 pointed bullet, formerly called the Spitzer bullet, was mutilated by drilling a $\frac{1}{10}$-inch hole $\frac{1}{16}$ inch deep through the metal case, on the side of bullet close to its base, removing 1.4 grains of metal. Several of the tests were made with an $\frac{1}{8}$-inch drill, removing 2.2 grains metal, but this mutilation proved excessive.

The primary object of these investigations was to determine the length of the air spiral or the time of one gyration, and also to determine the diameter of this spiral. We regret to report a failure. The motion of this sharp-pointed projectile from the Springfield rifle at our command was so changeable that, up to the date of closing the tests, no attempt was made to place screens at the proper distances for the special purpose of determining the length and diameter of the air spiral. This report in no way militates against the use of the sharp-pointed bullet. It is simply a report of our meager experiments setting forth the few results obtained. The bore of the particular rifle used was in good order, nearly new. The machine rest held the rifle as a whole, and in this way it was unlike the V-rest with its concentric action which had been invariably used heretofore. Numbering the 16 shots (made on six various dates) consecutively, the report is as follows: —

Shot 1, with $\frac{1}{10}$-inch drill hole removing 1.4 grains of metal, tipped at 4 feet from muzzle 7 degrees and partly straightened up at 7 feet. All of the 16 mutilated bullets were purposely flattened slightly at the point, causing each to make satisfactory prints in the strong linen paper used for screens, therefore no difficulty was experienced in accurately observing the tip.

Shot 2 tipped 7 degrees at 4 feet with almost perfect print at 7 feet.

Shot 3, with $\frac{1}{8}$-inch drill hole, tipped 11 degrees at 3 feet, straightened to 9 degrees at 7 feet, passed on with 9 degrees tip through 10 screens more and was lost. No bluff-pointed bullet with this mutilation, or any recorded tip, failed to perfectly straighten for each oscillation.

Shot 4, with $\frac{1}{10}$-inch hole, tipped 2 degrees at 3 feet with constantly increasing tip up to 7 degrees at 7 feet, passing on through all the screens (set 4 inches apart) with same tip, printing at 100 yards, 7 inches out of normal group. This is the first shot ever recorded on the range that showed no tendency to straighten up within 12 feet of muzzle. At 3 feet 8 inches the base stood true at 6 o'clock, at 9 feet true at 12 o'clock, making one-half of an oscillation and gyration combined in 5 feet 4 inches. At the last screen, at 12 feet 4 inches, the base stood at 4:30 o'clock, indicating that at 14 feet 4 inches, if there had been a screen, the base would have stood at 6 o'clock again, as it did 10 feet 8 inches previous or at the 3 feet 8 inch screen.

Shot 5, with $\frac{1}{10}$-inch hole, made a gradually increasing tip from the first screen 12 inches from muzzle to a 9-degree tip at 7 feet, printing at 100 yards 9 inches out from normal group. At 13 feet 4 inches it was exactly point on, and at 12 inches beyond this it had developed $1\frac{1}{2}$ degrees tip at 1 o'clock. At 12 inches from the muzzle it had developed a tip (from its no-tip position at the muzzle) of the same $1\frac{1}{2}$ degrees and the base stood at 3:30 o'clock. This indicates one gyration in about one and one-third oscillation of 13 feet 4 inches each. As far as this computation goes, one gyration would be made in 17 feet 8 inches. This is not conjecturing but imperfect experimentation.

Shot 6, similar in its action to shot 4 recorded above, is well illustrated by (Fig. 155), it being the second of the only two shots which have acted in this way on the homestead range. The screens were 4 inches apart from 1 foot from muzzle to 14 feet 8 inches, some of them being omitted from the lower row in the figure. After 2 feet 8 inches the prints have been penciled to aid the reader. Observe that the first five screens show as the bullet left them, and that the paper had wiped nearly all the black from the bullet at the fourth screen. The author flatters himself in obtaining such fine prints of a bullet as shown in (Fig. 155), and especially at so short a distance as 12 inches from the muzzle from a bullet traveling 2700 feet per second. The blast at the muzzle from the 50 grains of high-pressure powder and 24-inch barrel is terrific. Only steel rings and steel bolts were of any avail to protect the 12-inch screen. It will readily be seen in (Fig. 155) that the bullet obtained its maximum tip at about 6 feet and afterwards remained at this tip as far as the last screen. At 5 feet its base stands at 6 o'clock, at 9 feet 8 inches at 12 o'clock, thus making one half turn in 4 feet 8 inches, similar to shot 4 which made this half turn in 5 feet 4 inches. Shot 4 made a motion, having the shape of a complete gyration once in 13 feet 4 inches and shot 6 once in 10 feet 8 inches. Shot 6 printed at 100 yards 5 inches out.

Shot 7 was mutilated with $\frac{1}{10}$-inch hole, maximum tip 2 degrees at 4 feet, barely point on at 7 feet. The screens were 4 inches apart, as with previous shots, and we may conclude that this bullet did not entirely straighten up. It printed $7\frac{1}{2}$ inches out at 100 yards.

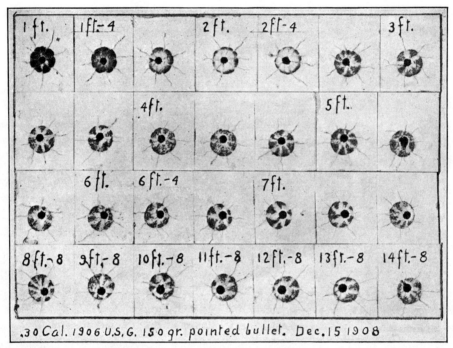

Fig. 155.

Shot 8 tipped 2 degrees at 4 feet, point on at 6 feet 8 inches, 13 feet 4 inches, 20 feet, and 26 feet. It printed 5 inches out at 100 yards.

Shot 9 tipped 2 degrees at 4 feet, exactly point on at 7 feet, 12 feet 8 inches, 19, and 26 feet. It printed $6\frac{1}{2}$ inches out at 100 yards. These shots, 7, 8, and 9, on same day, printed at 100 yards from $5\frac{1}{2}$ to $7\frac{1}{2}$ inches out from normal group, and in a direction relative to their o'clock tip at the 4-foot screen. This corresponds to the action of all bluff-pointed bullets with base mutilations. These three bullets only developed a slight tip in comparison with the preceding ones of the same kind containing the same mutilation. There were so many variations in the flight of these Spitzer bullets from day to day, that we were unable to complete our search during the time at our command.

Shot 10 with $\frac{1}{8}$-inch hole in base tipped 6 degrees at 4 feet, not so much as some of the shots with much less mutilation, point on at 8 feet, meeting target 10 inches out at 9 o'clock.

Shot 11, with same hole as shot 10, same tip, nearly straight at 7 feet, printing 7 inches out at 100 yards at 6 o'clock.

Shot 12 with $\frac{1}{10}$-inch hole tipped the same as number 10 and 11, perfectly straight at 7 feet, printing at 100 yards 12 inches out at 12 o'clock. Observe that 10 tipped at 4 feet at 1:30 o'clock, 11 at 4 feet at 10:30 o'clock, and 12 at 4 feet at 4:30 o'clock, all in the same relation as the direction of their prints were from the normal group at 100 yards. The same fact was noted after shots 7, 8, and 9. Shot 10 at 3 feet, that is 3 feet from its no-tip position at the muzzle, tipped at 12:15 o'clock. At 11 feet, 3 feet from its no-tip position at 8 feet, it stood at 2:30 o'clock, a movement in its gyration of $2\frac{1}{4}$ hours for one oscillation of 8 feet, making its computed gyration equal to 8 times $5\frac{1}{4}$ or 42 feet. This is quite different from the computed gyration of 17 feet 8 inches made from shot 5. In shot 4 where there was no apparent oscillation, its combined gyration and oscillation was positively shown by the screens to be 14 feet 4 inches.

This search, during the cold month of December, for the length and diameter of the spiral of this new bullet, was made with the hopes of throwing some light upon its reputed small wind deflection over long ranges for which there must be some cause. Here may be a good place to note the assertion made by W. E. Mann, who was present at the making of these mutilated shots, that "none tipped at the 100-yard butt." All were tippers at 4 feet, and the majority were excessive tippers, seven of which were caught through proper linen paper at 100 yards to show tip had it been present at this exact distance and all the other shots were caught on ordinary paper. Although the author admits no such conclusion, it was a fact that my brother would not change his assertion by anything obtained from close examination of all available 100-yard prints.

Shot 13, similar to some of the previous shots.

Shot 14 is finely represented by (Fig. 156), giving 14 screens in order to illustrate the perfect point-on position of one Spitzer bullet shot. It straightened up at 8 feet 4 inches and tipped at least 3 degrees at 4 feet. After the 6 feet 8 inch screen the prints were penciled in, the paper having wiped off the dirt on the bullet which makes so good a print at 3 and 4 feet. This care in

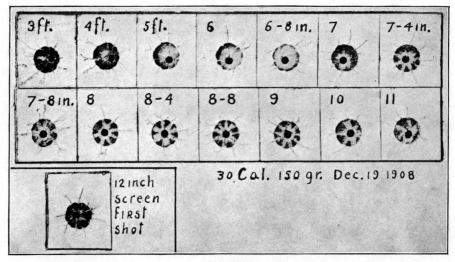

Fig. 156.

observing prints is for a purpose. Where did the black come from which the 3-foot screen took from the bullet in (Fig. 156) and the 1-foot screen in (Fig. 155)? It was suggested that the black came from the previous powder charge. To illustrate this: On December 19, the rifle was doubly cleaned before the first shot of the day. It had been carefully cleaned with concentrated ammonia after use on December 17. Again on December 19, ammonia was used faithfully until no trace of copper remained. After dry swabbing gasoline was used, followed by dry swabs. With the rifle in this condition the first shot was made through a screen 12 inches from muzzle. This screen is appended at the foot of (Fig. 156). After this test it was suggested by W. E. Mann that the outer surface of the bullet was burning hot because the heat had not been given time to be conducted into the metal of the jacket. But tests show that at 100 yards after the surface temperature is reduced the bullet makes the same colored print. The reader is at liberty to form conclusions upon the subject in question.

In this connection we beg to call attention to the so much discussed ball of air or globe of air which all rifle bullets have been supposed to carry along with or around them. If there was such a condition, or if there was a dense layer of air which could be carried many feet deep into water, or several inches into a clay bank, how could the screen paper wipe off the black so perfectly from a metal bullet traveling 2700 feet per second? How could the paper hug the jacket so closely that no air intervened? The screen paper must lie snugly against the metal in order to clean it so quickly.

It has also been suggested that the bullet in any rifle travels so rapidly that it cannot pick up the dirt from the bore, because the inertia of the dirt would hold the dirt back and the bullet would shoot through the dirt, leaving the latter behind. It will be admitted that the black substance, on the bullets here being tested, is traveling at about 2700 feet per second, and yet its inertia is arrested and the dirt held by the screen paper which is stationary. If any bullet does not pick up the dirt which it finds in the bore, would it be on account of the inertia of the dirt?

The last two shots, 15 and 16, made on the same day as number 14 require no particular description.

The Spitzer Bullet Straightens Up. The above observation of W. E. Mann, that "None of the mutilated 1906 pointed bullets were tippers at 100 yards," opened up a subject connected with this bullet that might well attract our attention.

The framework for six large linen paper screens was constructed for use at the farther end of the 100-yard range, and on December 28, 1908, the screens were properly placed 12 inches apart, commencing at the 100-yard butt and working towards the shooting table. Other screens were also placed as usual along the platform, beginning three feet from the muzzle. Six shots were made through these screens, each bullet being mutilated with a $\frac{1}{10}$-inch drill hole, and each tipped regularly at 4 feet from the muzzle and were point on at $7\frac{1}{2}$ and 8 feet. They formed a hollow group at the 100-yard butt about 15 inches in diameter, cutting the edges of the 11 × 16-inch screens there placed, grazing and smashing the frames to which screens were tacked. Not one bullet passed through enough papers to enable us to form any conclusions in regard to its oscillations or tip. The two or three prints that were obtained, however, determined clearly enough that each bullet was point on at the distance where it was caught. This day's work was practically without results.

Upon the following day, December 29, after repairing the screen frames, eight more shots, mutilated in the same manner, were made. The base of each cartridge was properly marked with paint and so loaded that the mutilated side of the bullet would issue at 9 o'clock at muzzle, which made the shot fly eight inches low at 100 yards. Changing elevation of the machine rest caused the group of mutilated shots, five in number, to print centrally in all the screens, making a 3.75-inch group at 100 yards. This group was considerably smaller than one of ten shots, carefully made at machine rest with normal cartridges from the Frankfort Arsenal, and the 1903 Model U. S. Rifle.

In this December 29 test eight shots were made, the mutilated bullets being so entered in the chamber as to print at 6 o'clock.

The location of the mutilation made by a drill in the bullet was indicated by a black mark on the base of shell, which could be seen after being placed in the chamber, and the first shot was made with this mark at 12 o'clock. It printed at 10 : 30 o'clock and nine inches out. From that shot it could be easily determined at what phase of the clock succeeding shots should be loaded so as to print at any phase desired. To print at 6 o'clock the marked shell occupied the 7 : 30 phase in the chamber.

The second shot missed the screen because the elevation was incorrectly changed to 16 inches instead of eight. Figure 157 shows the results of shots 3, 4, 6, 7, and 8; number 5 was lost, having struck the framework of the 7-foot screen, which was turned by the wind just before firing. The number of each shot is indicated in the figure over each column, and the distance from muzzle is noted on each screen. The upper row shows perfectly the maximum tips of each shot at any of the seven screens through which they severally passed at the end of the range. Each shot was exactly point on at some one of

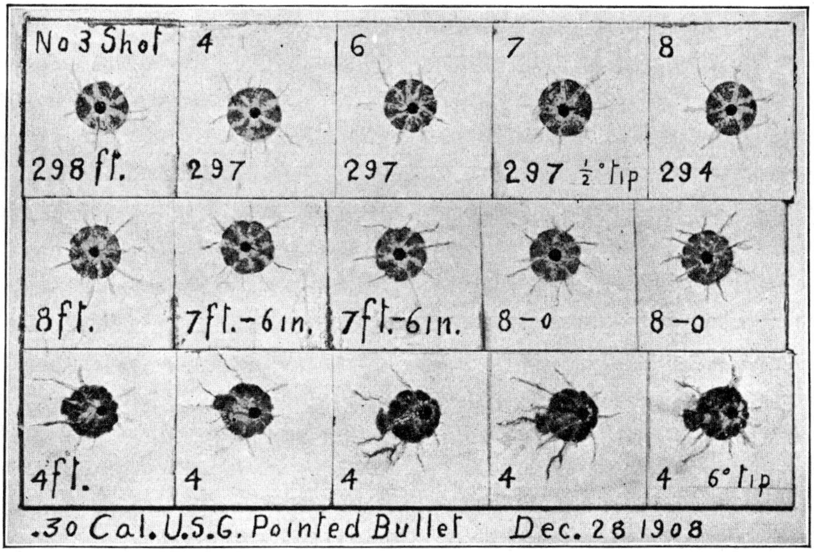

FIG. 157.

these screens. The center row gives the print and the distance at which each shot passed its no-tip position, and the lower row shows finely their respective maximum tips at four feet from rifle's muzzle.

It will be observed that the tip at this last-mentioned screen is in the same direction for all five shots, and it is because each cartridge was inserted uniformly with respect to the mutilated bullet which it contained.

It should also be noticed that the maximum tip of any shot, at 100 yards, was less than one degree, and as a whole all the prints could be called perfect under ordinary conditions of the usual observer. That they were excessive tippers at four feet from the muzzle and made a 3.75-inch group eight inches away from the center of a normal group at 100 yards, is indisputable.

This sharp-pointed bullet, at 50 yards, as shown by screens, has a much smaller tip than at the four-foot screen, showing that it was gradually straightening up. It has practically lost its tip when the 100-yard screen is reached. The flight of this bullet, however, is so rapid that its change from a tipper to point on is almost instantaneous, although not until the causes of the x and y errors have had ample time to exert their respective influences to produce the same error and in the same direction as all other bullets that we have recorded.

As stated at the beginning of this article, experiments were commenced with this Spitzer bullet and the new Government rifle, to discover if possible what peculiar property of the bullet caused its reputed wind deflection at 1000 yards, by determining the length and diameter of its air spiral, but without success.

W. E. Mann's assertion, however, that the bullet straightened up before reaching 100 yards and kept point on, switched us to this other test for proof, and it was found that he was correct. Of course, these experiments, although unsuccessful in demonstrating the cause of the reduced wind deflection of this pointed projectile over the older and much heavier bullets, have given rise to plausible theories respecting both wind deflection and that of its ballistic qualities of retaining its speed over long ranges. We consider it wise, however, as with many foregoing tests, to leave the subject right here for the reader to consider and build upon.

Kinetics of Spin. It may be of interest to determine how much force is required to produce the motion of rotation in a bullet as it is being driven through the rifle bore. The following discussion will be limited to that part of the bore at which the projectile is receiving its greatest acceleration, since this is the place at which the greatest force is being exerted upon the metal of the bullet by the leading sides of the lands, and therefore the place at which our interest centers. When this force, or pressure per square inch, upon the metal of the bullet is known for any particular case, we have added to our knowledge and may form a more correct judgment, if at any time we become interested in the question of the bullet stripping the grooves, or of the fusion of the metal which supports this pressure. By combining and simplifying several equations, each of which bears its part in the different steps of the problem before us, we have the following working formula: —

$$\frac{F}{A} = .00000697 \left(\frac{WrV^2}{PLdn}\right) \tag{7}$$

Where

F = total force in pounds exerted upon the bullet by the leading sides of the lands.

A = total area in square inches of the leading sides of all the lands taken by the bullet.

$\frac{F}{A}$ = force per square inch upon the leading sides of the lands.

W = weight of bullet in grains.

V assumed to be equal to the muzzle velocity of bullet by which assumption formula (7) becomes empirical.

P = pitch of rifling in inches.

n = number of grooves in rifle bore.

r = radius of bullet in inches.

L = length of bullet which takes the grooves in inches.

d = depth of rifling in inches.

If we assume V to be equal to the muzzle velocity and thus make formula (7) empirical, the above quantities are all known, except $\frac{F}{A}$, which we wish to determine.

EXAMPLE. Given the .25-cal. special ammunition, we have $W = 100$, $V = 2500$, $P = 12$, $n = 6$, $r = .125$, $L = .62$, $d = .0025$. Substituting in formula (7) we have

$$\frac{F}{A} = .00000697 \left(\frac{100 \times .125 \times (2500)^2}{12 \times .62 \times .0025 \times 6}\right)$$

Reducing: $\frac{F}{A}$, or the **force per square inch** required upon the leading sides of the lands to produce rotation, = 4900 pounds. If the acceleration for the first 12 inches of the bullet's flight from the chamber was uniform, and if its velocity was V at one foot from chamber, formula (7) would be mathematical. But since the acceleration is much greater at one and two inches from the chamber than at any other place, we may assume a value for V to be equal to the muzzle velocity. Constant experimentation with short barrels and with a 12-inch .32-cal. vs. a 32-inch .32-cal. leads us to make the above estimate. If this be correct, it would follow that if the acceleration of the bullet was constant for 12 inches and equal to its acceleration at one inch, its velocity at one foot would equal its muzzle velocity under normal conditions at 32 inches. The above supposition is simply put in to aid the reader in the use of the formula given, since in the present state of our knowledge some supposition must be decided upon. The following data and formulæ from which the working formula (7) was deduced may also be of interest. The factors and quantities entering these formulæ with their representative letters are as follows: —

W = weight of bullet in grains.

r = radius of bullet in inches.

V = velocity of bullet at the end of one foot of uniform acceleration in feet per second.

P = pitch of rifling in inches.

F = total force exerted on grooves of bullet or on rifling in pounds.

A = total area of leading sides of grooves in square inches.

n = number of grooves.

ω = angular velocity in radians per second.

α = angular acceleration in radians per second per second.

g = 32.2 feet per second per second, or the acceleration of gravity.

I = moment of inertia.

Then, $$A = L \times d \times n \tag{1}$$

Torque or twisting moment = moment of inertia $\times$ angular acceleration.

Then, $$I = \frac{1}{2} \times \frac{W}{7000} \times \frac{1}{g} \times \frac{r^2}{144} \tag{2}$$

$$\omega = 2\pi \times \frac{V}{\frac{P}{12}} \tag{3}$$

Assuming the acceleration to be constant, as we do,

$$a = \frac{\omega}{t} \tag{4}$$

Also under our supposition,

$$t = \frac{2}{V} \tag{5}$$

Hence,

$$a = 2\pi \times \frac{V}{\frac{P}{12}} \times \frac{V}{2} \tag{6}$$

in this case.

Torque = force × radius at which it is applied

$$= F \times \frac{r}{12}$$

Therefore,

$$F \times \frac{r}{12} = \frac{1}{2} \times \frac{W}{7000} \times \frac{1}{32.2} \times \frac{r^2}{144} \times 2\pi \times \frac{V}{\frac{P}{12}} \times \frac{V}{2}$$

Reducing this equation and combining, we find that

$$\frac{F}{A} = .00000697 \frac{r \times W \times V^2}{P \times L \times d \times n} \tag{7}$$

Stripping the Grooves. "Stripped the grooves" and "the bullet jumped the grooves" are phrases sometimes heard among riflemen, applied probably more often to the cast or lead bullet than to the more modern metal-cased bullet. Possibly this idea of stripping the grooves is more prevalent where attempts are made to speed up the non-jacketed bullet in modern sharp-twist rifles and with high-pressure powder, or in fact with black powder, while using excessive charges either experimentally or during systematic methods of producing an "express" load for cast bullets. The foregoing mathematics under the head of "Kinetics of Spin" were introduced to give us a better foundation upon which some judgment may be based about this "stripping the grooves" and raise our ideas about this subject slightly above those of pure conjecture. Following the plan of this book and following the mathematics of this subject, we append several tests, trusting still further to furnish a basis for more correct reasoning.

Test 181. — A .32–40–200-grain, 1 to 30 cast bullet was swaged lightly into a 3-inch piece of a new Winchester, .32-caliber barrel containing their regular rifling. Pushing this bullet partly out of the rifled barrel, left $\frac{1}{4}$ inch of the base end in the grooves. The point and $\frac{1}{2}$ inch of the body was left exposed. After grasping the point in a vise, the barrel was rotated. The $\frac{1}{4}$ inch of rifling on the base of the bullet held firmly, but the entire point was twisted off from the body of the bullet at the line of juncture.

Test 182. — This experiment was identical with the foregoing, except the vise was made to grasp the front end of the cylindrical part of bullet. In this case upon rotating the barrel, which contained $\frac{1}{4}$ inch of the base end of bullet, the body twisted within itself 45 degrees when the rotation of barrel was stopped and the bullet pushed out for examination. The lead of the bullet in the rifling had not yielded.

Test 183. — Under same conditions as the last test, the body of bullet being held in a vise, the barrel was slowly rotated. The twisting of the body which was not retained in the rifle grooves occurred as before, but the rotation was continued until the twisting of the body reached the muzzle and extended into it, after which the $\frac{1}{4}$ inch of the base turned round and round in the rifled bore. The $\frac{1}{4}$ inch of base of bullet which had been retained in the bore was found lengthened and reduced to bore diameter or .315 inch. The lead was not sheared off by the lands in the bore, since the grooves were found empty; but the part of the bullet which had been contained in the rifle grooves was crushed into the body of the lead, elongating this part of the bullet backwards into the rifle bore and forwards as the twisting process continued. It was like twisting a rope out of a tight hole.

Test 184. — In this test a similar bullet, whose body was $\frac{3}{4}$ inch long, was swaged in the rifled piece until it filled the grooves and left there. Before swaging a square plug $\frac{3}{8}$ inch long, one-half the length of the body of bullet, filed from a quarter-inch round bar of steel, had been placed into the base of this bullet. The swaging process caused a perfect fit of this square bar in the body of the lead. The corners of this square wrench filled the bore of the swage to within $\frac{1}{32}$ inch of either side. It was as large a square wrench as the diameter of the bullet would well admit. Rotating this wrench produced no movement of the outer surface of the bullet as it engaged the lands in the swage. The square wrench turned round and round in the bullet, forcing lead out at the base end and producing a hole in the bullet $\frac{1}{4}$ inch in diameter, but no stripping of the grooves.

In addition to the above tests, a well-known fact may be recalled, one which has been in print several times, and vouched for to me by Mr. Pope. If a regular .32-caliber smooth bore be placed in a rifling machine, and the rifling head be exchanged for a lead plug containing coarse emery, the slight scratches made are sufficient to properly rotate the normal projectile for this bore, and the rifle for a few shots will do fair work. Still further, we can look back over the period from 1900 to 1907, during which hundreds of different styles of bullets were recovered in snow or oiled sawdust, without mutilation. All were examined and usually with a glass, yet in no one case was there a suspicion of the grooves being stripped or even widened. One .38-caliber Babbitt metal bullet, bore diameter and 55-grain powder charge, upset so slightly that only the faintest trace of the land was left upon its surface, and yet, when taken from the snow, the land mark seemed normal as to width. This bullet had no base band which could engage the grooves. Two grains of lead were gas-cut from this bullet, and yet, as far as could be detected, it followed the grooves well.

Without conjecturing, more facts may yet be stated. In the test 183, just given, where the bullet was forced to the crushing point of the lands, and made to jump the grooves by mechanical means, the lead contained in the rifle grooves was free to move forward or backward, or in other words the bullet was free to elongate either forward or backward. After this lengthening had occurred, and not before, did the bullet jump the grooves. With the above in mind and considering a normal case where the bullet is being driven by the powder blast, the conditions are somewhat changed. At the time when the bullet is receiving its greatest acceleration, at the time when it has the strongest pressure on the leading side of the lands, which results in the maximum tendency to jump the grooves, the bullet is being forcibly shortened by the pressure which is driving it forward. The inertia of the point is a powerful obstruction to the lengthening of the body in a forward direction, while the powder blast is a still greater force to oppose the lengthening of the bullet backward. This powder pressure at the position of greatest acceleration, or about one inch in front of the chamber, is several times the crushing point of lead. It would seem, then, that tests 181 to 184, which indicate by crude mechanical means how firmly the bullet is grasped by normal rifling, in reality illustrates but one-third of this gripping power which the grooves have upon the bullet at the point of maximum acceleration under normal conditions in its flight through the rifle bore.

Let us consider the following forces, which oppose the acceleration of the motion of the bullet's rotation. First, the inertia of the bullet about its long axis. Second, the rotational friction of the ends of the bullet against the air in front and the powder gas in the rear. Third, the component of the friction of the bullet against the top of the lands in the plane of rotation.

Since these are the three forces opposing the motion of revolution, their sum represents the rotative force acting upon the bullet to produce its motion of spin. The component of this force, which is normal, or at right angles to the leading surface of the lands, is the shearing strain or the crushing strain which tends to strip the grooves. The first force, that of inertia, is mathematically computed by formula (7), page 369. The second force, of gas friction upon the ends of the bullet, is unappreciable and its consideration omitted, since at the position of greatest acceleration, or one inch from the chamber of the rifle, the rotation of the bullet is low in velocity. The third force may be exemplified by taking the case of revolving a tight-fitting bullet once around in a smooth bore, without carrying the bullet forward. This friction in the plane of rotation is noted above as the third force. Since the rifle grooves are themselves grooved by the imperfections of the tool that cuts them, they assist rather than oppose the action of the lands. The third force, as listed above, may in a new rifle barrel be confined as noted to the top of the lands.

The foregoing subjects, "Kinetics of Spin" and "Stripping the Grooves" are introduced in this place to arrest the attention of those who lightly talk of the bullet jumping the grooves. What happens when the bullet is said to have jumped the grooves, or when the condition of the rifle bore would lead the riflemen to such a conclusion, is not discussed here.

The foregoing mathematics may assist somewhat. Reference to the table on page 378 giving some of the properties of lead, tin, and copper, may also be of some assistance in forming conclusions upon this problem of stripping the grooves, such as are not reached by mere conjecture.

A LIST OF CONSTANTS

A list of constants gathered from time to time while making ready and while studying out the various tests performed during our rifle study, will be found interesting and useful to the rifleman.

RIFLEMAN'S TABLE OF MEASURES

3 feet make 1 yard.	INCHES THOUSANDTHS
1760 yds. make 1 mile.	$\frac{1}{16}'' = .064''$
200 yds. make $\frac{1}{8}$ mile approx.	$\frac{1}{32}'' = .032''$
1000 yds. make $\frac{5}{8}$ mile approx.	$\frac{1}{8}'' = .125''$
400 yds. make $\frac{1}{4}$ mile approx.	$\frac{1}{4}'' = .250''$
40 rds. make 220 yds.	$\frac{3}{8}'' = .375''$
1 rod makes $5\frac{1}{2}$ yds.	$\frac{5}{8}'' = .625''$
	$\frac{3}{4}'' = .750''$
	$\frac{7}{8}'' = .875''$

RIFLEMAN'S TABLE OF WEIGHTS

The grain weight is the same in all tables, Avoirdupois, Apothecary, and Troy.
1 pound Avois. = 16 Avois. oz. = 7000 grs.
1 pound Apoth. or Troy = 12 Troy oz. = 5760 grs.
1 ounce Avois. = 437.5 grs.
1 ounce Apoth. or Troy = 480 grs.
1 ounce of water by measure from a graduated glass = 1 oz. Avois. = 437.5 grs.
1 grain of lead = a cube $\frac{1}{16}''$ approx.
1 dram of black powder by measure from a powder flask = $\frac{1}{16}$ oz. = 1 dram Avois. by weight = 27.3 grs.

MISCELLANEOUS DATA

1 cu. foot of air weighs 525 grs.
1 cu. foot of water weighs 437,000 grs.
1 cu. foot of lead weighs 4,937,000 grs.
$\frac{1}{16}$ inch cube of lead = 1 gr. approx.
The .32 cal. bore 32'' long contains 1.01 grs. of air.
The .30 cal. bore 24'' long contains .63 grs. of air.
A bar of cast tin $\frac{1}{4}''$ square will suspend 220 lbs.
A bar of cast lead $\frac{1}{4}''$ square will suspend 100 lbs.
Lead by volume weighs 9300 times more than air.

A LIST OF CONSTANTS

DIAMETER OF DRILLS USED FOR MUTILATING BULLETS

No. 20 drill = .160″.
No. 30 drill = $\frac{1}{8}$″.
No. 39 drill = $\frac{1}{10}$″.
No. 50 drill = .07.″

THICKNESS OF METAL JACKETS

.25 cal. 117 gr. U. M. C. soft-nose bullet = .0165″.
.25 cal. 86 gr. smooth jacket soft nose = .015″.
.30 cal. U. M. C. 220 gr. full mantle = .019″.
.30 cal. Hudson 220 gr. = .021″.

LAFLIN & RAND SMOKELESS POWDER

AVERAGE MEASUREMENTS

Sharpshooter, thickness of disk = .018″.
Number of pellets to the gr. weight, 42.
Lightning, thickness of disk = .027″.
Number of pellets to the gr. weight, 22.
W. A., the thickness of disk = .050″.
Number of pellets to the gr. weight, 12.

DEGREE SCALE FOR BULLET TIP

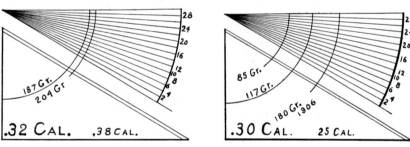

Fig. 158.

Figure 1 in the above cut is correctly drawn for .32 and .38 caliber bullets. Arcs are drawn on the scale for the 187 and 204 grain .32 caliber. Other arcs may be drawn by the experimenter for any weight bullet that he may be using. See instructions and directions for the use of this scale on page 232. One of the parallel lines in figure 1 is drawn for the .32 caliber and the other parallel line for the .38 caliber.

Figure 2 for the .25 and .30 caliber bullets explains itself when the working of figure 1 is understood.

VARIOUS PROPERTIES OF THE METALS AND GASES CONNECTED WITH THE BALLISTICS OF SMALL ARMS

	LEAD	TIN	ANTI-MONY	COPPER	NICKEL	GOV'T JACKET, 15% NICKEL 85% COPPER	WATER	AQUE-OUS VAPOR	AIR	HYDRO-GEN
Spec. Grav., water = 1	11.3	7.3	6.8 *7.8	8.9	8.8	8.9	1	.0006	.0013	.00009
Atomic Weight	206.9	119	120.2	63.6	58.7					
Melting point Fahr. scale	625°	440°	1160° *2500°	2000°	3000°	over 2000°	32°			
Tensile strength lbs. per sq. inch	cast 1600	cast 3500	*90000	drawn 60000		over †60000				
Specific heat, water = 1	.031	.056	.051	.095	.109		1	.480	.237	3.4
Heat or thermal resistance, copper = 1	8.6	5.0	*7	1		18				
Weight per cu. inch in grains	2880	1840	1710 *1960	2250	2200	2250	252	.15	.31	.02
per cu. ⅛ inch	5.6	3.6	3.3 *3.8	4.4	4.3	4.4	0.5	.0003	.0006	

* Steel.

† 1200 pounds ruptured the firmest attachment we were able to devise in many attempts to draw apart the empty government jacket of the 220-grain bullet. The jacket showed no permanent alteration under this stress. Its cross section is .02 sq. inch. 1200 ÷ .02 = 60000.

The above table was made up as an aid towards appreciating some of the interesting properties of the metals and gases which are concerned in rifle shooting. It is particularly connected with the two articles, "Kinetics of Spin" and "Stripping the Grooves," on pages 368 and 371, also with test 173, page 332. It can be used probably with advantage in connection with many experiments pertaining to rifle work. For accurate mathematical purposes some of the data given in the table would need elaboration and correction.

The U. S. five-cent coin, the metal case of the U. S. Government bullet, and German silver are very similar alloys of copper and nickel.

Attention is called to the high thermal resistance of the Government jacket metal.

INDEX

Air holes in bullets, 262.
   motion around bullet, 336, 338, 365.
   properties of, 378.
   space, excessive and group, 138.
   spiral, plotted from actual test, 271, 275, 281.
   tests, .28-cal., 137.
Ammonia *vs.* primer acid, 151.
Ammunition fixed, difficult, 132, 133, 152.
   for 7 mm., 153, 154.
   groups, 34.
   tests with, 33, 34, 117–120, 132.
   three powder group, 34.
   .28-cal., 117–120.
Ammunition, making special fixed, 153.
   .25-cal., Ill., 176.
Anthony, Professor, on recoil, 18.
Axis of air spiral is straight, 271, 275, 281.
Axis of gyration, 221.

Ball of air around bullet, 365.
Barrel, bore of, not straight, 95.
   gilt edge, 266.
   ringed, 15, 17, 74.
   See Caliber.
   shooting ringed, 16.
   36-in. *vs.* 20-in., 17.
Barrels, pipestem, 142.
Base band as gas check, 158.
Bell muzzle, 44.
Black powder discarded, 118.
   reflections on, 120.
Bore-diameter bullets discussed, 183.
   not advised, 184.
Bore diameter varies in size, 87.
Bore, permanent obstruction in, 111.
Brace, "bob sled" front, 28.
   Dr. Skinner's stock, 30.
   Horace Warner's front, 30.
Buckle, amount of, for .32-cal., 82.
Bullet mold, home-made steel, 3.
   plaster of paris, 2.
Bullet, burning hot, 365.
   force required to rotate, 368–371.

Bullet, force required to strip the grooves, 4
   tests, 371–375.
   molds, 261.
   press described, 51.
   Spitzer, 359–368.
   tip scale, 232, 233, 377.
   tip scale for actual use, 377.
   1906 U. S. Gov't., *see* Spitzer.
Bullets, air in, 262.
   and small twigs, 326.
   bore-diameter, discussed, 183.
   bore-diameter not advised, 184.
   bore-diameter cylindrical, 13, 14, 17, 31, 37, 41, 42.
   brass base, 95, 96.
   buzz-saw motion of, 108.
   compared to a boat, 238.
   cylinder or flat end, gyrate backwards, 323–325.
   centering, in bore, 3.
   cup base, 97, 98.
   cast, simulate putty, 120.
   deflected $\frac{1}{4}''$ in 22 ft., 238.
   deflected by paper plank, 336.
   deflected by whizzer, 117.
   deflected by flat surface, 205–211.
   experimental, Ill., 6.
   express, 120, 121.
   first two cylinder metal cased, 145, 147.
   flat end, plain cyl., 16, 36, 323, 325.
   flight of, 185.
   flight of, Ill., 237.
   flight of, Ill. by gyroscope, 222.
   front seating, without seater, 4, 140, 145, 174.
   gyrations of, through screens, 192–201, 245–277.
   hemisphere base, 97.
   hemisphere base, *vs.* flat, 99.
   in clean bore melt and throw lead, 332.
   jacketed, reflections upon, 149.
   lead *vs.* alloyed, 122.
   metal cased, reducing, 262.
   metal cased, upset, 146, 149, 157.
   motion of, in no tip position, 292–297.

380   INDEX

Bullets, motions of, summarized, 243.
    much heavier than air, 238.
    mutilated, that do not tip, 319.
    non-grooved, non-lubricated, 5, 12, 14.
    non-upsetable, 122.
    oblique base, shooting, 214, 215.
    oscillate and gyrate about center of gravity, 321.
    oscillating, Ill. by gyroscope, 222.
    oscillations of, through screens, 192, 196, 245–277.
    patterned after modern cannon, 22.
    pick up tacks, 327.
    plugged and swaged, 212–220.
    point plug, 212, 213.
    points of, mutilated, 212.
    rate of spin of, at 100 yd., 318.
    relative position of, at 6 ft., 284–289.
    six-degree tip, Ill., 229, 230.
    slightly mutilated, 212–215, 217. 219.
    speeding up, stops tipping, 148, 149, 174.
    spherical, 308, 343.
    spherical, do not gyrate, 310, 343.
    square end, 323–325.
    swaging, in bore, 28.
    the first two cylinder, 22.
    tip, 21-inch twist, 148.
    tipping, deceptive, 196, 198, 229, 231, 321.
    tipping, Ill., 229, 230, 237.
    two cylinder 7 mm., 157, 158.
    two cylinder, 22, 94.
    unbalanced, equations for, 354–357.
    unbalanced, how produced, 261.
    unbalanced in bore, 52, 53, 55, 264.
    very slightly mutilated, 217, 219.
    where deformed, 122.
    7 mm., cut down, 152.
    .25-cal. short barrel, 167–170.
    .25-cal., no base band, 165.
    .30-cal., no base band, 149.
    .32-cal., mutilated, through screens, 185–205.
    100-grain .25-cal., special, 174–176.
    100-grain miniature .30-cal., 312–315.
    220-grain U. S. Gov't., 311–315.

Caliber, 7mm. barrel, 152, 158.
    7 mm. 24-inch, 159.
    .25, screen shooting, 84, 86.
    .25–36 Marlin Factory, 164.
    .25–36 special 14 twist, 174–177.
    first .28 Pope, 31.
    .28, fixed ammunition, 117–119, 132.

Caliber, .28, first fixed ammunition, 117–119.
    .28, Herrick shell reduced, 132, 140. 142, 143.
    .28–9, 140.
    discarding two .28, 141.
    .30, 8-inch twist, 145.
    .30, 8-inch twist, 290.
    .30, 8-inch twist, 290.
    .30, 21-inch twist, 147.
    .30, covered by steel tube, 148.
    first .30–30, 145.
    .30–40 Krag, 18.
    .32, Pope 1902, 93.
    .32 rim-fire Wesson, 2.
    .32 Ross-Pope, 41, 82, 83.
    .32, smooth bore, 107.
    .32, Stevens taper chamber, 3.
    .32, vented, 109, 110.
    .32, Winchester 20 inches, 17.
    .32, Winchester 30 inches, 15.
    .32, Winchester 36 inches, 12.
    .32, 12 inches long, 104.
    .38, Pope special, 22.
    .236, Lee navy, 303.
    old .44, muzzle loading, 1.
Cannon, 358.
Chamber, adapting to high pressure powder, 118.
    auxiliary, removable, .33-cal., 26.
    for .28 fixed ammunition, 117, 118, 132.
    groove diameter, 22, 23, 133.
    necessity for special, 118.
    true taper for, 4.
Chase-patch, 13, 14, 16, 38, 68, 149, 183.
Cleaning rifles by ammonia, 151.
Competition *vs.* scientific investigation, 39.
Concentric rings on barrel, 48.
    action, Ill., 49.
Constants, a list of, for riflemen, 376–377.
Copper, properties of, 378.
Correcting measurements, 234.
Corrosion with lead bullets, 134, 136.
    rings, 134, 136.
Covered range, *see* Muslin cover.
Cross hairs on target, 181.
Crow rifle abandoned, 31.

Deflection at muzzle, first noticed, 1.
    by plank at muzzle, 44, 282.
Degree scale for bullet tip, 232, 377.
Diameter of spiral, Ill., 237.
Dirt in bore, inertia of, 365.
Dougherty's, Mr., 100-yd. groups, 25.
Dovetail for telescope mount, 129.
    sight block patented, 128, 152.
Drift of bullet, 239.

INDEX

381

Drift of, defined, 242.
Drills, diameter of, 377.
Dry sawdust, effect of, 59.

Equations (1) to (11), 351–358.
        for kinetics of spin, 369.
        for x and y error, 351–357.

Firing pin tests, 135.
Flight of a bullet, Ill., 237.
Forces acting on lands in bore, 374.
Fore stock, cast iron, 5.
Front seating without bullet seater, 133, 136, 140, 144, 146.
        .25-cal., 164, 165, 174.
        .28-cal., 136–141.

Gases, properties of, 378.
Gilt-edged chuck rifle at last, 160.
        rifle does not deform bullet, 266.
        shooting *vs.* 15,000 shots, 185.
Group, air space, 138.
        auxiliary chamber, 27, 28.
        brass base, 96.
        cannon ball, 24.
        Dougherty's 100-yd., 25.
        dry swab and smokeless, 139.
        fixed ammunition, on "Medicus" range, 34.
        hollow base *vs.* spherical, 97.
        seven shot oblique base, 58.
        Stevens taper chamber, 6.
        V-rest *vs.* muzzle rest, 100.
        vented *vs.* unvented, 110.
        12-inch barrel, 105.
        .25-cal. fixed ammunition, 165.
        .32-cal. non-grooved, non-lubricated, 13.
        .32-cal. rotation, 87, 95.
Groups, average spreading constant, 121, 260.
        fortunate ones not important, 12.
        proportional to distance from muzzle, 85, 202, 204.
Gyration, how produced, 221.
        Ill., 221, 224.
        of spherical shot, 310, 343.
        time of, 200, 273, 280, 297, 298, 313.
Gyroscope, 222.

Harwood, Reuben, removable chamber, 27.
Heavy ordnance, 358.
Helix, 350.
Herrick shell, 132, 133, 140, 143.
Homestead Range described and Ill., 46, 50, 268, 270.
House, shooting, disbanded, 12.

Inertia of dirt in bore, 365.

Jewell, N., and telescopes, 125, 130.

Kent, Perry E., venting system, 114.
Kinetics of spin, 368–371.

Lands, force exerted by, 368–371.
        gripping force on bullet, 371–375.
Lead, properties of, 378.
Lee Navy, .236 accuracy and other tests, 304–306.
Leopold, E. A., and fixed ammunition, 117.
        and oiled sawdust, 59.
        heavy ordnance, 358.
        hollow group, 214.
        in *Pacific Coast Journal*, 117.
        letter on venting barrel, 115.
        letter to, on tel. mounts, 131.
        on black powder, 162.
        on tip, 147.
        on x-error, 250, 347.
        states the trouble, 102.
        vents a barrel, 114.
        whizzer, 115.
        40, 84.
Line of fire, 103.
Line of sight with V-rest, 186.
Loading, speed of on hunt, 132.
Lubrication, kerosene and machine oil, 12.
        non-uniform, 36.

Machine rest for 1903 gov't rifle, 359.
Main spring tests, 136.
Mann, W. E., details discussed, 140.
        on friction, 139.
        on Spitzer bullet, 363, 365.
        putty plug theory, 65.
        shooting match, 160.
        square box for muzzle, 44.
        suggests a severe test, 218.
        vs. V-rest shooting, 100, 101.
Mathematics, 349, 357.
        of spin, 368–371.
Measures, rifleman's table of, 376.
Measuring tipping bullets, 232, 233, 234.
        wind drift, 239.
Mechanical rifle shooting defined, 39, 40.
"Medicus," *see* Skinner, Dr.
Metal jackets, ductility of, 265, 266.
        for woodchucks, 145.
        sizing down, 265.
        thickness of, 377.
        will upset, 146, 149, 157.
Metals, properties of, 378.
Mirage *vs.* telescopes, 177–180.

Miscellaneous data, 376.
Moss in bore, 134.
Motions of flying bullets, 243.
Muslin cover to range, 1st test, 83.
    Ill., 46, 50.
        Pope .32-cal. test, 86.
Mutilated bullets, Lee Navy, 306.
        shooting, 212–220, 276–345.
        spherical, 308.
        220 gr. U. S. gov't, 311.
Mutilations, equations for, 354–357.
    of barrel at muzzle, 105, 106, 107.
    of bullet very slight, effect of, 217, 219.
    three cases of, 354.
Muzzle, beveled, 105.
    blast effect, questioned, 216, 217.
    blast on hemisphere base, 98.
    blast tested, 43, 282.
    blast, 12-inch barrel, 104.
    burrs in, 105.
    defective, 105.
    loading, adapted to hunting, 38.
    Pope's false, two-part, 93.

Nickel, properties of, 378.

Oblique base, mathematics of, 357.
Oblique bases, proportional to group, 58.
    tabulated, 55.
    tested, 54.
O'clock tip scale, 233.
Off shot, defined by Dr. Skinner, 92.
    discussed, 89.
    one in five, 15.
Oiled sawdust discovered, 59.
    hunting for bullets in, 60.
    See Sawdust.
Oleo wad, see Wad.
Oscillating plank shot, gov't bullet, 334.
Oscillations, about center of gravity, 321.
    are not gyrations, 243.
    at 100 yd., 298, 299.
    form of, 292, 342.
    how produced, 221, 242, 345.
    Ill., 222, 224, 228, 340, 342.
    point on position of, 292–297, 340–342.
    time of, 197, 198, 200.
    vary in time, 292, 293, 340.
    vibration and swing, 242.

Paper plank shooting, 337.
Patent telescope mount, 126, 129.
Personal element, 39.

Pettit, H., on venting, 114.
Pipestem barrel, 30 cal., 149.
Plank and screen shooting, 205–211.
    at muzzle, 43, 282.
    eight-foot, groups, 206, 209.
Plank shooting, bullet deflected away, 316.
    round balls, 308.
    summarized, 210.
    with service rifle, 334.
    100-gr. miniature .30-cal., 313–315.
Plots of actual air spiral, 271, 275, 281.
Point deflection, 243.
Pope, H. M., compliments, 93.
    vents barrels, 109.
Pope's plank discovery verified, 205–211.
Powder, cannon, Ill., 125.
    old black abandoned, 162.
    sharpshooter first tested, 123.
    solvents fruitless, 134.
    vials for weighed charges, 182.
Primer acid, 135.
    residue, 136.
Primers, comparative strength of, 135.
    exploded by heat, 135.
    explosion pressure of, 135.
    failed to explode, 135.
    No. $9\tfrac{1}{2}$, 136.
    vs. smokeless powder, 135.
    $2\tfrac{1}{2}$ vs. $7\tfrac{1}{2}$, 135.
Prong in bore, 111.
Putty plug theory, 65, 122.

Questions, incidental, 10.

Rainbow of the riflemen, 258.
Range extended to 175 yd., 203.
    ready for screen shooting, 268.
Recoil, action and reaction equal, 18.
    Ill., 19.
    of rifle during discharge, 30.
    mathematics of, 18.
Records kept for 12 years, 12.
Reducing die for metal cased bullets, 265.
Reflections, 247.
    after tests with .28 cal., 37.
Rest, change of position of, 36.
    Dr. Baker's woodchuck, 163.
    muzzle vs. V-rest, 100, 101.
    target work, error in, 258.
Rifle, see Caliber.
    remodeling a .28-cal., 143.
Rifles, range vs. target, 131.
Rotation test, Lee Navy, 306.
    Pope, 1902, 95.

INDEX

Rotation test, Pope's personal .32-cal., 87.
    Ross-Pope, 83.
    Rotative force required to spin bullet in bore, 374.
    8-inch twist, .30-cal., 146.
    21-inch twist, .30-cal., 148, 150.
Round bullet, *see* Bullet, spherical.
Rust ring, 134, 136.

Sawdust, oiled, 7 mm. shooting, 156.
    .25, .30-cal. shooting, 156.
Scale for tipping bullets, 232, 233.
Scores, fortunate, discarded, 26.
Screen shooting and mutilated bullets, 185–205, 283–337.
    errors in, 186, 194.
    Lee Navy, .236-cal., 304–306.
    platform for, 268.
    spherical bullets, 308–310.
Shell, Herrick's, .28 reduced, 117, 132, 140, 143.
    .28 shortened, less air space, 143, 144.
    7 mm. has no flange, 152.
    .30–40 reduced to 7 mm., 152.
    .30–40 reduced to .25-cal., 166, 174.
    various .25-cal., Ill., 176.
Shells, Babbitt metal in base, 83.
    experimental, 22, 24, 143, 144, 145, 154, 164, 166.
    groove diameter, 22, 24, 143, 144, 145.
    special, Ideal, 4.
Shooting cans of water, 19.
    cylinder bullets, 323–325.
    from Baker's rest, 163.
    Gibraltar, Dr. Skinner, 45.
    lead blocks with Krag, 23.
    perpendicular, with Krag, 20.
    snow, 52.
    snow, oblique bases, 53.
    steam pipe, 19.
    the trouble in rifle, 102.
    two bullets in short barrel, 73.
    unbalanced bullets, 212–220, 276–345.
    under muslin cover, 83, 84.
    window glass, Ill., 21.
Shorkley, Maj. Geo. S., 91.
Short barrels and .32-cal. bullets, 64, 66, 70.
    and .38-cal. bullets, 68.
    and brass wad, 76.
    deductions, 82.
    described and Ill., 61, 62.
    discussion, H. Pettit, 77, 79.
    with high-pressure powder, 70.
    unsafe, 63.
    30-inch targets, 69.

Short barrels, .25-cal. tests, 166–170.
    .30-cal. tests, 170–173.
    .30-cal. target, 173.
Shot, *see* Bullet.
Sighting up 7 mm. for "chucks," 160.
Sights, soldering, 1.
Skinner, Dr., article on, 90.
    letter to Author, 92.
    letter to, 88.
    old reliable .38-cal., 141.
    shooting match, 99, 141.
    shooting table, 101.
Smokeless powder corrosion, 134.
Smooth bore tests, 107, 108.
Spearing, vented barrels, 113.
Spin of bullet determined by screens, 318, 339.
Spin, rate of at 100 yd., 318.
Spinning machine for bullets, 225.
Spiral, bore, to find diameter of, 355.
    developing the, 352.
    Ill., 237, 351.
    of flight, Ill., 237.
    x, 251, 252.
Spitzer bullets, fly irregularly, 360.
    from clean bore, 365.
    mutilated, 359–368.
    oscillate, 364.
    straighten up at 100 yd., 363, 365–367.
    tests, 359–368.
Steel shell, 133.
Stripping the grooves, 371–375.
Swage makes oblique base, 57.
Swages for metal jackets, 145, 265.
    for .28-cal., 155.
Swaging grooved bullets, 263.

Table of properties, metals, gases, 378.
Tacks, driving, with bullets, Ill., 326.
Tangent to x-spiral, 251, 252.
    to y-spiral, 254, 256.
Target, erroneous reading of, 231.
    35-yd., 6.
    100-yd. disdained, 202.
    100 *vs.* 200 yd., 202.
Telescope, cross hairs on target, 181.
    error in, detected, 11.
    estimated distances on target with 181.
    Mogg, 3.
    mounts described, 125.
    mounts, novel features, 130.
Telescopes and mirage, 177–180.
Tests, haphazard, 249.
Throating out .25-cal. Marlin, 166.

Tin, properties of, 378.
Tip, direction of, depends upon mutilation, 283, 286, 288, 290, 321.
    excessive, how produced disclosed, 344.
Tipping bullet and increased charge, 148, 174.
    deceptive, 196, 198, 229, 231.
Tops, are unbalanced, 10, 263.
    spinning, 11.
Trajectory, defined, 123.
    deflection discussed, 245.
    Lee Navy .236-cal., 308.
    lowered by loose bullet, 185.
    two-cylinder U. S. factory bullet, 164.
    7 mm. cal., 158, 159.
    .25–36 Marlin, 164, 165, 166, 174, 175.
    100 yd. .28-cal., 123, 133, 134.
Trough at muzzle, 43.
Tube, paper, shooting, 337.
Twist, determined by screens, 339.
    increased, enlarges x-error, 230, 290.
    increased, quickens oscillations, 290, 313.
    .30-cal. 8 and 21 inch, 145.
Twists of rifles, difficulties with, 267.
Two-cylinder bullet mold, 144.
    bullets, 22, 94, 147, 157, 158.

Upset of .30-cal. metal-cased, 149.
    where occurs, 81, 82.

V machine rest, 45, 48.
V-rest vs. any old rest, 162.
    Baker's rest, 162.
Velocity tested by rate of spin, 318, 339.
Vents, blast from, 113.
    how many required, 114.
Vented barrel, first conception of, 1.
    oblique base, 110.
    test, 109–113.
Vented barrels, E. A. Leopold, 114, 115.
Venting, second attempt, 113.
Verification of past work, 268.

Vials for carrying weighed charges, 182.

Wad, Leopold oleo, 56, 119.
Weights, rifleman's table of, 376.
Whizzer, the, Ill., 115, 116.
Wind drift experiment, 239–241.
Woodchuck, gilt-edge, rifle, 37.
    hunt with expert, 33.
    rifle, 31.
    size of target, Ill., 35.
    sufficient time to load, 33.
    .28–9 rifle, 140–141.
Woodchuck's, boiling out, head, 91.

x-error and north pole, 248.
    cause of, 122.
    cause of, disclosed, 249.
    cause of, Ill., 251, 252.
    computed at 6 ft., 346.
    diagram of, at target, 353.
    Ill. by tests 1 to test 182, 121.
    made apparent at 6 ft., 328–331.
    mathematics of, 351–353.
    not the whole error, 347–349.
    separated from y-error, 300.
    spiral, diameter of computed, 354–357.
    stands alone, 319.
    surmised, 137.
    what is the cause of, 121.
    why search for cause of, 248.
$x + y$ error equals 80 %, 258.
x and y spirals join, 254, 256.

y-error, cause of, disclosed, 253.
    diagram of, at target, 353.
    disclosed by plank shooting, 270.
    Ill., Tests 1 to 182, 121.
    separated from x-error, 300.
    stands alone, 322.
    $2\frac{1}{2}$ times x-error, 300.
y-spiral, actual diam. by test, 271, 275, 281.
    Ill., 254, 256.

Marginal Notes

The publisher has endeavored to decipher these notations by Harry Pope, F.W. Mann and others. In those notes where meaning was not clear, or where certain words could not be deciphered, publisher's comments appear in brackets. In cases where writing is clear and meaningful, the notations have been omitted in this section.

Page II
Born   1856. Died Tuesday, November 14, 1916. *[Day of the week questioned, possibly Sunday.]* My best friend, H.M. Pope. Last 25 barrels never fitted. Received from Mrs. Mann 2/28/22. Above was shipped a week before he died. Account of war delays *[?]* did not reach Milford for a week. He did not go to shop and died before he saw it.

Page XXIII
Prof. Anthony, page 18 — action and reaction energy.

Page XXV
Insert between pages 94 and 95 (in file of papers) — 7/8/42 — must be from Mann book. Put it there.

Page 3
I had 6 years before made my first muzzleloading barrel, the bullet of which fitted the bore forward and the back bands tapering, to fit and fill the grooves, had also discovered that such a bullet made the best breech loading shooting I had ever seen, as the forward portion entered the bore true and the rear would follow or could be seated true through a shell guide of proper size. Had also discovered and used nitro priming and to some extent the effects on accuracy of different loads and tempers.

This diameter and taper muzzleloading bullet of Mr. Pope seems to me as different in principle and action than the Stevens taper chamber bullet. *[Outside notation believed to be F.W. Mann handwriting.]*

Page 10
Mr. M might have saved much time and learned faster had he mixed with really expert shooters. H.M.

This statement maybe would be modified, if Mr. Pope or Mr. Mann knew what in reality Mr. Mann was trying to accomplish. (Frank)

Page 12
Okay if Mr. Mann is to be judged by what he has written — but Mr. Mann did not for 8 years make a single experiment with primers, powder, and not more than 1 or 2 experiments with wads — these were kept uniform for all cast bullet experiments.

Mr. M seems to have been ignorant of the great difference in accuracy caused solely by changes of primer, nitro priming, grain and make of powder, temper of bullets and grease. The oleo wad is often a decided source of inaccuracy.

Page 13
It has been well known for many years before this by me ('84 with my first 32/40 Ballard 20-inch twist) that a smooth bullet will shoot in a slower twist than a grooved 185 patch in

a bore printing perfectly while 185-grain keyholed flat.* Experiments to correct this show that a fine point bullet also showed much less tip due to less air resistance, a 175-grain nearly sharp pointed bullet was adopted, which tipped less than a 165-grain Marlin.
*[Apparently transposed words mean "185-grain flatpoint keyholed."]

Page 14
He could not possibly do it except by reaming and rifling together, though probably reasonably close.

Bore diameter bullets seated in breech, were used many years before this in Ballard rifles to my knowledge by some of our best shooters just tight enough to hold, light visible all around through grooves. *[Dovler ?]* Hayes — 18 years ago.

Page 15
Imperfect muzzle — Mann?
No choke in Winchester barrels often belled.
Winchester barrels were not choked except accidentally if grooves were choked, lands must have been given it conversely. Okay, M. *[Pope's choice of the word,* conversely, *makes the meaning of this comment unclear.]*

Page 17
Better muzzle. Just as likely to be an improvement due to gradual improvements in detail of which Mr. Mann was ignorant, also better swages or better methods in swaging bullets or many other details.

Page 18
S&F, October 26, '93, page 11, Horace Warner bullet and recoil velocity proportional to weight. *["S&F" believed to be Shooting and Fishing magazine.]*

Page 19
It might have been the plug.

Plate 2 — opposite Page 21
.30 Krag Govt. Load 2,000 ft. Velocity + (lighter bullet) at 20 feet, 1½ times size, square end bullet.

Page 30
Recoil 1/10 before bullet is out.

Page 41
Was not perfect, was breech worn.
Not experimented with to determine the accuracy of a Pope but to become conversant with its use. F.W.M.

Page 42
It apparently was not as no attempt was made, apparently, to discover best load, or trouble. O.K. Mann

Page 45
Anything loose decidedly affects accuracy. O.K. Mann
Concrete foundation or masonry with iron plate suggested by H.M. based on his at Hfd. Ct. Mann's rest at Walnut Hill angle iron in cement was so springy as to require extreme care in its use. The fixed V rest is Frank's. O.K. Mann

Page 52
[Comment in upper corner:] Frank was ill.
The first time I visited Frank about 1896 I took a lot of bullets recovered from snow to show him. It was then old. I don't know when I first heard of it. The first I ever saw were some I picked up on the range, shot in winter, after snow had melted.
Was starter correctly handled.

Page 56
A 7½ U.M.C. and no other charge, fired in shell with tube (bullet breech loaded) as described, will blow a hole in base of bullet every time. Lead is apparently melted away by hot flame of primer. Have done it many times. *[Words* good *and* doubted *are initialed FWM — Dr. Franklin W. Mann.]* No doubt, have done it many times with primers only. Stream of fire like blow pipe. HM

Page 58
Looks like two.

Page 59
I suggested with sawdust to Mr. M in Hfd. O.K. Mann

Page 61
I frequently recovered bullets from soft loam good enough to show rifling and upset in the '80s.
Had shown Frank snow bullets first Milford visit but he was so ill he could hardly talk. See note page 52.

Page 64
This upset and battering up of short barreled bullets suggests that there may be some such action with revolvers and pistols. Try it some time.
2-in double chambered 32/40
8-5/8-32 into 3-inch 40 caliber rifled

Page 94
[Allen Pope changes date of Pope's birth to the 15th.]

Page 110
A bunch of letters and papers in the Newton-Kent were in here. Mann-Pope argument. Are tied up in a bunch on book shelves. Pope 7/31/40

Page 117
The thin card and the bullet striking squarely through it, except for rotation, would alter the shape of a bullet practically nothing, while a lead bullet striking a hard piece of wood might have its point badly battered and that on one side, and could go almost anywhere. L's *[Leopold]* reasoning does not hold.

Page 118
Leopold wad would open group.
Priming a primer not mentioned. This 28/30 Herrick would shoot nothing but 2½ Peters Nitro primed, a very coarse powder, FG semi-smokeless and often cg and a quite hard bullet. This load was fine.
No temper mentioned probably lead, which would not shoot fine. Swaging would unbalance them. Best bullet is exactly as it is cast. No machine made bullet equals it.

Page 119
Most all fine Schuetzen men knew differently.
These are fairly common sizes for 200 yard shooting with proper loads and 10 shots — not 5.

Page 120
Wouldn't you think another load would be thought of.
Then why not a temper bullet when every shooter of ability was using them.

Page 121
No it doesn't. Usual velocity for Pope Schuetzen loads.
Not lead, but tin tempered.
Mann's very last work was solid steel bullets to over 6,000 feet the week before he died.

Page 122
Always one primer and powder load. No search for best combination. When at Milford January '14 I shot my number 150 — 33/220 1/25 muzzleloader — 35 shots at 100 yards into size of head of shell [about .4 ? ?]. Frank had never seen such shooting, V rest, concentric action.

Page 123
They do.

Page 125
Mistake Stevens made no scopes till about early '02.

Page 135
Primer intended to explode by blow not heat an entirely different matter.

Page 136
No residuim [?] of black powder is alkaline.

Page 145
Suggested 30/30 because when properly made and loaded it was a very fine and suitable load and I had the tools to make it.
1905 See page 151.
I told him that "I knew of no way in which as fine accuracy could be secured with bullets seated in shell as with metal patched bullets and smokeless powder." And showed him groups so shot. Suggested it for accuracy only with fixed ammunition. Showed him my .30 Special "Wimbleton" gun and 1902 — 220-grain [heel ?] bullets (his later two ?).

Page 150
Not steel, Winchester or Remington never made them.

Page 151
See note by page 145. Pope [heel ?] bullets and rifle.

Page 152
The last work at Stevens. I left there December 31.
Pope shell also shown him at shop on last visit.

Page 153
    Shown him also.

Page 159
    Probably shot a burr out (burred throating).

Page 162
    I had quit Stevens and had been burned out at San Francisco and was then in Los Angeles.

Page 176
    Why he lapped the bore and did not get it right.

Page 177
    I showed him this.

Page 184
    The 2 cylinder bullet.
    Sure does. A breech-loading gun kicks much more than muzzle-loader, same cartridge.

Page 185
    I doubt it.
    of course
    I used them in Hartford about '86-'87. Discovered spiral flight and length.

Page 186
    It got leaded, he thought its size changed unevenly and finally cut off 8 or 9 inches. He had also bruised muz[zle]. I refinished it in Jersey City in   ? but could not put the length back.

Page 189
    Old range 102 yards

Page 196
    Pope's birthday 41 years

Page 210
    I shot in Hartford 1/2" left of 12 foot plank 12 feet long Pope 32/200 bullet, it was deflected about 12" right at 200 yards.

Page 240
    10 drops, no wind
    [in target] This target turned 1/10 [?] drop.
    [left] dropped 41"; equal 200 yard drop
    [right] dropped 9 7/8" at 100 yards

Page 241
    Diagonal change of axis in flight mentioned S&F 11/14/89 page 8, vol. 7 [Shooting and Fishing magazine?] which mentions earlier article to same effect, by H. Warner. Look earlier article up. HM 3/6/41
    [Numbers 1460 behind bullet, in bullet path; 36 drop at target.]

Page 243
   Gyrate. To move spirally — to whirl around a central point.
   Ossilate *[sic]*. To swing, vibrate like a pendulum.

Page 245
   E.A.L. denies this in letter 5/24/10, "Never said so, knows better."

Page 253
   — started by Pope to test Glendale troubles. See page 205

Page 262
   These holes were not caused by included air, but by shrinkage as probably neither mould or lead were hot enough, and because ladle was not left on the sprew hole long enough to fill shrinkage hole. A bullet always cools from the outside to centre, contracting as it cools, leaving hole in centre.

Page 304
   No, nearly so, but shots are changing position considerably.

Plate 34 — Page 307
   (6) notched base corner .02 gr.
   (8) .06 dr*[illed]* near base, .7 grain. Lost at 100 *[yards ?]*.
   (9) do *[ditto]* 1/2 grain.
   (10) point filed .02 grain.

Page 340
   (No — 2000)

Page 343
   How can screen show tip of a spherical bullet, the hole would be circular no matter what direction the axis of spin took.

Page 368
   Time *[reference to different time-of-flight?]*.

Page 373
   Stripping

Page 242
   I wonder. Bullets dropped as illustrated would be drifted. Bullets shot from a rifle are moved by wind many times further in same time. This would be deflection and a curve varying with force and time.

Page 391
   December, 1942. A nearly reprint of this book just out by Herman P. Dean, Huntington, West Virginia. Published by Standard Printing and Publishing Company, October, 1942. Has a letter on Mann's home life by a daughter, now Mrs. Willard Lewis, Forest Hills, Augusta, Georgia, about Mrs. F. Mann's appreciation and her own help and appreciation to help *[with ?]* Franklin's work, (not strictly true).

The inscription on the back of the above photograph reads:
  "Harry M. Pope at lathe in shop 18 Morris Street Jersey City, N.J. about 1941?" Photo original in Library Of The Congress Washington, D.C. Photo by Arnold Genthé famed for his photographs of San Francisco fire 1906.